Student Solutions Manual

for

Calculus for the Managerial, Life, and Social Sciences

Seventh Edition

S. T. Tan
Stonehill College

THOMSON

BROOKS/COLE

Australia • Canada • Mexico • Singapore • Spain • United Kingdom • United States

For more information about our products, contact us at:
Thomson Learning Academic Resource Center
1-800-423-0563

For permission to use material from this text or product, submit a request online at
http://www.thomsonrights.com.
Any additional questions about permissions can be submitted by email to **thomsonrights@thomson.com.**

Thomson Higher Education
10 Davis Drive
Belmont, CA 94002-3098
USA

Asia (including India)
Thomson Learning
5 Shenton Way
#01-01 UIC Building
Singapore 068808

Australia/New Zealand
Thomson Learning Australia
102 Dodds Street
Southbank, Victoria 3006
Australia

Canada
Thomson Nelson
1120 Birchmount Road
Toronto, Ontario M1K 5G4
Canada

UK/Europe/Middle East/Africa
Thomson Learning
High Holborn House
50–51 Bedford Row
London WC1R 4LR
United Kingdom

Latin America
Thomson Learning
Seneca, 53
Colonia Polanco
11560 Mexico
D.F. Mexico

Spain (including Portugal)
Thomson Paraninfo
Calle Magallanes, 25
28015 Madrid, Spain

CONTENTS

CHAPTER 5   EXPONENTIAL AND LOGARITHMIC FUNCTIONS

CHAPTER 6   INTEGRATION

CHAPTER 7   ADDITIONAL TOPICS IN INTEGRATION

CHAPTER 8   CALCULUS OF SEVERAL VARIABLES

CHAPTER 1

EXERCISES 1.1, page 9

1. The statement is false because -3 is greater than -20. (See the number line that follows).

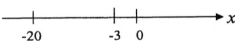

3. The statement is false because 2/3 [which is equal to (4/6)] is less than 5/6.

5. The interval (3,6) is shown on the number line that follows. Note that this is an open interval indicated by (and).

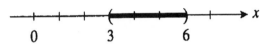

7. The interval [-1,4) is shown on the number line that follows. Note that this is a half-open interval indicated by [(closed) and) (open).

9. The infinite interval (0,∞) is shown on the number line that follows.

11. First, $2x + 4 < 8$
    Next, $2x < 4$ (Add -4 to each side of the inequality.)
    and $x < 2.$ (Multiply each side of the inequality by 1/2)
    We write this in interval notation as $(-\infty, 2)$.

1

1 Preliminaries

13. We are given the inequality $-4x \geq 20$.

Then $x \leq -5$. (Multiply both sides of the inequality by -1/4 and reverse the sign of the inequality.)

We write this in interval notation as $(-\infty, -5]$.

15. We are given the inequality $-6 < x - 2 < 4$.

First $\quad -6 + 2 < x < 4 + 2$ (Add +2 to each member of the inequality.)

and $\quad -4 < x < 6$,

so the solution set is the open interval $(-4, 6)$.

17. We want to find the values of x that satisfy the inequalities

$$x + 1 > 4 \text{ or } x + 2 < -1.$$

Adding -1 to both sides of the first inequality, we obtain

$$x + 1 - 1 > 4 - 1,$$

or $\quad\quad\quad x > 3$.

Similarly, adding -2 to both sides of the second inequality, we obtain

$$x + 2 - 2 < -1 - 2,$$

or $\quad\quad\quad x < -3$.

Therefore, the solution set is $(-\infty, -3) \cup (3, \infty)$.

19. We want to find the values of x that satisfy the inequalities

$$x + 3 > 1 \text{ and } x - 2 < 1.$$

Adding -3 to both sides of the first inequality, we obtain

$$x + 3 - 3 > 1 - 3,$$

or $\quad\quad\quad x > -2$.

Similarly, adding 2 to each side of the second inequality, we obtain

$$x - 2 + 2 < 1 + 2, \text{ or } x < 3.$$

Since both inequalities must be satisfied, the solution set is $(-2, 3)$.

21. $|-6 + 2| = 4$.

23. $\dfrac{|-12 + 4|}{|16 - 12|} = \dfrac{|-8|}{|4|} = 2$.

25. $\sqrt{3}|-2| + 3|-\sqrt{3}| = \sqrt{3}(2) + 3\sqrt{3} = 5\sqrt{3}$.

27. $|\pi - 1| + 2 = \pi - 1 + 2 = \pi + 1$.

29. $|\sqrt{2} - 1| + |3 - \sqrt{2}| = \sqrt{2} - 1 + 3 - \sqrt{2} = 2$.

31. False. If $a > b$, then $-a < -b$, $-a + b < -b + b$, and $b - a < 0$.

33. False. Let $a = -2$ and $b = -3$. Then $a^2 = 4$ and $b^2 = 9$, and $4 < 9$. Note that we only need to provide a counterexample to show that the statement is not always true.

35. True. There are three possible cases.

   Case 1 If $a > 0$, $b > 0$, then $a^3 > b^3$, since $a^3 - b^3 = (a - b)(a^2 + ab + b^2) > 0$.

   Case 2 If $a > 0$, $b < 0$, then $a^3 > 0$ and $b^3 < 0$ and it follows that $a^3 > b^3$.

   Case 3 If $a < 0$ and $b < 0$, then $a^3 - b^3 = (a - b)(a^2 + ab + b^2) > 0$, and we see that $a^3 > b^3$. (Note that $(a - b) > 0$ and $ab > 0$.)

37. False. Take $a = -2$, then $|-a| = |-(-2)| = |2| = 2 \neq a$.

39. True. If $a - 4 < 0$, then $|a - 4| = 4 - a = |4 - a|$. If $a - 4 > 0$, then
$$|4 - a| = a - 4 = |a - 4|.$$

41. False. Take $a = 3$, $b = -1$. Then $|a + b| = |3 - 1| = 2 \neq |a| + |b| = 3 + 1 = 4$.

43. $27^{2/3} = (3^3)^{2/3} = 3^2 = 9$.

45. $\left(\dfrac{1}{\sqrt{3}}\right)^0 = 1$. Recall that any number raised to the zero power is 1.

47. $\left[\left(\dfrac{1}{8}\right)^{1/3}\right]^{-2} = \left(\dfrac{1}{2}\right)^{-2} = (2^2) = 4$.

49. $\left(\dfrac{7^{-5} \cdot 7^2}{7^{-2}}\right)^{-1} = (7^{-5+2+2})^{-1} = (7^{-1})^{-1} = 7^1 = 7$.

51. $(125^{2/3})^{-1/2} = 125^{(2/3)(-1/2)} = 125^{-1/3} = \dfrac{1}{125^{1/3}} = \dfrac{1}{5}.$

53. $\dfrac{\sqrt{32}}{\sqrt{8}} = \sqrt{\dfrac{32}{8}} = \sqrt{4} = 2.$

55. $\dfrac{16^{5/8}16^{1/2}}{16^{7/8}} = 16^{(5/8+1/2-7/8)} = 16^{1/4} = 2.$

57. $16^{1/4} \cdot 8^{-1/3} = 2 \cdot \left(\dfrac{1}{8}\right)^{1/3} = 2 \cdot \dfrac{1}{2} = 1.$ 59. True.

61. False. $x^3 \times 2x^2 = 2x^{3+2} = 2x^5 \neq 2x^6.$ 63. False. $\dfrac{2^{4x}}{1^{3x}} = \dfrac{2^{4x}}{1} = 2^{4x}.$

65. False. $\dfrac{1}{4^{-3}} = 4^3 = 64.$ 67. False. $(1.2^{1/2})^{-1/2} = (1.2)^{-1/4} \neq 1.$

69. $(xy)^{-2} = \dfrac{1}{(xy)^2}.$ 71. $\dfrac{x^{-1/3}}{x^{1/2}} = x^{(-1/3)-(1/2)} = x^{-5/6} = \dfrac{1}{x^{5/6}}.$

73. $12^0(s+t)^{-3} = 1 \cdot \dfrac{1}{(s+t)^3} = \dfrac{1}{(s+t)^3}.$

75. $\dfrac{x^{7/3}}{x^{-2}} = x^{(7/3)+2} = x^{(7/3)+(6/3)} = x^{13/3}.$

77. $(x^2y^{-3})(x^{-5}y^3) = (x^{2-5}y^{-3+3}) = x^{-3}y^0 = x^{-3} = \dfrac{1}{x^3}.$

79. $\dfrac{x^{3/4}}{x^{-1/4}} = x^{(3/4)-(-1/4)} = x^{4/4} = x.$

81. $\left(\dfrac{x^3}{-27y^{-6}}\right)^{-2/3} = x^{3(-2/3)}\left(-\dfrac{1}{27}\right)^{-2/3}y^{6(-2/3)} = x^{-2}\left(-\dfrac{1}{3}\right)^{-2}y^{-4} = \dfrac{9}{x^2y^4}.$

83. $\left(\dfrac{x^{-3}}{y^{-2}}\right)^2\left(\dfrac{y}{x}\right)^4 = \dfrac{x^{-3(2)}y^4}{y^{-2(2)}x^4} = \left(\dfrac{y^{4+4}}{x^{4+6}}\right) = \dfrac{y^8}{x^{10}}.$

85. $\sqrt[3]{x^{-2}}\cdot\sqrt{4x^5} = x^{-2/3}\cdot 4^{1/2}\cdot x^{5/2} = x^{-(2/3)+(5/2)}\cdot 2 = 2x^{11/6}.$

87. $-\sqrt[4]{16x^4y^8} = -(16^{1/4}\cdot x^{4/4}\cdot y^{8/4}) = -2xy^2.$

89. $\sqrt[6]{64x^8y^3} = (64)^{1/6}\cdot x^{8/6}y^{3/6} = 2x^{4/3}y^{1/2}.$

91. $2^{3/2} = (2)(2^{1/2}) = 2(1.414) = 2.828.$

93. $9^{3/4} = (3^2)^{3/4} = 3^{6/4} = 3^{3/2} = 3\cdot 3^{1/2} = 3(1.732) = 5.196.$

95. $10^{3/2} = 10^{1/2}\cdot 10 = (3.162)(10) = 31.62.$

97. $10^{2.5} = 10^2\cdot 10^{1/2} = 100(3.162) = 316.2.$

99. $\dfrac{3}{2\sqrt{x}}\cdot\dfrac{\sqrt{x}}{\sqrt{x}} = \dfrac{3\sqrt{x}}{2x}.$

101. $\dfrac{2y}{\sqrt{3y}}\cdot\dfrac{\sqrt{3y}}{\sqrt{3y}} = \dfrac{2y\sqrt{3y}}{3y} = \dfrac{2}{3}\sqrt{3y}.$

103. $\dfrac{1}{\sqrt[3]{x}}\cdot\dfrac{\sqrt[3]{x^2}}{\sqrt[3]{x^2}} = \dfrac{\sqrt[3]{x^2}}{\sqrt[3]{x^3}} = \dfrac{\sqrt[3]{x^2}}{x}.$

105. $\dfrac{2\sqrt{x}}{3}\cdot\dfrac{\sqrt{x}}{\sqrt{x}} = \dfrac{2x}{3\sqrt{x}}.$

107. $\sqrt{\dfrac{2y}{x}} = \dfrac{\sqrt{2y}}{\sqrt{x}}\cdot\dfrac{\sqrt{2y}}{\sqrt{2y}} = \dfrac{2y}{\sqrt{2xy}}.$

109. $\dfrac{\sqrt[3]{x^2z}}{y}\cdot\dfrac{\sqrt[3]{xz^2}}{\sqrt[3]{xz^2}} = \dfrac{\sqrt[3]{x^3z^3}}{y\sqrt[3]{xz^2}} = \dfrac{xz}{y\sqrt[3]{xz^2}}.$

111. If the car is driven in the city, then it can be expected to cover

$$(18.1)(20) = 362 \qquad \text{(miles/gal} \cdot \text{gal)}$$

or 362 miles on a full tank. If the car is driven on the highway, then it can be expected to cover

$$(18.1)(27) = 488.7 \qquad \text{(miles/gal} \cdot \text{gal)}$$
or 488.7 miles on a full tank. Thus, the driving range of the car may be described by the interval [362, 488.7].

113. $6(P - 2500) \le 4(P + 2400)$
$6P - 15000 \le 4P + 9600$
$2P \le 24600, \quad \text{or} \quad P \le 12300.$
Therefore, the maximum profit is $12,300.

115. Let x represent the salesman's monthly sales in dollars. Then
$$0.15(x - 12000) \ge 3000$$
$$15(x - 12000) \ge 300000$$
$$15x - 180000 \ge 300000$$
$$15x \ge 480000$$
$$x \ge 32000.$$
We conclude that the salesman must earn at least $32,000 to reach his goal.

117. The rod is acceptable if $0.49 < x < 0.51$ or $-0.01 < x - 0.5 < 0.01$. This gives the required inequality $|x - 0.5| < 0.01$.

119. We want to solve the inequality
$$-6x^2 + 30x - 10 \ge 14. \qquad \text{(Remember } x \text{ is expressed in thousands.)}$$
Adding -14 to both sides of this inequality, we have
$$-6x^2 + 30x - 10 - 14 \ge 14 - 14,$$
or $- 6x^2 + 30x - 24 \ge 0.$
Dividing both sides of the inequality by -6 (which reverses the sign of the inequality), we have $x^2 - 5x + 4 \le 0$.
Factoring this last expression, we have $(x - 4)(x - 1) \le 0$.
From the following sign diagram,

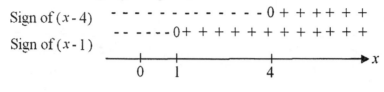

we see that x must lie between 1 and 4. (The inequality is only satisfied when the two factors have opposite signs.) Since x is expressed in thousands of units, we see that the manufacturer must produce between 1000 and 4000 units of the commodity.

121. False. Take $a = 1$, $b = 2$, and $c = 3$. Then $a < b$, but
$a - c = 1 - 3 = -2 \not> 2 - 3 = -1 = b - c$.

123. True. $|a - b| = |a + (-b)| \le |a| + |-b| = |a| + |b|$.

EXERCISES 1.2, page 21

1. $(7x^2 - 2x + 5) + (2x^2 + 5x - 4) = 7x^2 - 2x + 5 + 2x^2 + 5x - 4$
$$= 9x^2 + 3x + 1.$$

3. $(5y^2 - 2y + 1) - (y^2 - 3y - 7) = 5y^2 - 2y + 1 - y^2 + 3y + 7$
$$= 4y^2 + y + 8.$$

5. $x - \{2x - [-x - (1 - x)]\} = x - \{2x - [-x - 1 + x]\}$
$$= x - \{2x + 1\}$$
$$= x - 2x - 1$$
$$= -x - 1.$$

7. $(\frac{1}{3} - 1 + e) - (-\frac{1}{3} - 1 + e^{-1}) = \frac{1}{3} - 1 + e + \frac{1}{3} + 1 - \frac{1}{e}$
$$= \frac{2}{3} + e - \frac{1}{e}$$
$$= \frac{3e^2 + 2e - 3}{3e}.$$

9. $3\sqrt{8} + 8 - 2\sqrt{y} + \frac{1}{2}\sqrt{x} - \frac{3}{4}\sqrt{y} = 3\sqrt{4 \cdot 2} + 8 + \frac{1}{2}\sqrt{x} - \frac{11}{4}\sqrt{y}$
$$= 6\sqrt{2} + 8 + \frac{1}{2}\sqrt{x} - \frac{11}{4}\sqrt{y}.$$

11. $(x + 8)(x - 2) = x(x - 2) + 8(x - 2) = x^2 - 2x + 8x - 16 = x^2 + 6x - 16.$

13. $(a + 5)^2 = (a + 5)(a + 5) = a(a + 5) + 5(a + 5) = a^2 + 5a + 5a + 25$

7

$$= a^2 + 10a + 25.$$

15. $(x + 2y)^2 = (x + 2y)(x + 2y) = x(x + 2y) + 2y(x + 2y)$
$$= x^2 + 2xy + 2yx + 4y^2 = x^2 + 4xy + 4y^2.$$

17. $(2x + y)(2x - y) = 2x(2x - y) + y(2x - y) = 4x^2 - 2xy + 2xy - y^2$
$$= 4x^2 - y^2.$$

19. $(x^2 - 1)(2x) - x^2(2x) = 2x^3 - 2x - 2x^3 = -2x.$

21. $2(t + \sqrt{t})^2 - 2t^2 = 2(t + \sqrt{t})(t + \sqrt{t}) - 2t^2$
$$= 2(t^2 + 2t\sqrt{t} + t) - 2t^2$$
$$= 2t^2 + 4t\sqrt{t} + 2t - 2t^2$$
$$= 4t\sqrt{t} + 2t = 2t(2\sqrt{t} + 1).$$

23. $4x^5 - 12x^4 - 6x^3 = 2x^3(2x^2 - 6x - 3).$

25. $7a^4 - 42a^2b^2 + 49a^3b = 7a^2(a^2 + 7ab - 6b^2).$

27. $e^{-x} - xe^{-x} = e^{-x}(1 - x).$

29. $2x^{-5/2} - \frac{3}{2}x^{-3/2} = \frac{1}{2}x^{-5/2}(4 - 3x).$

31. $6ac + 3bc - 4ad - 2bd = 3c(2a + b) - 2d(2a + b) = (2a + b)(3c - 2d).$

33. $4a^2 - b^2 = (2a + b)(2a - b).$ [Difference of two squares]

35. $10 - 14x - 12x^2 = -2(6x^2 + 7x - 5) = -2(3x + 5)(2x - 1).$

37. $3x^2 - 6x - 24 = 3(x^2 - 2x - 8) = 3(x - 4)(x + 2).$

39. $12x^2 - 2x - 30 = 2(6x^2 - x - 15) = 2(3x - 5)(2x + 3).$

41. $9x^2 - 16y^2 = (3x)^2 - (4y)^2 = (3x - 4y)(3x + 4y).$

43. $x^6 + 125 = (x^2)^3 + (5)^3 = (x^2 + 5)(x^4 - 5x^2 + 25)$.

45. $(x^2 + y^2)x - xy(2y) = x^3 + xy^2 - 2xy^2 = x^3 - xy^2$.

47. $2(x - 1)(2x + 2)^3[4(x - 1) + (2x + 2)]$
$$= 2(x - 1)(2x + 2)^3[4x - 4 + 2x + 2]$$
$$= 2(x - 1)(2x + 2)^3[6x - 2]$$
$$= 4(x - 1)(3x - 1)(2x + 2)^3.$$

49. $4(x - 1)^2(2x + 2)^3(2) + (2x + 2)^4(2)(x - 1)$
$$= 2(x - 1)(2x + 2)^3[4(x - 1) + (2x + 2)] = 2(x - 1)(2x + 2)^3(6x - 2)$$
$$= 4(x - 1)(3x - 1)(2x + 2)^3.$$

51. $(x^2 + 2)^2[5(x^2 + 2)^2 - 3](2x) = (x^2 + 2)^2[5(x^4 + 4x^2 + 4) - 3](2x)$
$$= (2x)(x^2 + 2)^2(5x^4 + 20x^2 + 17).$$

53. $x^2 + x - 12 = 0$, or $(x + 4)(x - 3) = 0$, so that $x = -4$ or $x = 3$. We conclude that the roots are $x = -4$ and $x = 3$.

55. $4t^2 + 2t - 2 = (2t - 1)(2t + 2) = 0$. Thus, $t = 1/2$ and $t = -1$ are the roots.

57. $\frac{1}{4}x^2 - x + 1 = (\frac{1}{2}x - 1)(\frac{1}{2}x - 1) = 0$. Thus $\frac{1}{2}x = 1$, and $x = 2$ is a double root of the equation.

59. Here we use the quadratic formula to solve the equation $4x^2 + 5x - 6 = 0$. Then, $a = 4$, $b = 5$, and $c = -6$. Therefore,

$$x = \frac{-b \pm \sqrt{b^2 - 4ac}}{2a} = \frac{-(5) \pm \sqrt{(5)^2 - 4(4)(-6)}}{2(4)} = \frac{-5 \pm \sqrt{121}}{8}$$
$$= \frac{-5 \pm 11}{8}.$$

Thus, $x = -\frac{16}{8} = -2$ and $x = \frac{6}{8} = \frac{3}{4}$ are the roots of the equation.

61. We use the quadratic formula to solve the equation $8x^2 - 8x - 3 = 0$. Here $a = 8$, $b = -8$, and $c = -3$. Therefore,

1 Preliminaries

$$x = \frac{-b \pm \sqrt{b^2 - 4ac}}{2a} = \frac{-(-8) \pm \sqrt{(-8)^2 - 4(8)(-3)}}{2(8)} = \frac{8 \pm \sqrt{160}}{16}$$

$$= \frac{8 \pm 4\sqrt{10}}{16} = \frac{2 \pm \sqrt{10}}{4}.$$

Thus, $x = \frac{1}{2} + \frac{1}{4}\sqrt{10}$ and $x = \frac{1}{2} - \frac{1}{4}\sqrt{10}$ are the roots of the equation.

63. We use the quadratic formula to solve $2x^2 + 4x - 3 = 0$. Here, $a = 2$, $b = 4$, and $c = -3$. Therefore

$$x = \frac{-b \pm \sqrt{b^2 - 4ac}}{2a} = \frac{-(4) \pm \sqrt{(4)^2 - 4(2)(-3)}}{2(2)} = \frac{-4 \pm \sqrt{40}}{4}$$

$$= \frac{-4 \pm 2\sqrt{10}}{4} = \frac{-2 \pm \sqrt{10}}{2}.$$

Thus, $x = -1 + \frac{1}{2}\sqrt{10}$ and $x = -1 - \frac{1}{2}\sqrt{10}$ are the roots of the equation.

65. $\dfrac{x^2 + x - 2}{x^2 - 4} = \dfrac{(x+2)(x-1)}{(x+2)(x-2)} = \dfrac{x-1}{x-2}.$

67. $\dfrac{12t^2 + 12t + 3}{4t^2 - 1} = \dfrac{3(4t^2 + 4t + 1)}{4t^2 - 1} = \dfrac{3(2t+1)(2t+1)}{(2t+1)(2t-1)} = \dfrac{3(2t+1)}{2t-1}.$

69. $\dfrac{(4x-1)(3) - (3x+1)(4)}{(4x-1)^2} = \dfrac{12x - 3 - 12x - 4}{(4x-1)^2} = -\dfrac{7}{(4x-1)^2}.$

71. $\dfrac{2a^2 - 2b^2}{b-a} \cdot \dfrac{4a + 4b}{a^2 + 2ab + b^2} = \dfrac{2(a+b)(a-b)4(a+b)}{-(a-b)(a+b)(a+b)} = -8.$

73. $\dfrac{3x^2 + 2x - 1}{2x + 6} \div \dfrac{x^2 - 1}{x^2 + 2x - 3} = \dfrac{(3x-1)(x+1)}{2(x+3)} \cdot \dfrac{(x+3)(x-1)}{(x+1)(x-1)} = \dfrac{3x-1}{2}.$

75. $\dfrac{58}{3(3t+2)}+\dfrac{1}{3}=\dfrac{58+3t+2}{3(3t+2)}=\dfrac{3t+60}{3(3t+2)}=\dfrac{t+20}{3t+2}.$

77. $\dfrac{2x}{2x-1}-\dfrac{3x}{2x+5}=\dfrac{2x(2x+5)-3x(2x-1)}{(2x-1)(2x+5)}=\dfrac{4x^2+10x-6x^2+3x}{(2x-1)(2x+5)}$

$$=\dfrac{-2x^2+13x}{(2x-1)(2x+5)}=-\dfrac{x(2x-13)}{(2x-1)(2x+5)}.$$

79. $\dfrac{4}{x^2-9}-\dfrac{5}{x^2-6x+9}=\dfrac{4}{(x+3)(x-3)}-\dfrac{5}{(x-3)^2}$

$$=\dfrac{4(x-3)-5(x+3)}{(x-3)^2(x+3)}=-\dfrac{x+27}{(x-3)^2(x+3)}.$$

81. $\dfrac{1+\dfrac{1}{x}}{1-\dfrac{1}{x}}=\dfrac{\dfrac{x+1}{x}}{\dfrac{x-1}{x}}=\dfrac{x+1}{x}\cdot\dfrac{x}{x-1}=\dfrac{x+1}{x-1}.$

83. $\dfrac{4x^2}{2\sqrt{2x^2+7}}+\sqrt{2x^2+7}=\dfrac{4x^2+2\sqrt{2x^2+7}\sqrt{2x^2+7}}{2\sqrt{2x^2+7}}=\dfrac{4x^2+4x^2+14}{2\sqrt{2x^2+7}}$

$$=\dfrac{4x^2+7}{\sqrt{2x^2+7}}.$$

85. $\dfrac{2x(x+1)^{-1/2}-(x+1)^{1/2}}{x^2}=\dfrac{(x+1)^{-1/2}(2x-x-1)}{x^2}=\dfrac{(x+1)^{-1/2}(x-1)}{x^2}$

$$=\dfrac{x-1}{x^2\sqrt{x+1}}.$$

87.
$$\frac{(2x+1)^{1/2}-(x+2)(2x+1)^{-1/2}}{2x+1}=\frac{(2x+1)^{-1/2}(2x+1-x-2)}{2x+1}$$
$$=\frac{(2x+1)^{-1/2}(x-1)}{2x+1}=\frac{x-1}{(2x+1)^{3/2}}.$$

89. $\dfrac{1}{\sqrt{3}-1}\cdot\dfrac{\sqrt{3}+1}{\sqrt{3}+1}=\dfrac{\sqrt{3}+1}{3-1}=\dfrac{\sqrt{3}+1}{2}.$

91. $\dfrac{1}{\sqrt{x}-\sqrt{y}}\cdot\dfrac{\sqrt{x}+\sqrt{y}}{\sqrt{x}+\sqrt{y}}=\dfrac{\sqrt{x}+\sqrt{y}}{x-y}.$

93. $\dfrac{\sqrt{a}+\sqrt{b}}{\sqrt{a}-\sqrt{b}}\cdot\dfrac{\sqrt{a}+\sqrt{b}}{\sqrt{a}+\sqrt{b}}=\dfrac{(\sqrt{a}+\sqrt{b})^2}{a-b}.$

95. $\dfrac{\sqrt{x}}{3}\cdot\dfrac{\sqrt{x}}{\sqrt{x}}=\dfrac{x}{3\sqrt{x}}.$

97. $\dfrac{1-\sqrt{3}}{3}\cdot\dfrac{1+\sqrt{3}}{1+\sqrt{3}}=\dfrac{1^2-(\sqrt{3})^2}{3(1+\sqrt{3})}=-\dfrac{2}{3(1+\sqrt{3})}.$

99. $\dfrac{1+\sqrt{x+2}}{\sqrt{x+2}}\cdot\dfrac{1-\sqrt{x+2}}{1-\sqrt{x+2}}=\dfrac{1-(x+2)}{\sqrt{x+2}(1-\sqrt{x+2})}=-\dfrac{x+1}{\sqrt{x+2}(1-\sqrt{x+2})}.$

101. True. The two real roots are $\dfrac{-b\pm\sqrt{b^2-4ac}}{2a}.$

103. False. Take $a=2$, $b=3$, and $c=4$. Then

$\dfrac{a}{b+c}=\dfrac{2}{3+4}=\dfrac{2}{7}.$ But $\dfrac{a}{b}+\dfrac{a}{c}=\dfrac{2}{3}+\dfrac{3}{4}=\dfrac{8+9}{12}=\dfrac{17}{12}.$

1.3 Problem Solving Tips

Suppose you are asked to determine whether a given statement is true or false, and you are also asked to explain your answer. How would you answer the question?

If you think the statement is true, then prove it. On the other hand, if you think the statement is false, then give an example that disproves the statement. For example, the statement "If a and b are real numbers, then $a - b = b - a$" is false and an example that disproves it may be constructed by taking $a = 3$ and $b = 5$. For these values of a and b, we find $a - b = 3 - 5 = -2$ but $b - a = 5 - 3 = 2$ and this shows that $a - b \neq b - a$. Such an example is called a **counterexample**.

CONCEPT QUESTIONS, EXERCISES 1.3, page 28

1. a. $a < 0$ and $b > 0$;   b. $a < 0$ and $b < 0$ c. $a > 0$ and $b < 0$

EXERCISES 1.3, page 28

1. The coordinates of A are (3,3) and it is located in Quadrant I.

3. The coordinates of C are (2,-2) and it is located in Quadrant IV.

5. The coordinates of E are (-4,-6) and it is located in Quadrant III.

7. A   9. E, F, and G.   11. F

For Exercises 13-19, refer to the following figure.

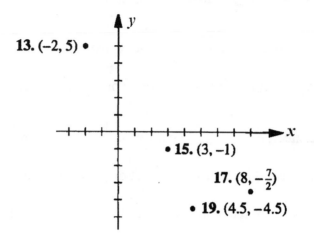

13. (−2, 5) •

• 15. (3, −1)

17. $(8, -\frac{7}{2})$

• 19. (4.5, −4.5)

21. Using the distance formula, we find that $\sqrt{(4-1)^2+(7-3)^2} = \sqrt{3^2+4^2} = \sqrt{25} = 5$.

23. Using the distance formula, we find that
$$\sqrt{(4-(-1))^2+(9-3)^2} = \sqrt{5^2+6^2} = \sqrt{25+36} = \sqrt{61}.$$

25. The coordinates of the points have the form $(x, -6)$. Since the points are 10 units away from the origin, we have
$$(x-0)^2+(-6-0)^2 = 10^2$$
$$x^2 = 64,$$
or $x = \pm 8$. Therefore, the required points are $(-8, -6)$ and $(8, -6)$.

27. The points are shown in the diagram that follows.

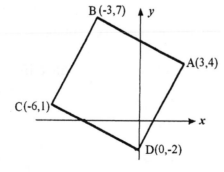

To show that the four sides are equal, we compute the following:

$$d(A,B) = \sqrt{(-3-3)^2 + (7-4)^2} = \sqrt{(-6)^2 + 3^2} = \sqrt{45}$$

$$d(B,C) = \sqrt{[(-6-(-3)]^2 + (1-7)^2} = \sqrt{(-3)^2 + (-6)^2} = \sqrt{45}$$

$$d(C,D) = \sqrt{[0-(-6)]^2 + [(-2)-1]^2} = \sqrt{(6)^2 + (-3)^2} = \sqrt{45}$$

$$d(A,D) = \sqrt{(0-3)^2 + (-2-4)^2} = \sqrt{(3)^2 + (-6)^2} = \sqrt{45}.$$

Next, to show that $\triangle ABC$ is a right triangle, we show that it satisfies the Pythagorean Theorem. Thus,

$$d(A,C) = \sqrt{(-6-3)^2 + (1-4)^2} = \sqrt{(-9)^2 + (-3)^2} = \sqrt{90} = 3\sqrt{10}$$

and $[d(A,B)]^2 + [d(B,C)]^2 = 90 = [d(A,C)]^2$. Similarly, $d(B,D) = \sqrt{90} = 3\sqrt{10}$, so $\triangle BAD$ is a right triangle as well. It follows that $\angle B$ and $\angle D$ are right angles, and we conclude that $ADCB$ is a square

29. The equation of the circle with radius 5 and center (2,-3) is given by
$$(x-2)^2 + [y-(-3)]^2 = 5^2, \text{ or } (x-2)^2 + (y+3)^2 = 25.$$

31. The equation of the circle with radius 5 and center (0, 0) is given by
$$(x-0)^2 + (y-0)^2 = 5^2, \text{ or } x^2 + y^2 = 25$$

33. The distance between the points (5,2) and (2,-3) is given by
$$d = \sqrt{(5-2)^2 + (2-(-3))^2} = \sqrt{3^2 + 5^2} = \sqrt{34}.$$

Therefore $r = \sqrt{34}$ and the equation of the circle passing through (5,2) and (2,-3) is
$$(x-2)^2 + [y-(-3)]^2 = 34, \text{ or } (x-2)^2 + (y+3)^2 = 34.$$

35. Referring to the diagram on page 29 of the text, we see that the distance from A to B is given by $d(A,B) = \sqrt{400^2 + 300^2} = \sqrt{250,000} = 500$. The distance from B to C is given by
$$d(B,C) = \sqrt{(-800-400)^2 + (800-300)^2} = \sqrt{(-1200)^2 + (500)^2}$$
$$= \sqrt{1,690,000} = 1300.$$

The distance from C to D is given by
$$d(C,D) = \sqrt{[-800-(-800)]^2 + (800-0)^2} = \sqrt{0+800^2} = 800.$$

The distance from D to A is given by

$$d(D, A) = \sqrt{[(-800) - 0]^2 + (0 - 0)} = \sqrt{640000} = 800.$$

Therefore, the total distance covered on the tour, is

$$d(A, B) + d(B, C) + d(C, D) + d(D, A) = 500 + 1300 + 800 + 800$$
$$= 3400, \quad \text{or 3400 miles.}$$

37. Referring to the following diagram,

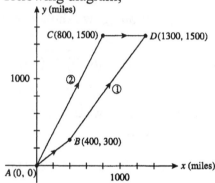

we see that the distance he would cover if he took Route (1) is given by

$$d(A, B) + d(B, D) = \sqrt{400^2 + 300^2} + \sqrt{(1300 - 400)^2 + (1500 - 300)^2}$$
$$= \sqrt{250,000} + \sqrt{2,250,000} = 500 + 1500 = 2000,$$

or 2000 miles. On the other hand, the distance he would cover if he took Route (2) is given by

$$d(A, C) + d(C, D) = \sqrt{800^2 + 1500^2} + \sqrt{(1300 - 800)^2}$$
$$= \sqrt{2,890,000} + \sqrt{250,000} = 1700 + 500 = 2200,$$

or 2200 miles. Comparing these results, we see that he should take Route (1).

39. Calculations to determine VHF requirements:

$$d = \sqrt{25^2 + 35^2} = \sqrt{625 + 1225} = \sqrt{1850} \approx 43.01.$$

Models B through D satisfy this requirement.

Calculations to determine UHF requirements:

$$d = \sqrt{20^2 + 32^2} = \sqrt{400 + 1024} = \sqrt{1424} = 37.74$$

Models C through D satisfy this requirement. Therefore, Model C will allow him to receive both channels at the least cost.

41. a. Let the position of ship A and ship B after t hours be $A(0, y)$ and $B(x, 0)$, respectively. Then $x = 30t$ and $y = 20t$. Therefore, the distance between the two

ships is $\qquad D = \sqrt{(30t)^2 + (20t)^2} = \sqrt{900t^2 + 400t^2} = 10\sqrt{13}t.$

b. The required distance is obtained by letting $t = 2$ giving $D = 10\sqrt{13}(2)$ or approximately 72.11 miles.

43. True. Plot the points.

45. False. The distance between $P_1(a,b)$ and $P_3(kc, kd)$ is

$d = \sqrt{(kc - a)^2 + (kd - b)^2}$

$\neq |k| D = |k| \sqrt{(c - a)^2 + (d - b)^2} = \sqrt{k^2(c - a)^2 + k^2(d - b)^2} = \sqrt{[k(c - a)]^2 + [k(d - b)]^2}$

47. Referring to the figure in the text, we see that the distance between the two points is given by the length of the hypotenuse of the right triangle. That is,

$$d = \sqrt{(x_2 - x_1)^2 + (y_2 - y_1)^2}$$

1.4 Problem Solving Tips

When you solve a problem in the exercises that follow each section, first read the problem. Then, before you start computing or writing out a solution, try to formulate a strategy for solving the problem. Then proceed by using your strategy to solve the problem.

Here we summarize some general problem-solving techniques that have been covered in this section.

1. **To show that two lines are parallel,** you need to show that the slopes of the two lines are equal or their slopes are undefined.

---

1 Preliminaries

2. To show that two lines L_1 and L_2 are perpendicular, you need to show that the slope m_1 of L_1 is the negative reciprocal of the slope m_2 of L_2; that is, $m_1 = -1/m_2$.

3. To find the equation of a line, you need the slope of the line and a point lying on the line. You can then find the equation of the line by using the point-slope form of the equation of a line: $(y - y_1) = m(x - x_1)$

1.4 CONCEPT QUESTIONS, page 40

1. The slope is $m = \dfrac{y_2 - y_1}{x_2 - x_1}$, where $P(x_1, y_1)$ and $P(x_2, y_2)$ are any two distinct points on the nonvertical line. The slope of a vertical line is undefined.

3. a. $m_1 = m_2$ b. $m_2 = -\dfrac{1}{m_1}$

EXERCISES 1.4, page 40

1. e   3. a   5. f

7. Referring to the figure shown in the text, we see that $m = \dfrac{2-0}{0-(-4)} = \dfrac{1}{2}$.

9. This is a vertical line, and hence its slope is undefined.

11. $m = \dfrac{y_2 - y_1}{x_2 - x_1} = \dfrac{8-3}{5-4} = 5.$

13. $m = \dfrac{y_2 - y_1}{x_2 - x_1} = \dfrac{8-3}{4-(-2)} = \dfrac{5}{6}.$

15. $m = \dfrac{y_2 - y_1}{x_2 - x_1} = \dfrac{d-b}{c-a}$ $(a \neq c).$

17. Since the equation is in the slope-intercept form, we read off the slope $m = 4$.
    a. If x increases by 1 unit, then y increases by 4 units.
    b. If x decreases by 2 units, y decreases by $4(-2) = -8$ units.

19. The slope of the line through A and B is $\dfrac{-10-(-2)}{-3-1}=\dfrac{-8}{-4}=2$.

The slope of the line through C and D is $\dfrac{1-5}{-1-1}=\dfrac{-4}{-2}=2$.

Since the slopes of these two lines are equal, the lines are parallel.

21. The slope of the line through A and B is $\dfrac{2-5}{4-(-2)}=-\dfrac{3}{6}=-\dfrac{1}{2}$.

The slope of the line through C and D is $\dfrac{6-(-2)}{3-(-1)}=\dfrac{8}{4}=2$. Since the slopes of these

two lines are the negative reciprocals of each other, the lines are perpendicular.

23. The slope of the line through the point $(1, a)$ and $(4,-2)$ is $m_1=\dfrac{-2-a}{4-1}$ and the

slope of the line through $(2,8)$ and $(-7, a+4)$ is $m_2=\dfrac{a+4-8}{-7-2}$. Since these two

lines are parallel, m_1 is equal to m_2. Therefore,

$$\frac{-2-a}{3}=\frac{a-4}{-9}$$
$$-9(-2-a)=3(a-4)$$
$$18+9a=3a-12$$
$$6a=-30 \quad \text{and} \quad a=-5.$$

25. An equation of a horizontal line is of the form $y=b$. In this case $b=-3$, so $y=-3$ is an equation of the line.

27. We use the point-slope form of an equation of a line with the point $(3,-4)$ and slope $m=2$. Thus $\quad y-y_1=m(x-x_1)$,

and
$$y-(-4)=2(x-3)$$
$$y+4=2x-6$$
$$y=2x-10.$$

29. Since the slope $m=0$, we know that the line is a horizontal line of the form $y=b$. Since the line passes through $(-3,2)$, we see that $b=2$, and an equation of the line is $y=2$.

31. We first compute the slope of the line joining the points (2,4) and (3,7). Thus,

$$m = \frac{7-4}{3-2} = 3.$$

Using the point-slope form of an equation of a line with the point (2,4) and slope $m = 3$, we find

$$y - 4 = 3(x - 2)$$
$$y = 3x - 2.$$

33. We first compute the slope of the line joining the points (1,2) and (–3,–2). Thus,

$$m = \frac{-2-2}{-3-1} = \frac{-4}{-4} = 1.$$

Using the point-slope form of an equation of a line with the point (1,2) and slope $m = 1$, we find

$$y - 2 = x - 1$$
$$y = x + 1.$$

35. We use the slope-intercept form of an equation of a line: $y = mx + b$. Since $m = 3$, and $b = 4$, the equation is $y = 3x + 4$.

37. We use the slope-intercept form of an equation of a line: $y = mx + b$. Since $m = 0$, and $b = 5$, the equation is $y = 5$.

39. We first write the given equation in the slope-intercept form:

$$x - 2y = 0$$
$$-2y = -x$$
$$y = \tfrac{1}{2}x \ .$$

From this equation, we see that $m = 1/2$ and $b = 0$.

41. We write the equation in slope-intercept form:

$$2x - 3y - 9 = 0$$
$$-3y = -2x + 9$$
$$y = \tfrac{2}{3}x - 3.$$

From this equation, we see that $m = 2/3$ and $b = -3$.

43. We write the equation in slope-intercept form:

$$2x + 4y = 14$$
$$4y = -2x + 14$$
$$y = -\tfrac{2}{4}x + \tfrac{14}{4} = -\tfrac{1}{2}x + \tfrac{7}{2}.$$

From this equation, we see that $m = -1/2$ and $b = 7/2$.

45. We first write the equation $2x - 4y - 8 = 0$ in slope- intercept form:
$$2x - 4y - 8 = 0$$
$$4y = 2x - 8$$
$$y = \tfrac{1}{2}x - 2$$

Now the required line is parallel to this line, and hence has the same slope. Using the point-slope form of an equation of a line with $m = 1/2$ and the point $(-2, 2)$, we have
$$y - 2 = \tfrac{1}{2}[x - (-2)]$$
$$y = \tfrac{1}{2}x + 3.$$

47. A line parallel to the x-axis has slope 0 and is of the form $y = b$. Since the line is 6 units below the axis, it passes through $(0, -6)$ and its equation is $y = -6$.

49. We use the point-slope form of an equation of a line to obtain
$$y - b = 0(x - a) \quad \text{or} \quad y = b.$$

51. Since the required line is parallel to the line joining $(-3, 2)$ and $(6, 8)$, it has slope
$$m = \frac{8 - 2}{6 - (-3)} = \frac{6}{9} = \frac{2}{3}.$$

We also know that the required line passes through $(-5, -4)$. Using the point-slope form of an equation of a line, we find
$$y - (-4) = \tfrac{2}{3}(x - (-5))$$

or $\qquad y = \tfrac{2}{3}x + \tfrac{10}{3} - 4;$ that is $\qquad y = \tfrac{2}{3}x - \tfrac{2}{3}$.

53. Since the point $(-3, 5)$ lies on the line $kx + 3y + 9 = 0$, it satisfies the equation. Substituting $x = -3$ and $y = 5$ into the equation gives
$$-3k + 15 + 9 = 0, \text{ or } k = 8.$$

55. $3x - 2y + 6 = 0$

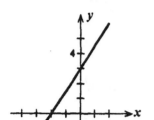

57. $x + 2y - 4 = 0$

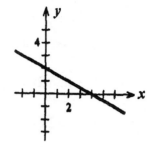

59. $y + 5 = 0$

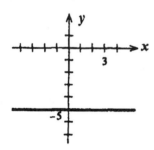

61. Since the line passes through the points $(a, 0)$ and $(0, b)$, its slope is

$m = \dfrac{b - 0}{0 - a} = -\dfrac{b}{a}$. Then, using the point-slope form of an equation of a line with the

point $(a, 0)$ we have

$$y - 0 = -\tfrac{b}{a}(x - a)$$

$$y = -\tfrac{b}{a}x + b$$

which may be written in the form $\tfrac{b}{a}x + y = b$.

Multiplying this last equation by $1/b$, we have $\dfrac{x}{a} + \dfrac{y}{b} = 1$.

63. Using the equation $\dfrac{x}{a} + \dfrac{y}{b} = 1$ with $a = -2$ and $b = -4$, we have $-\dfrac{x}{2} - \dfrac{y}{4} = 1$.

Then

$$-4x - 2y = 8$$
$$2y = -8 - 4x$$
$$y = -2x - 4.$$

65. Using the equation $\dfrac{x}{a} + \dfrac{y}{b} = 1$ with $a = 4$ and $b = -1/2$, we have

$$\dfrac{x}{4} + \dfrac{y}{-\frac{1}{2}} = 1$$
$$-\tfrac{1}{4}x + 2y = -1$$
$$2y = \tfrac{1}{4}x - 1$$
$$y = \tfrac{1}{8}x - \tfrac{1}{2}.$$

67. The slope of the line passing through A and B is $m = \dfrac{7-1}{1-(-2)} = \dfrac{6}{3} = 2$,

and the slope of the line passing through B and C is $m = \dfrac{13-7}{4-1} = \dfrac{6}{3} = 2$.

Since the slopes are equal, the points lie on the same line.

69. a.

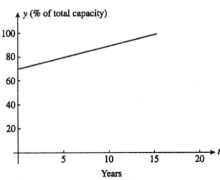

b. The slope is 1.9467 and the y-intercept is 70.082.

c. The output is increasing at the rate of 1.9467%/yr; the output at the beginning of 1990 was 70.082%.

d. We solve the equation $1.9467t + 70.082 = 100$ giving $t = 15.37$. We conclude that the plants will be generating at maximum capacity shortly after 2005.

71. a. $y = 0.55x$

b. Solving the equation $1100 = 0.55x$ for x, we have $x = \dfrac{1100}{0.55} = 2000$.

73. Using the points (0, 0.68) and (10, 0.80), we see that the slope of the required line

is $\qquad m = \dfrac{0.80 - 0.68}{10 - 0} = \dfrac{0.12}{10} = .012.$

Next, using the point-slope form of the equation of a line, we have
$$y - 0.68 = 0.012(t - 0)$$
or $\qquad\qquad y = 0.012t + 0.68.$

Therefore, when $t = 14$, we have $y = 0.012(14) + 0.68 = .848$, or 84.8%. That is, in 2004 women's wages are expected to be 84.8% of men's wages.

75. a. – b.

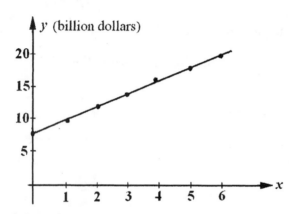

c. $m = \dfrac{18.8 - 7.9}{6 - 0} \approx 1.82$, $y - 7.9 = 1.82(x - 0)$, or $y = 1.82x + 7.9$.

d. $y = 1.82(5) + 7.9 \approx 17$ or $17 billion; This agrees with the actual data for that year.

77. True. The slope of the line is given by $-\dfrac{2}{4} = -\dfrac{1}{2}$.

79. False. Let the slope of L_1 be $m_1 > 0$. Then the slope of L_2 is $m_2 = -\dfrac{1}{m_1} < 0$.

81. True. Set $y = 0$ and we have $Ax + C = 0$ or $x = -C/A$ and this is where the line cuts the x-axis.

83. Writing each equation in the slope-intercept form, we have

$$y = -\frac{a_1}{b_1}x - \frac{c_1}{b_1} \quad (b_1 \neq 0) \quad \text{and} \quad y = -\frac{a_2}{b_2}x - \frac{c_2}{b_2} \quad (b_2 \neq 0)$$

Since two lines are parallel if and only if their slopes are equal, we see that the lines

are parallel if and only if $-\dfrac{a_1}{b_1} = -\dfrac{a_2}{b_2}$, or $a_1b_2 - b_1a_2 = 0$.

CHAPTER 1, CONCEPT REVIEW, page 46

1. ordered; abscissa (x-coordinate); ordinate (y-coordinate)

3. $\sqrt{(c-a)^2 + (d-b)^2}$

5. a. $\dfrac{y_2 - y_1}{x_2 - x_1}$ b. Undefined   c. 0   d. Positive

7. a. $y - y_1 = m(x - x_1)$;   point-slope form   b. $y = mx + b$; slope-intercept

CHAPTER 1, REVIEW EXERCISES, page 47

1. Adding x to both sides yields $3 \leq 3x + 9$ or $3x \geq -6$, and $x \geq -2$.
   We conclude that the solution set is $[-2, \infty)$.

3. The inequalities imply $x > 5$ or $x < -4$. So the solution set is
   $(-\infty, -4) \cup (5, \infty)$.

5. $|-5 + 7| + |-2| = |2| + |-2| = 2 + 2 = 4$.

7. $|2\pi - 6| - \pi = 2\pi - 6 - \pi = \pi - 6$.

9. $\left(\dfrac{9}{4}\right)^{3/2} = \dfrac{9^{3/2}}{4^{3/2}} = \dfrac{27}{8}$.

11. $(3 \cdot 4)^{-2} = 12^{-2} = \dfrac{1}{12^2} = \dfrac{1}{144}$.

13. $\dfrac{(3\cdot 2^{-3})(4\cdot 3^{5})}{2\cdot 9^{3}} = \dfrac{3\cdot 2^{-3}\cdot 2^{2}\cdot 3^{5}}{2\cdot (3^{2})^{3}} = \dfrac{2^{-1}\cdot 3^{6}}{2\cdot 3^{6}} = \dfrac{1}{4}.$

15. $\dfrac{4(x^{2}+y)^{3}}{x^{2}+y} = 4(x^{2}+y)^{2}.$

17. $\dfrac{\sqrt[4]{16x^{5}yz}}{\sqrt[4]{81xyz^{5}}} = \dfrac{(2^{4}x^{5}yz)^{1/4}}{(3^{4}xyz^{5})^{1/4}} = \dfrac{2x^{5/4}y^{1/4}z^{1/4}}{3x^{1/4}y^{1/4}z^{5/4}} = \dfrac{2x}{3z}.$

19. $\left(\dfrac{3xy^{2}}{4x^{3}y}\right)^{-2}\left(\dfrac{3xy^{3}}{2x^{2}}\right)^{3} = \left(\dfrac{3y}{4x^{2}}\right)^{-2}\left(\dfrac{3y^{3}}{2x}\right)^{3} = \left(\dfrac{4x^{2}}{3y}\right)^{2}\left(\dfrac{3y^{3}}{2x}\right)^{3} = \dfrac{(16x^{4})(27y^{9})}{(9y^{2})(8x^{3})} = 6xy^{7}.$

21. $-2\pi^{2}r^{3} + 100\pi r^{2} = -2\pi r^{2}(\pi r - 50).$

23. $16 - x^{2} = 4^{2} - x^{2} = (4 - x)(4 + x).$

25. $8x^{2} + 2x - 3 = (4x + 3)(2x - 1) = 0$ and $x = -3/4$ and $x = 1/2$ are the roots of the equation.

27. $-x^{3} - 2x^{2} + 3x = -x(x^{2} + 2x - 3) = -x(x + 3)(x - 1) = 0$ and the roots of the equation are $x = 0$, $x = -3$, and $x = 1$.

29. Here we use the quadratic formula to solve the equation $x^{2} - 2x - 5 = 0$. Then $a = 1$, $b = -2$, and $c = -5$. Thus,

$$x = \dfrac{-b\pm\sqrt{b^{2}-4ac}}{2a} = \dfrac{-(-2)\pm\sqrt{(-2)^{2}-4(1)(-5)}}{2(1)} = \dfrac{2\pm\sqrt{24}}{2} = 1\pm\sqrt{6}.$$

31. $\dfrac{(t+6)(60)-(60t+180)}{(t+6)^{2}} = \dfrac{60t+360-60t-180}{(t+6)^{2}} = \dfrac{180}{(t+6)^{2}}.$

33. $\dfrac{2}{3}\left(\dfrac{4x}{2x^{2}-1}\right)+3\left(\dfrac{3}{3x-1}\right) = \dfrac{8x}{3(2x^{2}-1)}+\dfrac{9}{3x-1} = \dfrac{8x(3x-1)+27(2x^{2}-1)}{3(2x^{2}-1)(3x-1)}$

$$= \frac{78x^2 - 8x - 27}{3(2x^2 - 1)(3x - 1)}.$$

35. $$\frac{\sqrt{x}-1}{x-1} = \frac{\sqrt{x}-1}{x-1} \cdot \frac{\sqrt{x}+1}{\sqrt{x}+1}$$

$$= \frac{(\sqrt{x})^2 - 1}{(x-1)(\sqrt{x}+1)} = \frac{x-1}{(x-1)(\sqrt{x}+1)} = \frac{1}{\sqrt{x}+1}.$$

37. The distance is

$$d = \sqrt{[1-(-2)]^2 + [-7-(-3)]^2} = \sqrt{3^2 + (-4)^2} = \sqrt{9+16} = \sqrt{25} = 5.$$

39. An equation is $x = -2$.

41. The slope of L is $m = \dfrac{\frac{7}{2}-4}{3-(-2)} = -\dfrac{1}{10}$ and an equation of L is

$y - 4 = -\frac{1}{10}[x - (-2)] = -\frac{1}{10}x - \frac{1}{5}$, or $y = -\frac{1}{10}x + \frac{19}{5}$. The general form of this equation is $x + 10y - 38 = 0$.

43. Writing the given equation in the form $y = \frac{5}{2}x - 3$, we see that the slope of the given line is 5/2. So a required equation is $y - 4 = \frac{5}{2}(x+2)$ or $y = \frac{5}{2}x + 9$
The general form of this equation is $5x - 2y + 18 = 0$.

45. Rewriting the given equation in the slope-intercept form, we have $4y = -3x + 8$ or $y = -\frac{3}{4}x + 2$ and conclude that the slope of the required line is $-3/4$. Using the point-slope form of the equation of a line with the point $(2,3)$ and slope $-3/4$, we obtain $y - 3 = -\frac{3}{4}(x-2)$, and so $y = -\frac{3}{4}x + \frac{6}{4} + 3 = -\frac{3}{4}x + \frac{9}{2}$. The general form of this equation is $3x + 4y - 18 = 0$.

47. The slope of the line passing through $(-2,-4)$ and $(1,5)$ is $m = \dfrac{5-(-4)}{1-(-2)} = \dfrac{9}{3} = 3$. So the required line is $y - (-2) = 3[x - (-3)]$ and $y + 2 = 3x + 9$, or $y = 3x + 7$.

49. Setting $x = 0$ gives $y = -6$ as the y-intercept. Setting $y = 0$ gives $x = 8$ as the

x-intercept. The graph of the equation $3x - 4y = 24$ follows:

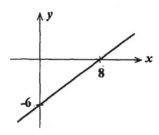

51. $2(1.5C + 80) \le 2(2.5C - 20) \Rightarrow 1.5C + 80 \le 2.5C - 20$, so $C \ge 100$ and the minimum cost is \$100.

53. a.

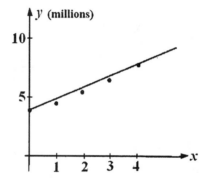

c. $P_1(0, 3.9)$ and $P_2(4, 7.8)$; $m = \dfrac{7.8 - 3.9}{4 - 0} = \dfrac{3.9}{4} = 0.975$

$y - 3.9 = 0.975(x - 0)$ or $y = 0.975x + 3.9$

d. If $x = 3$, then $y = 0.975(3) + 3.9 = 6.825$. So, the number of systems installed in 2005 ($x = 3$) is 6,825,000 which is close to the projection of 6.8 million.

CHAPTER 1, Before Moving On, page 48

1. a. $\left| \pi - 2\sqrt{3} \right| - \left| \sqrt{3} - \sqrt{2} \right| = -(\pi - 2\sqrt{3}) - (\sqrt{3} - \sqrt{2}) = \sqrt{3} + \sqrt{2} - \pi$.

   b. $\left[\left(-\tfrac{1}{3} \right)^{-3} \right]^{1/3} = \left(-\tfrac{1}{3} \right)^{(-3)(\tfrac{1}{3})} = \left(-\tfrac{1}{3} \right)^{-1} = -3$.

2. a. $\sqrt[3]{64x^6} \cdot \sqrt{9y^2x^6} = (4x^2)(3yx^3) = 12x^5y$

b. $\left(\dfrac{a^{-3}}{b^{-4}}\right)^2 \left(\dfrac{b}{a}\right)^{-3} = \dfrac{a^{-6}}{b^{-8}} \cdot \dfrac{b^{-3}}{a^{-3}} = \dfrac{b^8}{a^6} \cdot \dfrac{a^3}{b^3} = \dfrac{b^5}{a^3}$

3. a. $\dfrac{2x}{3\sqrt{y}} \cdot \dfrac{\sqrt{y}}{\sqrt{y}} = \dfrac{2x\sqrt{y}}{3y}$

b. $-\dfrac{3x}{\sqrt{x+2}} + 3\sqrt{x+2} = \dfrac{x}{\sqrt{x}-4} \cdot \dfrac{\sqrt{x}+4}{\sqrt{x}+4} = \dfrac{x(\sqrt{x}+4)}{x-16}$

4. a. $\dfrac{(x^2+1)(\frac{1}{2}x^{-1/2}) - x^{1/2}(2x)}{(x^2+1)^2} = \dfrac{\frac{1}{2}x^{-1/2}[(x^2+1)-4x^2]}{(x^2+1)^2} = \dfrac{1-3x^2}{2x^{1/2}(x^2+1)^2}$

b. $-\dfrac{3x+3(x+2)}{\sqrt{x+2}} = \dfrac{6}{\sqrt{x+2}} = \dfrac{6\sqrt{x+2}}{x+2}$

5. $\dfrac{\sqrt{x}+\sqrt{y}}{\sqrt{x}-\sqrt{y}} = \dfrac{\sqrt{x}+\sqrt{y}}{\sqrt{x}-\sqrt{y}} \cdot \dfrac{\sqrt{x}-\sqrt{y}}{\sqrt{x}-\sqrt{y}} = \dfrac{x-y}{(\sqrt{x}-\sqrt{y})^2}$

6. a. $12x^3 - 10x^2 - 12x = 2x(6x^2 - 5x - 6) = 2x(2x-3)(3x+2)$

b. $2bx - 2by + 3cx - 3cy = 2b(x-y) + 3c(x-y) = (2b+3c)(x-y)$

7. a. $12x^2 - 9x - 3 = 0$; $3(4x^2 - 3x - 1) = 0$; $3(4x+1)(x-1) = 0$ so $x = -\frac{1}{4}$ or $x = 1$.

b. $3x^2 - 5x + 1 = 0$. Using the quadratic equations, with $a = 3$, $b = -5$, and $c = 1$, we

have $x = \dfrac{-(-5) \pm \sqrt{25-12}}{2(3)} = \dfrac{5 \pm \sqrt{13}}{6}$

8. $d = \sqrt{[6-(-2)]^2 + (8-4)^2} = \sqrt{64+16} = \sqrt{80} = 4\sqrt{5}$

9. $m = \dfrac{5-(-2)}{4-(-1)} = \dfrac{7}{5}$; $y - (-2) = \dfrac{7}{5}(x-(-1))$;

29

$$y + 2 = \frac{7}{5}x + \frac{7}{5} \quad \text{or} \quad y = \frac{7}{5}x - \frac{3}{5}$$

10. $m = -\dfrac{1}{3}, \quad b = \dfrac{4}{3}; \quad y = -\dfrac{1}{3}x + \dfrac{4}{3}$

CHAPTER 2

2.1 Problem Solving Tips

New mathematical terms in each section appear in blue bold-faced type along with their definition or they are boxed (the green boxes). Each time you encounter a new term, read through the definition and then try to express the definition in your own words without looking at the book. Once you understand these definitions, it will be easier for you to work the exercise sets that follow each section.

Here are some hints for solving the problems in the exercises that follow:

1. **To find the domain of a function** $f(x)$, find all values of x for which $f(x)$ is a real number.

    a. *If the function involves a quotient*, check to see if there are any values of x at which the denominator is equal to zero. (Remember, division by zero is not allowed.) Then exclude those points from the domain.

    b. *If the function involves the root of a real number*, check to see if the root is an even or an odd root. If n is even, the nth root of a negative number is not defined and, consequently, those values of x yielding the nth root of a negative number must be excluded from the domain of f. For example, $\sqrt{x-1}$ is only defined for

$x \geq 1$, so the domain of $f(x) = \sqrt{x-1}$ is $[1, \infty)$.

2. **To evaluate a piecewise-defined function** $f(x)$ at a specific value of x, check to see which subdomain x lies in. Then evaluate the function using the rule for that subdomain.

3. **To determine whether a curve is the graph of a function**, use the vertical-line test. If you can draw a vertical line through the curve that intersects the curve in more than one point, then the curve is *not* a function.

CONCEPT QUESTIONS, EXERCISES 2.1, page 58

1. a. A function is a rule that associates with each element in a set A exactly one element in a set B.
    b. The domain of a function f is the set of all elements x in the set such that $f(x)$ is an element in B. The range of f is the set of all elements $f(x)$ whenever x is an element in its domain.
    c. An independent variable is a variable in the domain of a function f. The dependent variable is $y = f(x)$.

EXERCISES 2.1, page 58

1. $f(x) = 5x + 6$. Therefore $f(3) = 5(3) + 6 = 21$; $f(-3) = 5(-3) + 6 = -9$;
    $f(a) = 5(a) + 6 = 5a + 6$; $f(-a) = 5(-a) + 6 = -5a + 6$; and
    $f(a + 3) = 5(a + 3) + 6 = 5a + 15 + 6 = 5a + 21$.

3. $g(x) = 3x^2 - 6x - 3$; $g(0) = 3(0) - 6(0) - 3 = -3$;
    $g(-1) = 3(-1)^2 - 6(-1) - 3 = 3 + 6 - 3 = 6$; $g(a) = 3(a)^2 - 6(a) - 3 = 3a^2 - 6a - 3$

$$g(x+1) = 3(x+1)^2 - 6(x+1) - 3 = 3(x^2 + 2x + 1) - 6x - 6 - 3$$
$$= 3x^2 + 6x + 3 - 6x - 9 = 3x^2 - 6.$$

5. $f(x) = 2x + 5$; $f(a+h) = 2(a+h) + 5 = 2a + 2h + 5$. $f(-a) = 2(-a) + 5 = -2a + 5$
$f(a^2) = 2(a^2) + 5 = 2a^2 + 5$; $f(a - 2h) = 2(a - 2h) + 5 = 2a - 4h + 5$
$f(2a - h) = 2(2a - h) + 5 = 4a - 2h + 5$

7. $s(t) = \dfrac{2t}{t^2 - 1}$. Therefore, $s(4) = \dfrac{2(4)}{(4)^2 - 1} = \dfrac{8}{15}$. $s(0) = \dfrac{2(0)}{0^2 - 1} = 0$

$s(a) = \dfrac{2(a)}{a^2 - 1} = \dfrac{2a}{a^2 - 1}$; $s(2+a) = \dfrac{2(2+a)}{(2+a)^2 - 1} = \dfrac{2(2+a)}{a^2 + 4a + 4 - 1} = \dfrac{2(2+a)}{a^2 + 4a + 3}$

$s(t+1) = \dfrac{2(t+1)}{(t+1)^2 - 1} = \dfrac{2(t+1)}{t^2 + 2t + 1 - 1} = \dfrac{2(t+1)}{t(t+2)}$.

9. $f(t) = \dfrac{2t^2}{\sqrt{t-1}}$. Therefore, $f(2) = \dfrac{2(2^2)}{\sqrt{2}-1} = 8$; $f(a) = \dfrac{2a^2}{\sqrt{a}-1}$;

$f(x+1) = \dfrac{2(x+1)^2}{\sqrt{(x+1)-1}} = \dfrac{2(x+1)^2}{\sqrt{x}}$; $f(x-1) = \dfrac{2(x-1)^2}{\sqrt{(x-1)-1}} = \dfrac{2(x-1)^2}{\sqrt{x-2}}$.

11. Since $x = -2 \le 0$, we see that $f(-2) = (-2)^2 + 1 = 4 + 1 = 5$. Since $x = 0 \le 0$, we see that $f(0) = (0)^2 + 1 = 1$. Since $x = 1 > 0$, we see that $f(1) = \sqrt{1} = 1$.

13. Since $x = -1 < 1$, $f(-1) = -\frac{1}{2}(-1)^2 + 3 = \frac{5}{2}$. Since $x = 0 < 1$,

$f(0) = -\frac{1}{2}(0)^2 + 3 = 3$. Since $x = 1 \ge 1$, $f(1) = 2(1^2) + 1 = 3$.

Since $x = 2 \ge 1, f(2) = 2(2^2) + 1 = 9$.

15. a. $f(0) = -2$ b. (i) $f(x) = 3$ when $x \approx 2$ (ii) $f(x) = 0$ when $x = 1$
    c. $[0,6]$ d. $[-2, 6]$

17. $g(2) = \sqrt{2^2 - 1} = \sqrt{3}$ and the point $(2, \sqrt{3})$ lies on the graph of g.

19. $f(-2) = \dfrac{|-2-1|}{-2+1} = \dfrac{|-3|}{-1} = -3$ and the point $(-2,-3)$ does lie on the graph of f.

21. Since $f(x)$ is a real number for any value of x, the domain of f is $(-\infty, \infty)$.

23. $f(x)$ is not defined at $x = 0$ and so the domain of f is $(-\infty,0) \cup (0,\infty)$.

25. $f(x)$ is a real number for all values of x. Note that $x^2 + 1 \geq 1$ for all x. Therefore, the domain of f is $(-\infty, \infty)$.

27. Since the square root of a number is defined for all real numbers greater than or equal to zero, we have $5 - x \geq 0$, or $-x \geq -5$ and so $x \leq 5$. (Recall that multiplying by -1 reverses the sign of an inequality.) Therefore, the domain of g is $(-\infty,5]$.

29. The denominator of f is zero when $x^2 - 1 = 0$ or $x = \pm 1$. Therefore, the domain of f is $(-\infty,-1) \cup (-1, 1) \cup (1,\infty)$.

31. f is defined when $x + 3 \geq 0$, that is, when $x \geq -3$. Therefore, the domain of f is $[-3,\infty)$.

33. The numerator is defined when $1 - x \geq 0$, $-x \geq -1$ or $x \leq 1$. Furthermore, the denominator is zero when $x = \pm 2$. Therefore, the domain is the set of all real numbers in $(-\infty,-2) \cup (-2, 1]$.

35. a. The domain of f is the set of all real numbers.
    b. $f(x) = x^2 - x - 6$. Therefore,
    $f(-3) = (-3)^2 - (-3) - 6 = 9 + 3 - 6 = 6$; $f(-2) = (-2)^2 - (-2) - 6 = 4 + 2 - 6 = 0$.

$f(-1) = (-1)^2 - (-1) - 6 = 1 + 1 - 6 = -4; \quad f(0) = (0)^2 - (0) - 6 = -6.$

$f\left(\frac{1}{2}\right) = \left(\frac{1}{2}\right)^2 - \left(\frac{1}{2}\right) - 6 = \frac{1}{4} - \frac{2}{4} - \frac{24}{4} = -\frac{25}{4}; \quad f(1) = (1)^2 - 1 - 6 = -6.$

$f(2) = (2)^2 - 2 - 6 = 4 - 2 - 6 = -4; \quad f(3) = (3)^2 - 3 - 6 = 9 - 3 - 6 = 0.$

c.

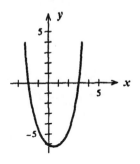

37. $f(x) = 2x^2 + 1$;

x	-3	-2	-1	0	1	2	3
$f(x)$	19	9	3	1	3	9	19

$(-\infty, \infty); \quad [1, \infty)$

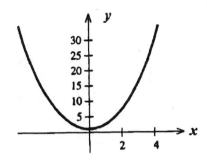

39. $f(x) = 2 + \sqrt{x}$ $[0, \infty); [2, \infty)$

x	0	1	2	4	9	16
$f(x)$	2	3	3.41	4	5	6

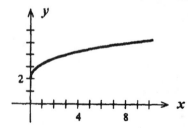

41. $f(x) = \sqrt{1-x}$

x	0	-1	-3	-8	-15
$f(x)$	1	1.4	2	3	4

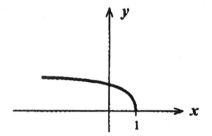

$(-\infty, 1]; [0, \infty)$

43. $f(x) = |x| - 1$ $(-\infty, \infty); [-1, \infty)$

x	-3	-2	-1	0	1	2	3
$f(x)$	2	1	0	-1	0	1	2

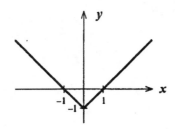

45. $f(x) = \begin{cases} x & \text{if } x < 0 \\ 2x+1 & \text{if } x \geq 0 \end{cases}$

x	-3	-2	-1	0	1	2	3
$f(x)$	-3	-2	-1	1	3	5	7

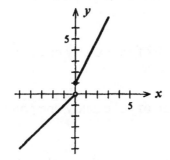

$(-\infty, \infty);\ (-\infty, 0) \cup [1, \infty)$

47. If $x \leq 1$, the graph of f is the half-line $y = -x + 1$. For $x > 1$, use the table

x	2	3	4
$f(x)$	3	8	15

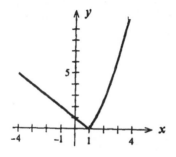

$(-\infty, \infty); [0, \infty)$

49. Each vertical line cuts the given graph at exactly one point, and so the graph represents y as a function of x.

51. Since there is a vertical line that intersects the graph at three points, the graph does not represent y as a function of x.

53. Each vertical line intersects the graph of f at exactly one point, and so the graph represents y as a function of x.

55. Each vertical line intersects the graph of f at exactly one point, and so the graph represents y as a function of x.

57. The circumference of a circle with a 5-inch radius is given by
$$C(5) = 2\pi(5) = 10\pi, \text{ or } 10\pi \text{ inches.}$$

59. $\dfrac{4}{3}(\pi)(2r)^3 = \dfrac{4}{3}\pi 8r^3 = 8(\dfrac{4}{3}\pi r^3)$. Therefore, the volume of the tumor is increased by a factor of 8.

61. a. From $t = 0$ to $t = 5$, the graph for cassettes lies above that for CDs so from 1985 to 1990, sales of prerecorded cassettes were greater than that of CDs.
    b. Sales of prerecorded CDs were greater than that of prerecorded cassettes from 1990 on.
    c. The graphs intersect at the point with coordinates $x = 5$ and $y \approx 3.5$, and this tells us that the sales of the two formats were the same in 1990 with the level of sales at approximately $3.5 billion.

63. a. The slope of the straight line passing through the points (0, 0.58) and (20, 0.95)

is $m = \dfrac{0.95 - 0.58}{20 - 0} = 0.0185$, and so an equation of the straight line passing through

these two points is

$$y - 0.58 = 0.0185(t - 0) \quad \text{or} \quad y = 0.0185t + 0.58$$

Next, the slope of the straight line passing through the points (20, 0.95) and

(30, 1.1) is $m = \dfrac{1.1 - 0.95}{30 - 20} = 0.015$, and so an equation of the straight line passing

through the two points is

$$y - 0.95 = 0.015(t - 20) \quad \text{or} \quad y = 0.015t + 0.65.$$

Therefore, the rule for f is

$$f(t) = \begin{cases} 0.0185t + 0.58 & 0 \le t \le 20 \\ 0.015t + 0.65 & 20 < t \le 30 \end{cases}$$

b. The ratios were changing at the rates of 0.0185/yr and 0.015/yr from 1960 through 1980, and from 1980 through 1990, respectively.

c. The ratio was 1 when $t \approx 20.3$. This shows that the number of bachelor's degrees earned by women equaled the number earned by men for the first time around 1983.

65. a. $T(x) = 0.06x$

b. $T(200) = 0.06(200) = 12$, or \$12.00; $\quad T(5.65) = 0.06(5.65) = 0.34$, or \$0.34.

67. The child should receive $D(4) = \frac{2}{25}(500)(4) = 160$, or 160 mg.

69. a. Take $m = 7.5$ and $b = 20$, then $f(t) = 7.5t + 20 \quad (0 \le t \le 6)$.

b. $f(6) = 7.5(6) + 20 = 65$, or 65 million households.

71. a. The graph of the function is a straight line passing through (0, 120,000) and

(10,0). Its slope is $m = \dfrac{0 - 120{,}000}{10 - 0} = -12{,}000$. The required equation is

$$V = -12{,}000n + 120{,}000.$$

b.

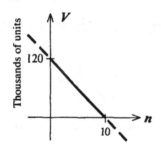

c. $V = -12{,}000(6) + 120{,}000 = 48{,}000$, or \$48,000.
d. This is given by the slope, that is, \$12,000 per year.

73. The domain of the function f is the set of all real positive numbers where $V \neq 0$; that is, $(0,\infty)$. The graph of f follows.

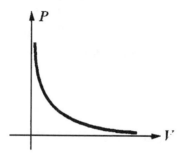

75. a. $N(0) = 3.6$ or 3.6 million people;

$N(25) = 0.0031(25)^2 + 0.16(25) + 3.6 = 9.5375$, or approximately 9.5 million people.

b. $N(30) = 0.0031(30)^2 + 0.16(30) + 3.6 = 11.19$, or approximately 11.2 million people.

77. When the proportion of popular votes won by the Democratic presidential candidate is 0.60, the proportion of seats in the House of Representatives won by Democratic candidates is given by

$$s(0.6) = \frac{(0.6)^3}{(0.6)^3 + (1-0.6)^3} = \frac{0.216}{0.216 + 0.064} = \frac{0.216}{0.280} \approx 0.77.$$

79. $N(t) = -0.0014t^3 + 0.027t^2 - 0.008t + 4.1$

    a. $N(0) = 4.1$, or 4.1 million.

    b. $N(12) = -0.0014(12)^3 + 0.027(12)^2 - 0.008(12) + 4.1 = 5.4728$, or 5.47 million.

81. a. The amount of solids discharged in 1989 ($t = 0$) was 130 tons/day; in 1992 ($t = 3$), it was 100 tons/day; and in 1996 ($t = 7$), it was
$f(7) = 1.25(7)^2 - 26.25(7) + 162.5 = 40$, or 40 tons/day.

    b.

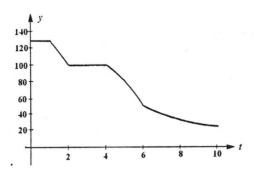

83. True, by definition of a function (page 50).

85. False. Let $f(x) = x^2$, then take $a = 1$, and $b = 2$. Then $f(a) = f(1) = 1$ and $f(b) = f(2) = 4$ and $f(a) + f(b) = 1 + 4 \neq f(a+b) = f(3) = 9$.

USING TECHNOLOGY EXERCISES 2.1, page 67

1.

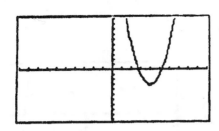

3.

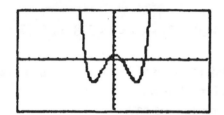

5. a.

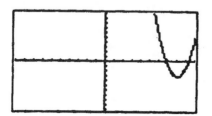

b.

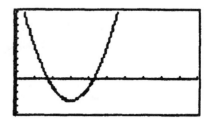

7. a.

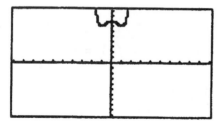

b.

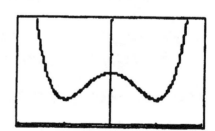

9. a.

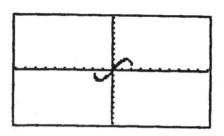

b.

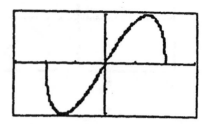

11.

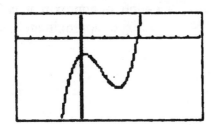

13.

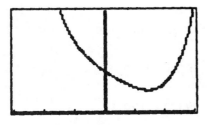

15.

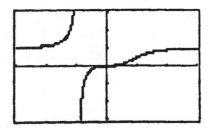

17. 18; $f(-1) = -3(-1)^3 + 5(-1)^2 - 2(-1) + 8 = 3 + 5 + 2 + 8 = 18$.

19. 2; $f(1) = \dfrac{(1)^4 - 3(1)^2}{1-2} = \dfrac{1-3}{-1} = 2$.

21. $f(2.145) \approx 18.5505$

23. $f(2.41) \approx 4.1616$

25. a.

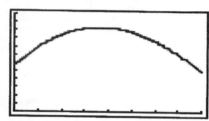

b. $f(2) \approx 9.4066$, or approximately 9.41%/yr
   $f(4) \approx 8.7062$, or approximately 8.71 %/yr.

27. a.

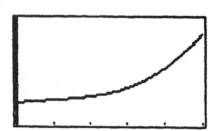

b. $f(6) = 44.7$;

   $f(8) = 52.7$;

   $f(11) = 129.2$.

2.2 Problem Solving Tips

When you come across new notation, make sure that you understand that notation. If

you can't express the notation verbally, you haven't yet grasped its use. For example, in

this section we introduced the notation $g \circ f$, read "g circle f." We use this notation to

describe the composition of the functions g and f. You should also note that $g \circ f$ is

different from $f \circ g$.

Here are some hints for solving the problems in the exercises that follow:

1. If f and g are functions with domains A and B, respectively, then **the domain of**

$f + g, f - g,$ **and** fg is $A \cap B$. The **domain of the quotient** f / g is $A \cap B$ excluding

all numbers x such that $g(x) = 0$.

2. **To find the rule for the composite function** $g \circ f$, evaluate the function g at $f(x)$.

Similarly, to find $f \circ g$ evaluate the function f at $g(x)$.

2.2 CONCEPT QUESTIONS, page 73

1. a. $(f + g)(x) = f(x) + g(x);$ $(f - g)(x) = f(x) - g(x);$ $(fg)(x) = f(x)g(x);$ all
    with domain $A \cap B$. $(f / g)(x) = \dfrac{f(x)}{g(x)}$, domain $A \cap B$ excluding $x \in A \cap B$ such
    that $g(x) = 0$.

    b. $(f + g)(2) = f(2) + g(2) = 3 + (-2) = 1; (f - g)(2) = f(2) - g(2) = 3 - (-2) = 5; ;$
    $(fg)(2) = f(2)g(2) = 3(-2) = -6; (f / g)(2) = \dfrac{f(2)}{g(2)} = \dfrac{3}{-2} = -\dfrac{3}{2}.$

EXERCISES 2.2, page 73

1. $(f + g)(x) = f(x) + g(x) = (x^3 + 5) + (x^2 - 2) = x^3 + x^2 + 3.$

3. $fg(x) = f(x)g(x) = (x^3 + 5)(x^2 - 2) = x^5 - 2x^3 + 5x^2 - 10.$

5. $\dfrac{f}{g}(x) = \dfrac{f(x)}{g(x)} = \dfrac{x^3 + 5}{x^2 - 2}.$

7. $\dfrac{fg}{h}(x) = \dfrac{f(x)g(x)}{h(x)} = \dfrac{(x^3+5)(x^2-2)}{2x+4} = \dfrac{x^5 - 2x^3 + 5x^2 - 10}{2x+4}$

9. $(f+g)(x) = f(x) + g(x) = x - 1 + \sqrt{x+1}.$

11. $(fg)(x) = f(x)g(x) = (x-1)\sqrt{x+1}$

13. $\dfrac{g}{h}(x) = \dfrac{g(x)}{h(x)} = \dfrac{\sqrt{x+1}}{2x^3-1}.$

15. $\dfrac{fg}{h}(x) = \dfrac{(x-1)(\sqrt{x+1})}{2x^3-1}$

17. $\dfrac{f-h}{g}(x) = \dfrac{x-1-(2x^3-1)}{\sqrt{x+1}} = \dfrac{x-2x^3}{\sqrt{x+1}}.$

19. $(f+g)(x) = x^2 + 5 + \sqrt{x} - 2 = x^2 + \sqrt{x} + 3.$

   $(f-g)(x) = x^2 + 5 - (\sqrt{x} - 2) = x^2 - \sqrt{x} + 7.$

   $(fg)(x) = (x^2 + 5)(\sqrt{x} - 2); \; (\dfrac{f}{g})(x) = \dfrac{x^2+5}{\sqrt{x}-2}.$

21. $(f+g)(x) = \sqrt{x+3} + \dfrac{1}{x-1} = \dfrac{(x-1)\sqrt{x+3}+1}{x-1}.$

   $(f-g)(x) = \sqrt{x+3} - \dfrac{1}{x-1} = \dfrac{(x-1)\sqrt{x+3}-1}{x-1}.$

   $(f\,g)(x) = \sqrt{x+3}\left(\dfrac{1}{x-1}\right) = \dfrac{\sqrt{x+3}}{x-1}.$

   $(\dfrac{f}{g}) = \sqrt{x+3}(x-1).$

23. $(f+g)(x) = \dfrac{x+1}{x-1} + \dfrac{x+2}{x-2} = \dfrac{(x+1)(x-2)+(x+2)(x-1)}{(x-1)(x-2)}$

$= \dfrac{x^2-x-2+x^2+x-2}{(x-1)(x-2)} = \dfrac{2x^2-4}{(x-1)(x-2)} = \dfrac{2(x^2-2)}{(x-1)(x-2)}.$

$(f-g)(x) = \dfrac{x+1}{x-1} - \dfrac{x+2}{x-2} = \dfrac{(x+1)(x-2)-(x+2)(x-1)}{(x-1)(x-2)}$

$= \dfrac{x^2-x-2-x^2-x+2}{(x-1)(x-2)} = \dfrac{-2x}{(x-1)(x-2)}.$

$(fg)(x) = \dfrac{(x+1)(x+2)}{(x-1)(x-2)}; \ \left(\dfrac{f}{g}\right) = \dfrac{(x+1)(x-2)}{(x-1)(x+2)}.$

25. $(f\circ g)(x) = f(g(x)) = f(x^2) = (x^2)^2 + x^2 + 1 = x^4 + x^2 + 1.$
$(g\circ f)(x) = g(f(x)) = g(x^2+x+1) = (x^2+x+1)^2.$

27. $(f\circ g)(x) = f(g(x)) = f(x^2-1) = \sqrt{x^2-1}+1.$
$(g\circ f)(x) = g(f(x)) = g(\sqrt{x}+1) = (\sqrt{x}+1)^2 - 1 = x+2\sqrt{x}+1-1 = x+2\sqrt{x}.$

29. $(f\circ g)(x) = f(g(x)) = f\left(\dfrac{1}{x}\right) = \dfrac{1}{x} \div \left(\dfrac{1}{x^2}+1\right) = \dfrac{1}{x}\cdot\dfrac{x^2}{x^2+1} = \dfrac{x}{x^2+1}.$

$(g\circ f)(x) = g(f(x)) = g\left(\dfrac{x}{x^2+1}\right) = \dfrac{x^2+1}{x}.$

31. $h(2) = g[f(2)]$. But $f(2) = 4+2+1 = 7$, so $h(2) = g(7) = 49$.

33. $h(2) = g[f(2)]$. But $f(2) = \dfrac{1}{2(2)+1} = \dfrac{1}{5}$, so $h(2) = g(\dfrac{1}{5}) = \dfrac{1}{\sqrt{5}} = \dfrac{\sqrt{5}}{5}.$

35. $f(x) = 2x^3 + x^2 + 1, \ g(x) = x^5.$

37. $f(x) = x^2 - 1, \ g(x) = \sqrt{x}.$

39. $f(x) = x^2 - 1$, $g(x) = \dfrac{1}{x}$.

41. $f(x) = 3x^2 + 2$, $g(x) = \dfrac{1}{x^{3/2}}$.

43. $f(a+h) - f(a) = [3(a+h)+4] - (3a+4) = 3a + 3h + 4 - 3a - 4 = 3h$.

45. $f(a+h) - f(a) = 4 - (a+h)^2 - (4 - a^2)$
$$= 4 - a^2 - 2ah - h^2 - 4 + a^2 = -2ah - h^2 = -h(2a + h).$$

47. $\dfrac{f(a+h) - f(a)}{h} = \dfrac{[(a+h)^2 + 1] - (a^2 + 1)}{h} = \dfrac{a^2 + 2ah + h^2 + 1 - a^2 - 1}{h} = \dfrac{2ah + h^2}{h}$
$$= \dfrac{h(2a+h)}{h} = 2a + h.$$

49. $\dfrac{f(a+h) - f(a)}{h} = \dfrac{[(a+h)^3 - (a+h)] - (a^3 - a)}{h}$
$$= \dfrac{a^3 + 3a^2h + 3ah^2 + h^3 - a - h - a^3 + a}{h}$$
$$= \dfrac{3a^2h + 3ah^2 + h^3 - h}{h} = 3a^2 + 3ah + h^2 - 1.$$

51. $\dfrac{f(a+h) - f(a)}{h} = \dfrac{\dfrac{1}{a+h} - \dfrac{1}{a}}{h} = \dfrac{\dfrac{a - (a+h)}{a(a+h)}}{h} = -\dfrac{1}{a(a+h)}.$

53. $F(t)$ represents the total revenue for the two restaurants at time t.

55. $f(t)g(t)$ represents the (dollar) value of Nancy's holdings at time t.

57. $g \circ f$ is the function giving the amount of carbon monoxide pollution at time t.

59. $C(x) = 0.6x + 12,100$.

61. a. $f(t) = 267$; $g(t) = 2t^2 + 46t + 733$
    b. $h(t) = (f + g)(t) = f(t) + g(t) = 267 + (2t^2 + 46t + 733) = 2t^2 + 46t + 1000$
    c. $h(13) = 2(13)^2 + 46(13) + 1000 = 1936$, or 1936 tons.

63. a. $P(x) = R(x) - C(x)$
       $= -0.1x^2 + 500x - (0.000003x^3 - 0.03x^2 + 200x + 100,000)$
       $= -0.000003x^3 - 0.07x^2 + 300x - 100,000$.
    b. $P(1500) = -0.000003(1500)^3 - 0.07(1500)^2 + 300(1500) - 100,000$
       $= 182,375$ or $182,375.

65. a. The gap is
       $$G(t) - C(t) = (3.5t^2 + 26.7t + 436.2) - (24.3t + 365)$$
       $$= 3.5t^2 + 2.4t + 71.2.$$
    b. At the beginning of 1983, the gap was
       $$G(0) = 3.5(0)^2 + 2.4(0) + 71.2 = 71.2, \text{ or } 71,200.$$
    At the beginning of 1986, the gap was
       $$G(3) = 3.5(3)^2 + 2.4(3) + 71.2 = 109.9, \text{ or } 109,900.$$

67. a. The occupancy rate at the beginning of January is
       $$r(0) = \frac{10}{81}(0)^3 - \frac{10}{3}(0)^2 + \frac{200}{9}(0) + 55 = 55, \text{ or } 55 \text{ percent.}$$
       $$r(5) = \frac{10}{81}(5)^3 - \frac{10}{3}(5)^2 + \frac{200}{9}(5) + 55 = 98.2, \text{ or } 98.2 \text{ percent.}$$

    b. The monthly revenue at the beginning of January is
       $$R(55) = -\frac{3}{5000}(55)^3 + \frac{9}{50}(55)^2 = 444.68, \text{ or } \$444,700.$$

    The monthly revenue at the beginning of June is
       $$R(98.2) = -\frac{3}{5000}(98.2)^3 + \frac{9}{50}(98.2)^2 = 1167.6, \text{ or } \$1,167,600.$$

69. True. $(f+g)(x)=f(x)+g(x)=g(x)+f(x)=(g+f)(x)$.

71. False. Take $f(x)=\sqrt{x}$ and $g(x)=x+1$. Then $(g\circ f)(x)=\sqrt{x}+1$, but $(f\circ g)(x)=\sqrt{x+1}$.

2.3 Problem Solving Tips

When you solve a problem involving a function, it is helpful to identify the type of function you are working with. For example, if you wish to find the domain of a *polynomial function* you know that there are no restrictions on the domain, since a polynomial is defined for all real numbers. If you want to find the domain of a *rational function*, you know that you have to check to see if there are any values for which the denominator is equal to 0.

Here are some hints for solving the problems in the exercises that follow:

1. **To find the market equilibrium of a commodity**, find the point of intersection of the supply and demand equations for the commodity. (Market equilibrium prevails when the quantity produced is equal to the quantity demanded.).

2. **To construct a mathematical model,** follow the guidelines given in the text on page 84. First try solving Examples 5 and 6 in the text without looking at the solution. Then go on to try a few similar problems (#72-80, on page 90-91 of the text).

2.3 CONCEPT QUESTIONS, page 85

1. See page 77. (Answers will vary).

3. a. A demand function $p = D(x)$ gives the relationship between the unit price of a commodity, p, and the quantity, x, demanded. A supply function $p = S(x)$ gives the relationship between the unit price of a commodity, p, and the quantity, x, the supplier will make available in the market place. Market equilibrium occurs when the quantity produced is equal to the quantity demanded. To find the market equilibrium, we solve the equations $p = D(x)$ and $p = S(x)$ simultaneously.

EXERCISES 2.3, page 85

1. Yes. $2x + 3y = 6$ and so $y = -\frac{2}{3}x + 2$.

3. Yes. $2y = x + 4$ and so $y = \frac{1}{2}x + 2$.

5. Yes. $4y = 2x + 9$ and so $y = \frac{1}{2}x + \frac{9}{4}$.   7. No, because of the term x^2.

9. f is a polynomial function in x of degree 6.

11. Expanding $G(x) = 2(x^2 - 3)^3$, we have $G(x) = 2x^6 - 18x^4 + 54x^2 - 54$, and we conclude that G is a polynomial function in x of degree 6.

13. f is neither a polynomial nor a rational function.

15. $f(0) = 2$ gives $f(0) = m(0) + b = b = 2$. Next, $f(3) = -1$ gives $f(3) = m(3) + b = -1$. Substituting $b = 2$ in this last equation, we have $3m + 2 = -1$, and $3m = -3$, or $m = -1$. So $m = -1$ and $b = 2$.

17. a. $C(x) = 8x + 40,000$ b. $R(x) = 12x$
    c. $P(x) = R(x) - C(x) = 12x - (8x + 40,000) = 4x - 40,000$.
    d. $P(8000) = 4(8000) - 40,000 = -8000$, or a loss of $8000.
    $P(12,000) = 4(12,000) - 40,000 = 8000$, or a profit of $8000.

19. The individual's disposable income is $D = (1 - 0.28)60{,}000 = 43{,}200$, or \$43,200.

21. The child should receive $D(4) = \left(\dfrac{4+1}{24}\right)(500) = 104.17$, or 104 mg.

23. $P(28) = -\tfrac{1}{8}(28)^2 + 7(28) + 30 = 128$, or \$128,000.

25. $S(6) = 0.73(6)^2 + 15.8(6) + 2.7 = 123.78$, or 123.78 million kw-hr.
    $S(8) = 0.73(8)^2 + 15.8(8) + 2.7 = 175.82$, or 175.82 million kw-hr.

27. $N(0) = 0.7$, or 0.7 per 100 million vehicle miles driven.
    $N(7) = 0.0336(7)^3 - 0.118(7)^2 + 0.215(7) + 0.7 = 7.9478$, or 7.95 per 100 million vehicle miles driven.

29. a. $N(0) = 0.32$ or 320,000
    b. $N(4) = -0.0675(4)^4 + 0.5083(4)^3 - 0.893(4)^2 + 0.66(4) + 0.32 = 3.9232$
       or 3,923,200.

31. $N(5) = 0.0018425(10)^{2.5} \approx 0.58265$, or approximately 0.583 million.
    $N(10) = 0.0018425(15)^{2.5} \approx 1.6056$, or approximately 1.606 million.

33. a.

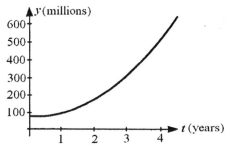

    b. $f(4) = 38.57(4^2) - 24.29(4) + 79.14 = 599.1$, or 599,100,000.

35. a. The given data implies that $R(40) = 50$, that is,
$$\frac{100(40)}{b+40} = 50$$

$$50(b+40) = 4000, \quad \text{or} \quad b = 40.$$

Therefore, the required response function is $R(x) = \dfrac{100x}{40+x}$.

b. The response will be $R(60) = \dfrac{100(60)}{40+60} = 60$, or approximately 60 percent.

37. a. $f(0) = 6.85$, $g(0) = 16.58$. Since $g(0) > f(0)$, we see that more film cameras were sold in 2001 ($t = 0$).

 b. We solve the equation $f(t) = g(t)$, that is,

$$3.05t + 6.85 = -1.85t + 16.58$$

$$4.9t = 9.73$$

$$t = 1.99 \approx 2$$

So sales of digital cameras first exceed those of film cameras in approximately 2003.

39. The slope of the line is $m = \dfrac{S-C}{n}$.

Therefore, an equation of the line is $y - C = \dfrac{S-C}{n}(t-0)$.

Letting $y = V(t)$, we have $V(t) = C - \dfrac{(C-S)}{n}t$.

41. The average U.S. credit card debt at the beginning of 1994 was
$$D(0) = 4.77(1+0)^{0.2676} = 4.77 \quad \text{or} \quad \$4770.$$
At the beginning of 1996, it was $D(2) = 4.77(1+2)^{0.2676} = 6.400$ or $6400. At the beginning of 1999, it was
$$D(5) = 5.6423(5^{0.1818}) \approx 7.560 \quad \text{or} \quad \$7560.$$

43. a. $A(0) = 16.4$, or $16.4 billion.

 $A(1) = 16.4(1+1)^{0.1} \approx 17.58$, or $17.58 billion.

 $A(2) = 16.4(2+1)^{0.1} \approx 18.30$, or $18.3 billion.

 $A(3) = 16.4(3+1)^{0.1} \approx 18.84$, ot $18.84 billion.

$$A(4) = 16.4(4+1)^{0.1} \approx 19.26, \text{ or } \$19.26 \text{ billion.}$$

The nutritional market has been growing over the years from 1999 through 2003.

b.

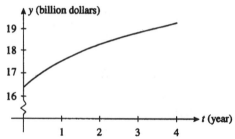

45. a.

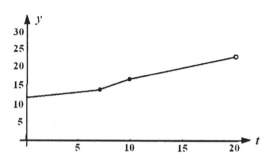

b. $f(5) = \frac{2}{7}(5) + 12 = \frac{10}{7} + 12 \approx 13.43$, or 13.43 percent

$f(15) = \frac{3}{5}(15) + 11 = 20$, or 20 percent.

47. $h(t) = f(t) - g(t) = \dfrac{110}{\frac{1}{2}t + 1} - 26(\frac{1}{4}t^2 - 1)^2 - 52.$

$h(0) = f(0) - g(0) = \dfrac{110}{\frac{1}{2}(0) + 1} - 26\left[\frac{1}{4}(0)^2 - 1\right]^2 - 52 = 110 - 26 - 52 = 32, \text{ or } \$32.$

$h(1) = f(1) - g(1) = \dfrac{110}{\frac{1}{2}(1) + 1} - 26\left[\frac{1}{4}(1)^2 - 1\right]^2 - 52 = 6.71, \text{ or } \$6.71.$

$h(2) = f(2) - g(2) = \dfrac{110}{\frac{1}{2}(2) + 1} - 26\left[\frac{1}{4}(2)^2 - 1\right]^2 - 52 = 3, \text{ or } \$3.$

We conclude that the price gap was narrowing.

49. a. $P(0) = 59.8$; $P(1) = 58.9$; $P(2) = 59.2$; $P(3) = 60.7$; $P(4) = 61.7$

b.

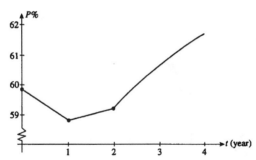

c. $P(3) = 60.7$, or 60.7%.

51. a.

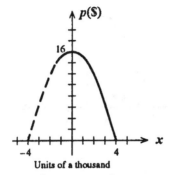

Units of a thousand

b. If $p = 7$, we have $7 = -x^2 + 16$, or $x^2 = 9$, so that $x = \pm 3$. Therefore, the quantity demanded when the unit price is $7 is 3000 units.

53. a.

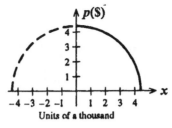

Units of a thousand

b. If $p = 3$, then $3 = \sqrt{18 - x^2}$, and $9 = 18 - x^2$, so that $x^2 = 9$ and $x = \pm 3$.

Therefore, the quantity demanded when the unit price is $3 is 3000 units.

55. a.

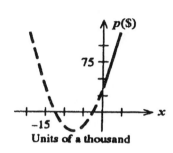

b. If $x = 2$, then $p = 2^2 + 16(2) + 40 = 76$, or $76.

57. a.

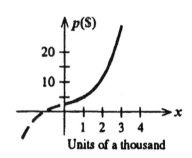

b. $p = 2^3 + 2(2) + 3 = 15$, or $15.

59. The slope of L_2 is greater than that of L_1. This means that for each drop of a dollar in the price of a clock radio, the quantity demanded of model B clock radios is greater than that of model A clock radios.

61. Substituting $x = 10$ into the demand function, we have
$$p = \frac{30}{0.02(10)^2 + 1} = \frac{30}{3} = 10, \text{ or } p = \$10.$$

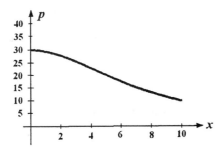

63.

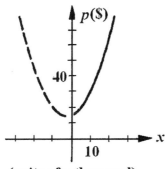

(units of a thousand)

If $x = 5$, then $p = 0.1(5)^2 + 0.5(5) + 15 = 20$, or \$20.

65. a. We solve the system of equations $p = cx + d$ and $p = ax + b$. Substituting the first equation in the second gives

$$cx + d = ax + b$$
$$(c - a)x = b - d$$

or
$$x = \frac{b - d}{c - a}.$$

Since $a < 0$ and $c > 0$, $c - a \neq 0$ and x is well-defined. Substituting this value of x into the second equation, we obtain

$$p = a\left(\frac{b - d}{c - a}\right) + b = \frac{ab - ad + bc - ab}{c - a} = \frac{bc - ad}{c - a}.$$

Therefore, the equilibrium quantity is $\dfrac{b - d}{c - a}$ and the equilibrium price is $\dfrac{bc - ad}{c - a}$.

2 Functions, Limits, and the Derivative

b. If c is increased, the denominator in the expression for x increases and so x gets smaller. At the same time, the first term in the first equation for p decreases and so p gets larger. This analysis shows that if the unit price for producing the product is increased then the equilibrium quantity decreases while the equilibrium price increases.

c. If b is decreased, the numerator of the expression for x decreases while the denominator stays the same. Therefore, x decreases. The expression for p also shows that p decreases. This analysis shows that if the (theoretical) upper bound for the unit price of a commodity is lowered, then both the equilibrium quantity and the equilibrium price drop.

67. We solve the equation $-2x^2 + 80 = 15x + 30$, or $2x^2 + 15x - 50 = 0$ for x. Thus, $(2x - 5)(x + 10) = 0$, or $x = 5/2$ or $x = -10$. Rejecting the negative root, we have $x = 5/2$. The corresponding value of p is $p = -2(\frac{5}{2})^2 + 80 = 67.5$. We conclude that the equilibrium quantity is 2500 and the equilibrium price is $67.50.

69. Solving both equations for x, we have $x = -(11/3)p + 22$ and $x = 2p^2 + p - 10$. Equating these two equations, we have
$$-\tfrac{11}{3}p + 22 = 2p^2 + p - 10,$$
or $\qquad -11p + 66 = 6p^2 + 3p - 30$
and $\qquad 6p^2 + 14p - 96 = 0.$
Dividing this last equation by 2 and then factoring, we have
$$(3p + 16)(p - 3) = 0,$$
or $p = 3$. The corresponding value of x is $2(3)^2 + 3 - 10 = 11$. We conclude that the equilibrium quantity is 11,000 and the equilibrium price is $3.

71. Equating the two equations, we have
$$144 - x^2 = 48 + \tfrac{1}{2}x^2$$
$$288 - 2x^2 = 96 + x^2$$
$$3x^2 = 192; \quad x^2 = 64,$$
or $x = \pm 8$. We take $x = 8$, and the corresponding value of p is $144 - 8^2 = 80$. We conclude that the equilibrium quantity is 8000 tires and the equilibrium price is $80.

73. The area of Juanita's garden is 250 sq ft. Therefore $xy = 250$ and $y = \dfrac{250}{x}$.

The amount of fencing needed is given by $2x + 2y$.

Therefore, $f = 2x + 2\left(\dfrac{250}{x}\right) = 2x + \dfrac{500}{x}$. The domain of f is $x > 0$.

75. Since the volume of the box is given by
$$V = (\text{area of the base}) \times \text{the height of the box}$$
$$= x^2 y = 20,$$

we have $y = \dfrac{20}{x^2}$. Next, the amount of material used in constructing the box is

given by the area of the base of the box, plus the area of the 4 sides, plus the area of the top of the box, or $x^2 + 4xy + x^2$. Then, the cost of constructing the box is

given by $f(x) = 0.30x^2 + 0.40x \cdot \dfrac{20}{x^2} + .20x^2 = 0.5x^2 + \dfrac{8}{x}$.

77. The average yield of the apple orchard is 36 bushels/tree when the density is 22 trees/acre. Let $x =$ the unit increase in tree density beyond 22. Then the yield of the apple orchard in bushels/acre is given by $(22 + x)(36 - 2x)$.

79. a. Let x denote the number of bottles sold beyond 10,000 bottles. Then
$$P(x) = (10,000 + x)(5 - 0.0002x)$$
$$= -0.0002x^2 + 3x + 50,000$$
b. He can expect a profit
$$P(6000) = -0.0002(6000^2) + 3(6000) + 50,000 = 60,800$$
or $60,800.

81. False. $f(x) = 3x^{3/4} + x^{1/2} + 1$ is not a polynomial function. The powers in x must be nonnegative integers.

83. False. $f(x) = x^{1/2}$ is not defined for negative values of x or $x = 0$.

1. (-3.0414, 0.1503); (3.0414, 7.4497)

3. (-2.3371, 2.4117); (6.0514, -2.5015)

5. (-1.0219, -6.3461); (1.2414, -1.5931), and (5.7805, 7.9391)

7. a.   b.   438 wall clocks; $40.92

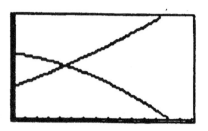

9. a. $y = 0.1375t^2 + 0.675t + 3.1$

    b.   c. 3.1; 3.9; 5; 6.4; 8; 9.9

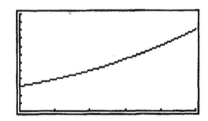

11. a. $y = -0.02028t^3 + 0.31393t^2 + 0.40873t + 0.66024$

    b.

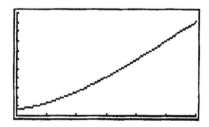

c. 0.66; 1.36; 2.57; 4.16; 6.02; 8.02; 10.03

13. a. $f(t) = 0.0012t^3 - 0.053t^2 + 0.497t + 2.55$ $(0 \le t \le 20)$

    b.

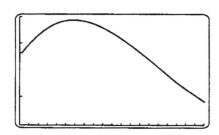

    c.

T	0	7	10	20
$f(t)$ trillion dollars	2.55	3.84	3.42	0.89

15. a. $y = 0.05833t^3 - 0.325t^2 + 1.8881t + 5.07143$

    b.   c. 6.7; 8.0; 9.4; 11.2; 13.7

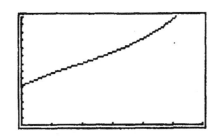

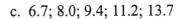

17. a. $y = 0.0125t^4 - 0.01389t^3 + 0.55417t^2 + 0.53294t + 4.95238$ $(0 \le t \le 5)$

    b.   c. 5.0; 6.0; 8.3; 12.2; 18.3; 27.5

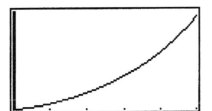

2.4 Problem Solving Tips

In this section, an important theorem was introduced (page 102). After you read Theorem 1, try to express the theorem in your own words. While you will not usually be required to prove these theorems in this course, you will be asked to understand the results of the theorem. For example, Theorem 1 gives us the properties of limits that allow us to evaluate the sum, difference, product, quotient, powers of functions and constant multiple of functions at a specified value with certain restrictions. You should be able to use the limit notation to write out each of these properties. You should also be able to use these properties to evaluate the limits of functions.

Here are some hints for solving the problems in the exercises that follow:

1. **To find the limit of a function** $f(x)$ **as** $x \to a$, where a is a real number, substitute a for x in the rule for f and simplify the result.

2. **To evaluate the limit of a quotient that has the indeterminate form 0/0:**

   a. First, replace the given function with an appropriate one that takes on the same values as the original function everywhere except at $x = a$.

   b. Next, evaluate the limit of this function as x approaches a.

2.4 CONCEPT QUESTIONS, page 111

1. The values of $f(x)$ can be made as close to 3 as we please by taking x sufficiently close to $x = 2$.

3. a. $\lim_{x \to 4} \sqrt{x}(2x^2 + 1) = \lim_{x \to 4}(\sqrt{x})\lim_{x \to 4}(2x^2 + 1)$ (Rule 4)

$$= \sqrt{4}\left[2(4)^2 + 1\right]$$ (Rules 1 and 3)

$$= 66$$

b. $\lim_{x \to 1}\left(\dfrac{2x^2 + x + 5}{x^4 + 1}\right) = \left(\lim_{x \to 1}\dfrac{2x^2 + x + 5}{x^4 + 1}\right)^{3/2}$ (Rule 1)

$$= \left(\dfrac{2 + 1 + 5}{1 + 1}\right)^{3/2}$$ (Rules 2, 3, and 5)

$$= 4^{3/2} = 8$$

5. $\lim_{x \to \infty} f(x) = L$ means $f(x)$ can be made as close to L as we please by taking x sufficiently large. $\lim_{x \to -\infty} f(x) = M$ means $f(x)$ can be made as close to M as we please by taking x as large as please in absolute value but negative.

EXERCISES 2.4, page 111

1. $\lim_{x \to -2} f(x) = 3.$ 3. $\lim_{x \to 3} f(x) = 3.$ 5. $\lim_{x \to -2} f(x) = 3.$

7. The limit does not exist. If we consider any value of x to the right of $x = -2$, $f(x) \le 2$. If we consider values of x to the left of $x = -2$, $f(x) \ge -2$. Since $f(x)$ does not approach any one number as x approaches $x = -2$, we conclude that the limit does not exist.

9. $\lim_{x \to 2}(x^2 + 1) = 5.$

x	1.9	1.99	1.999	2.001	2.01	2.1
$f(x)$	4.61	4.9601	4.9960	5.004	5.0401	5.41

11.

x	-0.1	-0.01	-0.001	0.001	0.01	0.1
$f(x)$	-1	-1	-1	1	1	1

The limit does not exist.

13.

x	0.9	0.99	0.999	1.001	1.01	1.1
$f(x)$	100	10,000	1,000,000	1,000,000	10,000	100

The limit does not exist.

15.

x	0.9	0.99	0.999	1.001	1.01	1.1
$f(x)$	2.9	2.99	2.999	3.001	3.01	3.1

$$\lim_{x \to 1} \frac{x^2 + x - 2}{x - 1} = 3.$$

17.

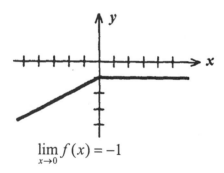

$$\lim_{x \to 0} f(x) = -1$$

19.

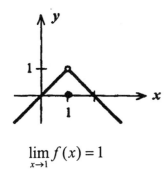

$$\lim_{x \to 1} f(x) = 1$$

21.

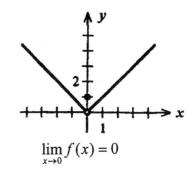

$$\lim_{x \to 0} f(x) = 0$$

23. $\lim\limits_{x \to 2} 3 = 3$

25. $\lim\limits_{x \to 3} x = 3$

27. $\lim\limits_{x \to 1} (1 - 2x^2) = 1 - 2(1)^2 = -1$

29. $\lim\limits_{x \to 1} (2x^3 - 3x^2 + x + 2) = 2(1)^3 - 3(1)^2 + 1 + 2 = 2.$

31. $\lim\limits_{s \to 0} (2s^2 - 1)(2s + 4) = (-1)(4) = -4.$

33. $\lim\limits_{x \to 2} \dfrac{2x + 1}{x + 2} = \dfrac{2(2) + 1}{2 + 2} = \dfrac{5}{4}.$

35. $\lim\limits_{x \to 2} \sqrt{x + 2} = \sqrt{2 + 2} = 2.$

37. $\lim\limits_{x \to -3} \sqrt{2x^4 + x^2} = \sqrt{2(-3)^4 + (-3)^2} = \sqrt{162 + 9} = \sqrt{171} = 3\sqrt{19}.$

39. $\lim\limits_{x \to -1} \dfrac{\sqrt{x^2 + 8}}{2x + 4} = \dfrac{\sqrt{(-1)^2 + 8}}{2(-1) + 4} = \dfrac{\sqrt{9}}{2} = \dfrac{3}{2}.$

41. $\lim\limits_{x\to a}[f(x)-g(x)]=\lim\limits_{x\to a}f(x)-\lim\limits_{x\to a}g(x)=3-4=-1.$

43. $\lim\limits_{x\to a}[2f(x)-3g(x)]=\lim\limits_{x\to a}2f(x)-\lim\limits_{x\to a}3g(x)=2(3)-3(4)=-6.$

45. $\lim\limits_{x\to a}\sqrt{g(x)}=\lim\limits_{x\to a}\sqrt{4}=2.$

47. $\lim\limits_{x\to a}\dfrac{2f(x)-g(x)}{f(x)g(x)}=\dfrac{2(3)-(4)}{(3)(4)}=\dfrac{2}{12}=\dfrac{1}{6}.$

49. $\lim\limits_{x\to1}\dfrac{x^2-1}{x-1}=\lim\limits_{x\to1}\dfrac{(x-1)(x+1)}{x-1}=\lim\limits_{x\to1}(x+1)=1+1=2.$

51. $\lim\limits_{x\to0}\dfrac{x^2-x}{x}=\lim\limits_{x\to0}\dfrac{x(x-1)}{x}=\lim\limits_{x\to0}(x-1)=0-1=-1.$

53. $\lim\limits_{x\to-5}\dfrac{x^2-25}{x+5}=\lim\limits_{x\to-5}\dfrac{(x+5)(x-5)}{x+5}=\lim\limits_{x\to-5}(x-5)=-10.$

55. $\lim\limits_{x\to1}\dfrac{x}{x-1}$ does not exist.

57. $\lim\limits_{x\to-2}\dfrac{x^2-x-6}{x^2+x-2}=\lim\limits_{x\to-2}\dfrac{(x-3)(x+2)}{(x+2)(x-1)}=\lim\limits_{x\to-2}\dfrac{x-3}{x-1}=\dfrac{-2-3}{-2-1}=\dfrac{5}{3}.$

59. $\lim\limits_{x\to1}\dfrac{\sqrt{x}-1}{x-1}=\lim\limits_{x\to1}\dfrac{\sqrt{x}-1}{x-1}\cdot\dfrac{\sqrt{x}+1}{\sqrt{x}+1}=\lim\limits_{x\to1}\dfrac{x-1}{(x-1)(\sqrt{x}+1)}=\lim\limits_{x\to1}\dfrac{1}{\sqrt{x}+1}=\dfrac{1}{2}.$

61. $\lim\limits_{x\to1}\dfrac{x-1}{x^3+x^2-2x}=\lim\limits_{x\to1}\dfrac{x-1}{x(x-1)(x+2)}=\lim\limits_{x\to1}\dfrac{1}{x(x+2)}=\dfrac{1}{3}.$

63. $\lim\limits_{x\to\infty}f(x)=\infty$ (does not exist) and $\lim\limits_{x\to-\infty}f(x)=\infty$ (does not exist).

65. $\lim_{x \to \infty} f(x) = 0$ and $\lim_{x \to -\infty} f(x) = 0$.

67. $\lim_{x \to \infty} f(x) = -\infty$ (does not exist) and $\lim_{x \to -\infty} f(x) = -\infty$ (does not exist).

69.

x	1	10	100	1000
$f(x)$	0.5	0.009901	0.0001	0.000001

x	-1	-10	-100	-1000
$f(x)$	0.5	0.009901	0.0001	0.000001

$\lim_{x \to \infty} f(x) = 0$ and $\lim_{x \to -\infty} f(x) = 0$

71.

x	1	5	10	100	1000
$f(x)$	12	360	2910	2.99×10^6	2.999×10^9

x	-1	-5	-10	-100	-1000
$f(x)$	6	-390	-3090	-3.01×10^6	-3.0×10^9

$\lim_{x \to \infty} f(x) = \infty$ (does not exist) and $\lim_{x \to -\infty} f(x) = -\infty$ (does not exist).

73. $\lim_{x \to \infty} \dfrac{3x + 2}{x - 5} = \lim_{x \to \infty} \dfrac{3 + \dfrac{2}{x}}{1 - \dfrac{5}{x}} = \dfrac{3}{1} = 3.$

75. $\displaystyle\lim_{x\to-\infty}\frac{3x^3+x^2+1}{x^3+1}=\lim_{x\to-\infty}\frac{3+\dfrac{1}{x}+\dfrac{1}{x^3}}{1+\dfrac{1}{x^3}}=3.$

77. $\displaystyle\lim_{x\to-\infty}\frac{x^4+1}{x^3-1}=\lim_{x\to-\infty}\frac{x+\dfrac{1}{x^3}}{1-\dfrac{1}{x^3}}=-\infty$; that is, the limit does not exist.

79. $\displaystyle\lim_{x\to\infty}\frac{x^5-x^3+x-1}{x^6+2x^2+1}=\lim_{x\to\infty}\frac{\dfrac{1}{x}-\dfrac{1}{x^3}+\dfrac{1}{x^5}-\dfrac{1}{x^6}}{1+\dfrac{2}{x^4}+\dfrac{1}{x^6}}=0.$

81. a. The cost of removing 50 percent of the pollutant is

$$C(50)=\frac{0.5(50)}{100-50}=0.5\text{, or }\$500,000.$$

Similarly, we find that the cost of removing 60, 70, 80, 90, and 95 percent of the

pollutants is $750,000; $1,166,667; $2,000,000, $4,500,000, and $9,500,000, respectively.

b. $\displaystyle\lim_{x\to100}\frac{0.5x}{100-x}=\infty$, which means that the cost of removing the pollutant increases

astronomically if we wish to remove almost all of the pollutant.

83. $\displaystyle\lim_{x\to\infty}\overline{C}(x)=\lim_{x\to\infty}2.2+\frac{2500}{x}=2.2$, or $2.20 per DVD.

In the long-run, the average cost of producing x DVDs will approach
$2.20/disc.

85. a. $T(1)=\dfrac{120}{1+4}=24$, or $24 million. $\qquad T(2)=\dfrac{120(4)}{8}=60$, $60 million.

$T(3)=\dfrac{120(9)}{13}=83.1$, or $83.1 million.

b. In the long run, the movie will gross

$$\lim_{x \to \infty} \frac{120x^2}{x^2+4} = \lim_{x \to \infty} \frac{120}{1+\dfrac{4}{x^2}} = 120, \text{ or } \$120 \text{ million.}$$

87. a. The average cost of driving 5000 miles per year is
$$C(5) = \frac{2010}{5^{2.2}} + 17.80 = 76.07,$$

or 76.1 cents per mile. Similarly, we see that the average cost of driving 10,000 miles per year; 15,000 miles per year; 20,000 miles per year; and 25,000 miles per year is 30.5, 23; 20.6, and 19.5 cents per mile, respectively.

  b.

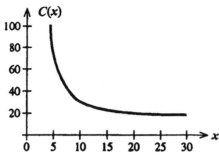

  c. It approaches 17.80 cents per mile.

89. False. Let $f(x) = \begin{cases} -1 & \text{if } x < 0 \\ 1 & \text{if } x > 0 \end{cases}$. Then $\lim_{x \to 0} f(x) = 1$, but $f(1)$ is not defined.

91. True. Division by zero is not permitted.

93. True. Each limit in the sum exists. Therefore,
$$\lim_{x \to 2} \left(\frac{x}{x+1} + \frac{3}{x-1} \right) = \lim_{x \to 2} \frac{x}{x+1} + \lim_{x \to 2} \frac{3}{x-1} = \frac{2}{3} + \frac{3}{1} = \frac{11}{3}.$$

95. $\lim_{x \to \infty} \dfrac{ax}{x+b} = \lim_{x \to \infty} \dfrac{a}{1+\frac{b}{x}} = a.$ As the amount of substrate becomes very large, the initial speed approaches the constant a moles per liter per second.

97. Consider the functions $f(x) = \begin{cases} -1 & \text{if } x < 0 \\ 1 & \text{if } x \geq 0 \end{cases}$ and $g(x) = \begin{cases} 1 & \text{if } x < 0 \\ -1 & \text{if } x \geq 0 \end{cases}$.

Then $\lim\limits_{x \to 0} f(x)$ and $\lim\limits_{x \to 0} g(x)$ do not exist, but $\lim\limits_{x \to 0} [f(x)g(x)] = \lim\limits_{x \to 0} (-1) = -1$.

This example does not contradict Theorem 1 because the hypothesis of Theorem 1 says that if $\lim\limits_{x \to 0} f(x)$ and $\lim\limits_{x \to 0} g(x)$ both exist, then the limit of the product of f and g also exists. It does not say that if the former do not exist, then the latter might not exist.

USING TECHNOLOGY EXERCISES 2.4, page 118

1. 5   3. 3   5. $\dfrac{2}{3}$ 7. $\dfrac{1}{2}$

9. e^2, or 7.38906

11. From the graph we see that $f(x)$ does not
    approach any finite number as x approaches 3.

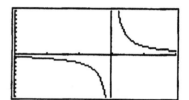

13. a.

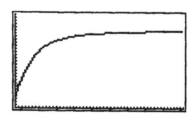

b. $\lim\limits_{t \to \infty} \dfrac{25t^2 + 125t + 200}{t^2 + 5t + 40} = 25$, so in the long run the population will approach 25,000.

2.5 Problem Solving Tips

The problem-solving skills that you learned in earlier sections are building-blocks for the rest of the course. You can't skip a section or a concept and "hope" that you will understand the material in a new section. It just won't work. If you don't build a strong foundation, you won't be able to understand the later concepts. For example, in this section we discussed one-sided limits. You need to understand the definition of a limit before you can understand what is meant by one-sided limits. That means you should be able to express the definition of a limit in your own words. If you can't grasp a new concept, it may well be that you still don't understand a previous concept. If so, you need to go back and review the earlier section before you go on. As another example, we also discussed the continuity of polynomial and rational functions on page 123. If you don't remember how to identify polynomial and rational functions, go back to Section 2.3 and review this material.

Here are some hints for solving the problems in the exercises that follow:

1. To evaluate the limit of a piecewise-defined function at a real number a, follow the same procedure that you used to evaluate a piecewise-defined function. First find the subdomain that a lies in. Next , use the rule for that subdomain to find the limit of f at a.

2. To determine the values of x at which a function is continuous check to see if the function is a polynomial or rational function. A polynomial function $y = P(x)$ is continuous at every value of x and a rational function is continuous at every value of x where $q(x) \neq 0$.

2.5 CONCEPT QUESTIONS, page 127

1. $\lim_{x \to 3^-} f(x) = 2$ means $f(x)$ can be made as close to 2 as we please by taking x sufficiently close to but to the left of $x = 3$. $\lim_{x \to 3^+} f(x) = 4$ means $f(x)$ can be made as close to 4 as we please by taking x sufficiently close to but to the right of $x = 3$.

3. a. f is continuous at a if $\lim_{x \to a} f(x) = f(a)$.

   b. f is continuous on an interval I if f is continuous at each point in I.

5. Refer to page 124. Answers will vary.

EXERCISES 2.5, page 127

1. $\lim_{x \to 2^-} f(x) = 3$, $\lim_{x \to 2^+} f(x) = 2$, $\lim_{x \to 2} f(x)$ does not exist.

3. $\lim_{x \to -1^-} f(x) = \infty$, $\lim_{x \to -1^+} f(x) = 2$. Therefore $\lim_{x \to -1} f(x)$ does not exist.

5. $\lim_{x \to 1^-} f(x) = 0$, $\lim_{x \to 1^+} f(x) = 2$, $\lim_{x \to 1} f(x)$ does not exist.

7. $\lim_{x \to 0^-} f(x) = -2$, $\lim_{x \to 0^+} f(x) = 2$, $\lim_{x \to 0} f(x)$ does not exist.

9. True   11. True   13. False

15. True   17. False   19. True

21. $\lim\limits_{x\to 1^+} (2x+4) = 6.$

23. $\lim\limits_{x\to 2^-} \dfrac{x-3}{x+2} = \dfrac{2-3}{2+2} = -\dfrac{1}{4}.$

25. $\lim\limits_{x\to 0^+} \dfrac{1}{x}$ does not exist because $1/x \to \infty$ as $x \to 0$ from the right..

27. $\lim\limits_{x\to 0^+} \dfrac{x-1}{x^2+1} = \dfrac{-1}{1} = -1.$

29. $\lim\limits_{x\to 0^+} \sqrt{x} = \sqrt{\lim\limits_{x\to 0^+} x} = 0.$

31. $\lim\limits_{x\to -2^+} (2x+\sqrt{2+x}) = \lim\limits_{x\to -2^+} 2x + \lim\limits_{x\to -2^+} \sqrt{2+x} = -4+0 = -4.$

33. $\lim\limits_{x\to 1^-} \dfrac{1+x}{1-x} = \infty$, that is, the limit does not exist.

35. $\lim\limits_{x\to 2^-} \dfrac{x^2-4}{x-2} = \lim\limits_{x\to 2^-} \dfrac{(x+2)(x-2)}{x-2} = \lim\limits_{x\to 2^-} (x+2) = 4.$

37. $\lim\limits_{x\to 0^+} f(x) = \lim\limits_{x\to 0^+} x^2 = 0,\ \lim\limits_{x\to 0^-} f(x) = \lim\limits_{x\to 0^-} 2x = 0$

39. The function is discontinuous at $x = 0$. Conditions 2 and 3 are violated.

41. The function is continuous everywhere.

43. The function is discontinuous at $x = 0$. Condition 3 is violated.

45. f is continuous for all values of x.

47. f is continuous for all values of x. Note that $x^2 + 1 \geq 1 > 0$.

49. f is discontinuous at $x = 1/2$, where the denominator is 0.

51. Observe that $x^2 + x - 2 = (x + 2)(x - 1) = 0$ if $x = -2$ or $x = 1$. So, f is discontinuous at these values of x.

53. f is continuous everywhere since all three conditions are satisfied.

55. f is continuous everywhere since all three conditions are satisfied.

57. Since the denominator $x^2 - 1 = (x - 1)(x + 1) = 0$ if $x = -1$ or 1, we see that f is discontinuous at these points.

59. Since $x^2 - 3x + 2 = (x - 2)(x - 1) = 0$ if $x = 1$ or 2, we see that the denominator is zero at these points and so f is discontinuous at these points.

61. The function f is discontinuous at $x = 1, 2, 3, ..., 11$ because the limit of f does not exist at these points.

63. Having made steady progress up to $x = x_1$, Michael's progress came to a standstill. Then at $x = x_2$ a sudden break-through occurs and he then continues to successfully complete the solution to the problem.

65. Conditions 2 and 3 are not satisfied at each of these points.

67. The graph of f follows.

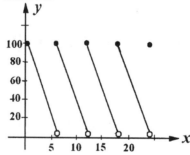

f is discontinuous at $x = 6, 12, 18, 24$.

69.

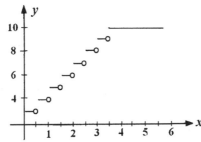

f is discontinuous at $x = \frac{1}{2}, 1, 1\frac{1}{2}, \ldots, 4$.

71. a. $\displaystyle\lim_{t \to 0^+} S(t) = \lim_{t \to 0^+} \frac{a}{t} + b = \infty$. As the time taken to excite the tissue is made smaller and smaller, the strength of the electric current gets stronger and stronger.

b. $\displaystyle\lim_{t \to \infty} \frac{a}{t} + b = b$. As the time taken to excite the tissue is made larger and larger, the strength of the electric current gets smaller and smaller and approaches b.

73. We require that $f(1) = 1 + 2 = 3 = \displaystyle\lim_{x \to 1^+} kx^2 = k$, or $k = 3$.

75. a. Yes, because if $f + g$ were continuous at a, then $g = (f + g) - f$ would be continuous (the difference of two continuous functions is continuous), and this would imply that g is continuous, a contradiction.
b. No. Consider the functions f and g defined by

$$f(x) = \begin{cases} -1 & \text{if } x < 0 \\ 1 & \text{if } x \geq 0 \end{cases} \quad \text{and} \quad g(x) = \begin{cases} 1 & \text{if } x < 0 \\ -1 & \text{if } x \geq 0 \end{cases}.$$

Both f and g are discontinuous at $x = 0$, but $f + g$ is continuous everywhere.

77. a. f is a polynomial of degree 2 and is therefore continuous everywhere, and in particular in $[1,3]$.
b. $f(1) = 3$ and $f(3) = -1$ and so f must have at least one zero in $(1,3)$.

79. f is a polynomial and is therefore continuous on $[-1,1]$.

$$f(-1) = (-1)^3 - 2(-1)^2 + 3(-1) + 2 = -1 - 2 - 3 + 2 = -4.$$
$$f(1) = 1 - 2 + 3 + 2 = 4.$$

Since $f(-1)$ and $f(1)$ have opposite signs, we see that f has at least one zero in $(-1,1)$.

81. $f(0) = 6$ and $f(3) = 3$ and f is continuous on $[0,3]$. So the Intermediate Value Theorem guarantees that there is at least one value of x for which $f(x) = 4$. Solving $f(x) = x^2 - 4x + 6 = 4$, we find $x^2 - 4x + 2 = 0$. Using the quadratic formula, we find that $x = 2 \pm \sqrt{2}$. Since $2 \pm \sqrt{2}$ does not lie in $[0,3]$, we see that $x = 2 - \sqrt{2} \approx 0.59$.

83. $x^5 + 2x - 7 = 0$

Step	Root of f(x) = 0 lies in
1	(1,2)
2	(1,1.5)
3	(1.25,1.5)
4	(1.25,1.375)
5	(1.3125,1.375)
6	(1.3125,1.34375)
7	(1.328125,1.34375)
8	(1.3359375,1.34375)
9	(1.33984375,1.34375)

We see that the required root is approximately 1.34.

85. a. $h(t) = 4 + 64(0) - 16(0) = 4$, and $h(2) = 4 + 64(2) - 16(4) = 68$.

b. The function h is continuous on $[0,2]$. Furthermore, the number 32 lies between 4 and 68. Therefore, the Intermediate Value Theorem guarantees that there is at least one value of t such that $h(t) = 32$, that is, Joan must see the ball at least once during the time the ball is in the air.

c. We solve

$$h(t) = 4 + 64t - 16t^2 = 32$$

or

$$16t^2 - 64t + 28 = 0$$
$$4t^2 - 16t + 7 = 0$$
$$(2t - 1)(2t - 7) = 0$$

giving $t = \frac{1}{2}$ or $t = \frac{7}{2}$. Joan sees the ball on its way up half a second after it was thrown and again $3\frac{1}{2}$ seconds later when it is on its way down.

87. False. Take
$$f(x) = \begin{cases} -1 & \text{if } x < 2 \\ 4 & \text{if } x = 2 \\ 1 & \text{if } x > 2 \end{cases}$$

Then $f(2) = 4$ but $\lim_{x \to 2}$ does not exist.

89. False. Consider the function $f(x) = x^2 - 1$ on the interval $[-2, 2]$. Here, $f(-2) = f(2) = 3$, but f has zeros at $x = -1$ and $x = 1$.

91. False. Let $f(x) = \begin{cases} x & \text{if } x \neq 0 \\ 1 & \text{if } x = 0 \end{cases}$. Then $\lim_{x \to 0^+} f(x) = \lim_{x \to 0^-} f(x)$, but $f(0) = 1$.

93. False. Take $f(x) = \begin{cases} \frac{1}{x} & \text{if } x \neq 0 \\ 0 & \text{if } x = 0 \end{cases}$. Then f is continuous for all $x \neq 0$ but

$\lim_{x \to 0} f(x)$ does not exist.

95. a. Both $g(x) = x$ and $h(x) = \sqrt{1 - x^2}$ are continuous on $[-1,1]$ and so
$f(x) = x - \sqrt{1 - x^2}$ is continuous on $[-1,1]$.
b. $f(-1) = -1$ and $f(1) = 1$ and so f has at least one zero in $(-1,1)$.
c. Solving $f(x) = 0$, we have $x = \sqrt{1 - x^2}$, $x^2 = 1 - x^2$, $2x^2 = 1$, or $x = \frac{\pm\sqrt{2}}{2}$.

97. a. (i). Repeated use of Property 3 shows that $g(x) = x^n = x \cdot x \cdots x$ (n times) is a continuous function since $f(x) = x$ is continuous by Property 1.
(ii). Properties 1 and 5 combine to show that $c \cdot x^n$ is continuous using the results of (a).
(iii). Each of the terms of $p(x) = a_0 x^n + a_1 x^{n-1} + \cdots + a_n$ is continuous and so Property 4 implies that p is continuous.

b. Property 6 now shows that $R(x) = \dfrac{p(x)}{q(x)}$ is continuous if $q(a) \neq 0$ since p

and q are continuous at $x = a$.

1. $x = 0, 1$ 3. $x = 2$ 5. $x = 0, \frac{1}{2}$

7. $x = -\frac{1}{2}, 2$ 9. $x = -2, 1$

11.

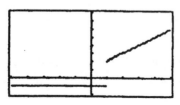

13.

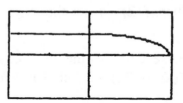

15.

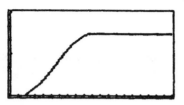

2.6 Problem Solving Tips

When you solve an applied problem, it is important to understand the question in

mathematical terms. For example, if you are given a function $f(t)$ describing the size

of a country's population at time t and asked to find the *rate of change* of that country's

population at any time t, this means that you need to find the derivative of the given

function; that is, find $f'(t)$. If you are then asked to find the population of the country at a specified time, say $t = 2$, you simply need to evaluate the function at that value of t. On the other hand, if you are asked to find the *rate of change* of the population at time $t = 2$, then you need to evaluate the derivative of the function at the value $t = 2$, that is find $f'(2)$. Here again, the key is to be familiar with the terminology and notation introduced in the chapter.

Here are some hints for solving the problems in the exercises that follow:

1. **To find the slope of the tangent line to the graph of a function at *any* point on the graph of that function, find the derivative of f.**

2. **To find the slope of the tangent line to the graph of a function at a *given* point (x_0, y_0) on the graph of that function, find f' and then evaluate f' at x_0.**

SECTION 2.6 CONCEPT QUESTIONS, page 147

1. a. $m = \dfrac{f(2+h) - f(2)}{h}$

    b. Slope of the tangent line is $\lim\limits_{h \to 0} \dfrac{f(2+h) - f(2)}{h}$.

3. a. It gives (i) the slope of the secant line passing through the points $(x, f(x))$ and $(x+h, f(x+h))$ and (ii) the average rate of change of f over the interval $[x, x+h]$.

    b. It gives (i) the slope of the tangent line to the graph of f at the point $(x, f(x))$ and (ii) the instantaneous rate of change of f at x.

EXERCISES 2.6, page 147

1. The rate of change of the average infant's weight when $t = 3$ is $(7.5)/5$, or 1.5 lb/month. The rate of change of the average infant's weight when $t = 18$ is $(3.5)/6$, or approximately 0.6 lb/month. The average rate of change over the infant's first year of life is $(22.5 - 7.5)/(12)$, or 1.25 lb/month.

3. The rate of change of the percentage of households watching television at 4 P.M. is $(12.3)/4$, or approximately 3.1 percent per hour. The rate at 11 P.M. is $(-42.3)/2 = -21.15$; that is, it is dropping off at the rate of 21.15 percent per hour.

5. a. Car A is travelling faster than Car B at t_1 because the slope of the tangent line to the graph of f is greater than the slope of the tangent line to the graph of g at t_1.
   b. Their speed is the same because the slope of the tangent lines are the same at t_2.
   c. Car B is travelling faster than Car A.
   d. They have both covered the same distance and are once again side by side at t_3.

7. a. P_2 is decreasing faster at t_1 because the slope of the tangent line to the graph of g at t_1 is more negative than the slope of the tangent line to the graph of f at t_1.
   b. P_1 is decreasing faster than P_2 at t_2.
   c. Bactericide B is more effective in the short run, but bactericide A is more effective in the long run.

9. $f(x) = 13$

   Step 1 $f(x + h) = 13$
   Step 2 $f(x + h) - f(x) = 13 - 13 = 0$
   Step 3 $\dfrac{f(x+h) - f(x)}{h} = \dfrac{0}{h} = 0$
   Step 4 $f'(x) = \lim\limits_{h \to 0} \dfrac{f(x+h) - f(x)}{h} = \lim\limits_{h \to 0} 0 = 0$

11. $f(x) = 2x + 7$

   Step 1 $f(x + h) = 2(x + h) + 7$
   Step 2 $f(x + h) - f(x) = 2(x + h) + 7 - (2x + 7) = 2h$
   Step 3 $\dfrac{f(x+h) - f(x)}{h} = \dfrac{2h}{h} = 2$

Step 4 $f'(x) = \lim\limits_{h \to 0} \dfrac{f(x+h) - f(x)}{h} = \lim\limits_{h \to 0} 2 = 2$

13. $f(x) = 3x^2$

   Step 1 $f(x+h) = 3(x+h)^2 = 3x^2 + 6xh + 3h^2$

   Step 2 $f(x+h) - f(x) = (3x^2 + 6xh + 3h^2) - 3x^2 = 6xh + 3h^2 = h(6x + 3h)$

   Step 3 $\dfrac{f(x+h) - f(x)}{h} = \dfrac{h(6x + 3h)}{h} = 6x + 3h$

   Step 4 $f'(x) = \lim\limits_{h \to 0} \dfrac{f(x+h) - f(x)}{h} = \lim\limits_{h \to 0} (6x + 3h) = 6x.$

15. $f(x) = -x^2 + 3x$

   Step 1 $f(x+h) = -(x+h)^2 + 3(x+h) = -x^2 - 2xh - h^2 + 3x + 3h$

   Step 2 $f(x+h) - f(x) = (-x^2 - 2xh - h^2 + 3x + 3h) - (-x^2 + 3x)$
   $$= -2xh - h^2 + 3h = h(-2x - h + 3)$$

   Step 3 $\dfrac{f(x+h) - f(x)}{h} = \dfrac{h(-2x - h + 3)}{h} = -2x - h + 3$

   Step 4 $f'(x) = \lim\limits_{h \to 0} \dfrac{f(x+h) - f(x)}{h} = \lim\limits_{h \to 0} (-2x - h + 3) = -2x + 3.$

17. $f(x) = 2x + 7$. Using the four-step process,

   Step 1 $f(x+h) = 2(x+h) + 7 = 2x + 2h + 7$

   Step 2 $f(x+h) - f(x) = 2x + 2h + 7 - 2x - 7 = 2h$

   Step 3 $\dfrac{f(x+h) - f(x)}{h} = \dfrac{2h}{h} = 2$

   Step 4 $f'(x) = \lim\limits_{h \to 0} \dfrac{f(x+h) - f(x)}{h} = \lim\limits_{h \to 0} 2 = 2$

   we find that $f'(x) = 2$. In particular, the slope at $x = 2$ is also 2. Therefore, a
   required equation is $y - 11 = 2(x - 2)$ or $y = 2x + 7$.

19. $f(x) = 3x^2$. We first compute $f'(x) = 6x$ (see Problem 13). Since the slope of the

   tangent line is $f'(1) = 6$, we use the point-slope form of the equation of a line and find that a required equation is $y - 3 = 6(x - 1)$, or $y = 6x - 3$.

21. $f(x) = -1/x$. We first compute $f'(x)$ using the four-step process.

Step 1 $f(x + h) = -\dfrac{1}{x + h}$

Step 2 $f(x + h) - f(x) = -\dfrac{1}{x + h} + \dfrac{1}{x} = \dfrac{-x + (x + h)}{x(x + h)} = \dfrac{h}{x(x + h)}$

Step 3 $\dfrac{f(x + h) - f(x)}{h} = \dfrac{\frac{h}{x(x + h)}}{h} = \dfrac{1}{x(x + h)}$

Step 4 $f'(x) = \lim\limits_{h \to 0} \dfrac{f(x + h) - f(x)}{h} = \lim\limits_{h \to 0} \dfrac{1}{x(x + h)} = \dfrac{1}{x^2}$.

The slope of the tangent line is $f'(3) = 1/9$. Therefore, a required equation is

$$y - (-\tfrac{1}{3}) = \tfrac{1}{9}(x - 3) \quad \text{or} \quad y = \tfrac{1}{9}x - \tfrac{2}{3}.$$

23. a. $f(x) = 2x^2 + 1$. We use the four-step process.

Step 1 $f(x + h) = 2(x + h)^2 + 1 = 2x^2 + 4xh + 2h^2 + 1$

Step 2 $f(x + h) - f(x) = (2x^2 + 4xh + 2h^2 + 1) - (2x^2 + 1) = 4xh + 2h^2$
$$= h(4x + 2h)$$

Step 3 $\dfrac{f(x + h) - f(x)}{h} = \dfrac{h(4x + 2h)}{h} = 4x + 2h$

Step 4 $f'(x) = \lim\limits_{h \to 0} \dfrac{f(x + h) - f(x)}{h} = \lim\limits_{h \to 0} (4x + 2h) = 4x$

b. The slope of the tangent line is $f'(1) = 4(1) = 4$. Therefore, an equation is
$y - 3 = 4(x - 1)$ or $y = 4x - 1$.

c.

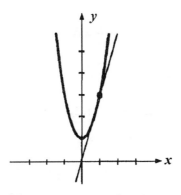

25. a. $f(x) = x^2 - 2x + 1$. We use the four-step process:

Step 1 $f(x + h) = (x + h)^2 - 2(x + h) + 1 = x^2 + 2xh + h^2 - 2x - 2h + 1$

Step 2 $f(x + h) - f(x) = (x^2 + 2xh + h^2 - 2x - 2h + 1) - (x^2 - 2x + 1)]$
$$= 2xh + h^2 - 2h = h(2x + h - 2)$$

Step 3 $\dfrac{f(x+h) - f(x)}{h} = \dfrac{h(2x + h - 2)}{h} = 2x + h - 2$

Step 4 $f'(x) = \lim_{h \to 0} \dfrac{f(x+h) - f(x)}{h} = \lim_{h \to 0} (2x + h - 2) = 2x - 2.$

b. At a point on the graph of f where the tangent line to the curve is horizontal, $f'(x) = 0$. Then $2x - 2 = 0$, or $x = 1$. Since $f(1) = 1 - 2 + 1 = 0$, we see that the required point is $(1,0)$.

c.

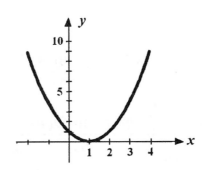

d. It is changing at the rate of 0 units per unit change in x.

27. a. $f(x) = x^2 + x$

$$\frac{f(3)-f(2)}{3-2} = \frac{(3^2+3)-(2^2+2)}{1} = 6$$

$$\frac{f(2.5)-f(2)}{2.5-2} = \frac{(2.5^2+2.5)-(2^2+2)}{0.5} = 5.5$$

$$\frac{f(2.1)-f(2)}{2.1-2} = \frac{(2.1^2+2.1)-(2^2+2)}{0.1} = 5.1$$

b. We first compute $f'(x)$ using the four-step process.

Step 1 $f(x+h) = (x+h)^2 + (x+h) = x^2 + 2xh + h^2 + x + h$

Step 2 $f(x+h)-f(x) = (x^2+2xh+h^2+x+h)-(x^2+x)]$
$$= 2xh + h^2 + h = h(2x+h+1)$$

Step 3 $\dfrac{f(x+h)-f(x)}{h} = \dfrac{h(2x+h+1)}{h} = 2x+h+1$

Step 4 $f'(x) = \lim\limits_{h \to 0} \dfrac{f(x+h)-f(x)}{h} = \lim\limits_{h \to 0}(2x+h+1) = 2x+1.$

The instantaneous rate of change of y at $x = 2$ is $f'(2) = 5$ or 5 units per unit change in x.

c. The results in (a) suggest that the average rates of change of f at $x = 2$ approach 5 as the interval $[2, 2+h]$ gets smaller and smaller ($h = 1, 0.5,$ and 0.1). This number is the instantaneous rate of change of f at $x = 2$ as computed in (b).

29. a. $f(t) = 2t^2 + 48t$. The average velocity of the car over $[20,21]$ is

$$\frac{f(21)-f(20)}{21-20} = \frac{[2(21)^2+48(21)]-[2(20)^2+48(20)]}{1} = 130 \text{ ft/sec}$$

Its average velocity over $[20,20.1]$ is

$$\frac{f(20.1)-f(20)}{20.1-20} = \frac{[2(20.1)^2+48(20.1)]-[2(20)^2+48(20)]}{0.1} = 128.2 \text{ ft/sec}$$

Its average velocity over $[20,20.01]$

$$\frac{f(20.01)-f(20)}{20.01-20} = \frac{[2(20.01)^2+48(20.01)]-[2(20)^2+48(20)]}{0.01} = 128.02 \text{ ft/sec}$$

b. We first compute $f'(t)$ using the four-step process.

Step 1 $f(t + h) = 2(t + h)^2 + 48(t + h) = 2t^2 + 4th + 2h^2 + 48t + 48h$

Step 2 $f(t + h) - f(t) = (2t^2 + 4th + 2h^2 + 48t + 48h) - (2t^2 + 48t)]$

$$= 4th + 2h^2 + 48h = h(4t + 2h + 48).$$

Step 3 $\dfrac{f(t + h) - f(t)}{h} = \dfrac{h(4t + 2h + 48)}{h} = 4t + 2h + 48$

Step 4 $f'(t) = \lim\limits_{t \to 0} \dfrac{f(t + h) - f(t)}{h} = \lim\limits_{t \to 0} 4t + 2h + 48 = 4t + 48$

The instantaneous velocity of the car at $t = 20$ is $f'(20) = 4(20) + 48$, or 128 ft/sec.

c. Our results shows that the average velocities do approach the instantaneous velocity as the intervals over which they are computed decreases.

31. a. We solve the equation $16t^2 = 400$ obtaining $t = 5$ which is the time it takes the screw driver to reach the ground.

b. The average velocity over the time [0,5] is

$$\frac{f(5) - f(0)}{5 - 0} = \frac{16(25) - 0}{5} = 80, \text{ or 80 ft/sec.} \quad [\text{Let s = f(t) = } 16t^2.]$$

c. The velocity of the screwdriver at time t is

$$v(t) = \lim\limits_{h \to 0} \frac{f(t + h) - f(t)}{h} = \lim\limits_{h \to 0} \frac{16(t + h)^2 - 16t^2}{h}$$

$$= \lim\limits_{h \to 0} \frac{16t^2 + 32th + 16h^2 - 16t^2}{h} = \lim\limits_{h \to 0} \frac{(32t + 16h)h}{h} = 32t.$$

In particular, the velocity of the screwdriver when it hits the ground (at $t = 5$) is

$$v(5) = 32(5) = 160, \text{ or 160 ft/sec.}$$

33. a. $V = \dfrac{1}{p}$. The average rate of change of V is

$$\frac{f(3) - f(2)}{3 - 2} = \frac{\frac{1}{3} - \frac{1}{2}}{1} = -\frac{1}{6}, \qquad [\text{Write } V = f(p) = \frac{1}{p}.]$$

or a decrease of $\frac{1}{6}$ liter/atmosphere.

b.

$$V'(t) = \lim_{h \to 0} \frac{\dfrac{f(p+h) - f(p)}{h}}{h} = \lim_{h \to 0} \frac{\dfrac{1}{p+h} - \dfrac{1}{p}}{h}$$

$$= \lim_{h \to 0} \frac{p - (p+h)}{hp(p+h)} = \lim_{h \to 0} -\frac{1}{p(p+h)} = -\frac{1}{p^2}.$$

In particular, the rate of change of V when $p = 2$ is

$$V'(2) = -\frac{1}{2^2}, \text{ or a decrease of } \tfrac{1}{4} \text{ liter/atmosphere}$$

35. a. Using the four-step process, we find that

$$P'(x) = \lim_{h \to 0} \frac{P(x+h) - P(x)}{h}$$

$$= \lim_{h \to 0} \frac{-\tfrac{1}{3}(x^2 + 2xh + h^2) + 7x + 7h + 30 - (-\tfrac{1}{3}x^2 + 7x + 30)}{h}$$

$$P'(x) = \lim_{h \to 0} \frac{P(x+h) - P(x)}{h}$$

$$= \lim_{h \to 0} \frac{-\tfrac{2}{3}xh - \tfrac{1}{3}h + 7h}{h} = \lim_{h \to 0}(-\tfrac{2}{3}x - \tfrac{1}{3}h + 7) = -\tfrac{2}{3}x + 7.$$

b. $P'(10) = -\tfrac{2}{3}(10) + 7 \approx 0.333$, or $333 per quarter.

$P'(30) = -\tfrac{2}{3}(30) + 7 \approx -13$, or a decrease of $13,000 per quarter.

37. $N(t) = t^2 + 2t + 50$. We first compute $N'(t)$ using the four–step process.

Step 1 $N(t + h) = (t + h)^2 + 2(t + h) + 50$
$$= t^2 + 2th + h^2 + 2t + 2h + 50$$

Step 2 $N(t + h) - N(t)$
$$= (t^2 + 2th + h^2 + 2t + 2h + 50) - (t^2 + 2t + 50)$$
$$= 2th + h^2 + 2h = h(2t + h + 2).$$

Step 3 $\dfrac{N(t+h) - N(t)}{h} = 2t + h + 2.$

Step 4 $N'(t) = \lim_{h \to 0}(2t + h + 2) = 2t + 2.$

The rate of change of the country's GNP two years from now will be $N'(2) = 6$, or $6 billion/yr. The rate of change four years from now will be $N'(4) = 10$, or $10 billion/yr.

39. $\dfrac{f(a+h)-f(a)}{h}$ gives the average rate of change of the seal population over the time interval $[a, a + h]$.

$\displaystyle\lim_{h\to 0}\dfrac{f(a+h)-f(a)}{h}$ gives the instantaneous rate of change of the seal population at $x = a$.

41. $\dfrac{f(a+h)-f(a)}{h}$ gives the average rate of change of the country's industrial production over the time interval $[a, a + h]$.

$\displaystyle\lim_{h\to 0}\dfrac{f(a+h)-f(a)}{h}$ gives the instantaneous rate of change of the country's industrial production at $x = a$.

43. $\dfrac{f(a+h)-f(a)}{h}$ gives the average rate of change of the atmospheric pressure over the altitudes $[a, a + h]$. $\displaystyle\lim_{h\to 0}\dfrac{f(a+h)-f(a)}{h}$ gives the instantaneous rate of change of the atmospheric pressure at $x = a$.

45. a. f has a limit at $x = a$.
   b. f is not continuous at $x = a$ because $f(a)$ is not defined.
   c. f is not differentiable at $x = a$ because it is not continuous there.

47. a. f has a limit at $x = a$.   b. f is continuous at $x = a$.
   c. f is not differentiable at $x = a$ because f has a kink at the point $x = a$.

49. a. f does not have a limit at $x = a$ because it is unbounded in the neighborhood of a.
   b. f is not continuous at $x = a$.
   c. f is not differentiable at $x = a$ because it is not continuous there.

51. Our computations yield the following results:
   32.1, 30.939, 30.814, 30.8014, 30.8001, 30.8000.
   The motorcycle's instantaneous velocity at $t = 2$ is approximately 30.8 ft/sec.

53. False. Let $f(x) = |x|$. Then f is continuous at $x = 0$, but is not differentiable there.

55. Observe that the graph of f has a kink at $x = -1$. We have

$$\frac{f(-1+h) - f(-1)}{h} = 1 \text{ if } h > 0, \text{ and } -1 \text{ if } h < 0,$$

so that $\lim_{h \to 0} \dfrac{f(-1+h) - f(-1)}{h}$ does not exist.

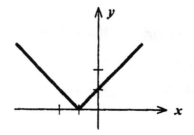

57. For continuity, we require that $f(1) = 1 = \lim_{x \to 1^+}(ax + b) = a + b$, or $a + b = 1$.

In order that the derivative exist at $x = 1$, we require that $\lim_{x \to 1^-} 2x = \lim_{x \to 1^+} a$, or $2 = a$.

Therefore, $b = -1$ and so $f(x) = \begin{cases} x^2 & \text{if } x \le 1 \\ 2x - 1 & \text{if } x > 1 \end{cases}$. The graph of f follows.

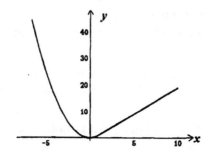

59. We have $f(x) = x$ if $x > 0$ and $f(x) = -x$ if $x < 0$. Therefore, when $x > 0$

$$f'(x) = \lim_{h \to 0}\frac{f(x+h) - f(x)}{h} = \lim_{h \to 0}\frac{x + h - x}{h} = \lim_{h \to 0}\frac{h}{h} = 1,$$

and when $x < 0$

$$f'(x) = \lim_{h \to 0} \frac{f(x+h) - f(x)}{h} = \lim_{h \to 0} \frac{-x - h - (-x)}{h} = \lim_{h \to 0} \frac{-h}{h} = -1.$$

Since the right–hand limit does not equal the left–hand limit, we conclude that $\lim_{h \to 0} f(x)$ does not exist.

USING TECHNOLOGY EXERCISES 2.6, page 154

1. a. $y = 4x - 3$
    b.

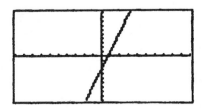

3. a. $y = 9x - 11$
    b.

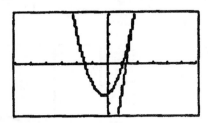

5. a. $y = \frac{1}{4}x + 1$
    b.

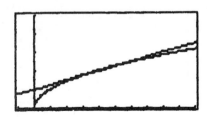

7. a. 4   b. $y = 4x - 1$
    c.

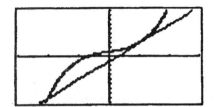

9. a. $\frac{3}{4}$ b. $y = \frac{3}{4}x - 1$
    c.

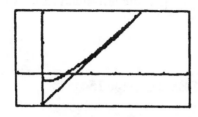

11. a. 4.02   b. $y = 4.02x - 3.57$
    c.

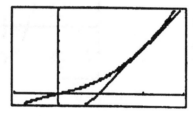

13. a.

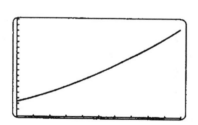

    b. 41.22 cents/mile
    c. 1.22 cents/mile/yr

CHAPTER 2 CONCEPT REVIEW, page 155

1. Domain; range; B

3. $f(x) \pm g(x)$; $f(x)g(x)$; $\dfrac{f(x)}{g(x)}$; $A \cap B$; $A \cap B$; zero

5. a. $P(x) = a_0 x^n + a_1 x^{n-1} + \cdots + a_{n-1}x + a_n$ ($a_0 \neq 0$, n, a positive integer)
   b. Linear; quadratic; cubic

c. Quotient; polynomials   d. x^r, r, a real number

7. a. L' b. $L \pm M$ c. LM d. $\dfrac{L}{M}$; $M \neq 0$

9. a. Right   b. Left   c. L; L

11. a. a; a: $g(a)$ b. Everywhere   c. $Q(x)$

13. a. $f'(a)$ b. $y - f(a) = m(x - a)$

CHAPTER 2 REVIEW, page 156

1. a. $9 - x \geq 0$ gives $x \leq 9$ and the domain is $(-\infty, 9]$.
    b. $2x^2 - x - 3 = (2x - 3)(x + 1)$, and $x = 3/2$ or $x = -1$.
    Since the denominator of the given expression is zero at these points, we see that
    the domain of f cannot include these points and so the domain of f is
    $(-\infty, -1) \cup (-1, \tfrac{3}{2}) \cup (\tfrac{3}{2}, \infty)$.

3. a.

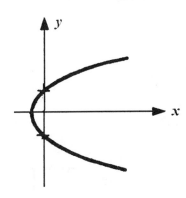

    b. For each value of $x > 0$, there are two values of y. We conclude that y is not a
    function of x. Equivalently, the function fails the vertical line test.
    c. Yes. For each value of y, there is only 1 value of x.

5. a. $f(x)g(x) = \dfrac{2x + 3}{x}$

    b. $\dfrac{f(x)}{g(x)} = \dfrac{1}{x(2x + 3)}$

    c. $f(g(x)) = \dfrac{1}{2x + 3}$.

    d. $g(f(x)) = 2\left(\dfrac{1}{x}\right) + 3 = \dfrac{2}{x} + 3$.

7. $\lim_{x \to 1}(x^2+1) = [(1)^2 + 1] = 1 + 1 = 2.$

9. $\lim_{x \to 3}\dfrac{x-3}{x+4} = \dfrac{3-3}{3+4} = 0.$

11. $\lim_{x \to -2}\dfrac{x^2-2x-3}{x^2+5x+6}$ does not exist. (The denominator is 0 at $x = -2$.)

13. $\lim_{x \to 3}\dfrac{4x-3}{\sqrt{x+1}} = \dfrac{12-3}{\sqrt{4}} = \dfrac{9}{2}.$

15. $\lim_{x \to 1^-}\dfrac{\sqrt{x}-1}{x-1} = \lim_{x \to 1^-}\dfrac{(\sqrt{x}-1)(\sqrt{x}+1)}{(x-1)(\sqrt{x}+1)} = \lim_{x \to 1^-}\dfrac{x-1}{(x-1)(\sqrt{x}+1)} = \lim_{x \to 1^-}\dfrac{1}{\sqrt{x}+1} = \dfrac{1}{2}.$

17. $\lim_{x \to -\infty}\dfrac{x+1}{x} = \lim_{x \to -\infty}\left(1+\dfrac{1}{x}\right) = 1.$

19. $\lim_{x \to -\infty}\dfrac{x^2}{x+1} = \lim_{x \to -\infty} x \cdot \dfrac{1}{1+\dfrac{1}{x}} = -\infty$, so the limit does not exist.

21. $\lim_{x \to 2^+} f(x) = \lim_{x \to 2^+}(x+2) = 4;$

$\lim_{x \to 2^-} f(x) = \lim_{x \to 2^-}(4-x) = 2.$

Therefore, $\lim_{x \to 2} f(x)$ does not exist.

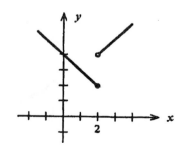

23. Since the denominator
$$4x^2 - 2x - 2 = 2(2x^2 - x - 1) = 2(2x+1)(x-1) = 0$$
if $x = -1/2$ or 1, we see that f is discontinuous at these points.

25. The function is discontinuous at $x = 0$.

27. $f(x) = 3x+5.$ Using the four-step process, we find

Step 1 $\quad f(x+h) = 3(x+h) + 5 = 3x + 3h + 5$

Step 2 $f(x+h)-f(x) = 3x+3h+5-3x-5 = 3h$

Step 3 $\dfrac{f(x+h)-f(x)}{h} = \dfrac{3h}{h} = 3.$

Step 4 $f'(x) = \lim\limits_{h \to 0} \dfrac{f(x+h)-f(x)}{h} = \lim\limits_{h \to 0}(3) = 3.$

29. $f(x) = \frac{3}{2}x+5.$ We use the four-step process to obtain

Step 1 $f(x+h) = \frac{3}{2}(x+h)+5 = \frac{3}{2}x+\frac{3}{2}h+5.$

Step 2 $f(x+h)-f(x) = \frac{3}{2}x+\frac{3}{2}h+5-\frac{3}{2}x-5 = \frac{3}{2}h.$

Step 3 $\dfrac{f(x+h)-f(x)}{h} = \dfrac{3}{2}.$

Step 4 $f'(x) = \lim\limits_{h \to 0}\dfrac{f(x+h)-f(x)}{h} = \lim\limits_{h \to 0}\dfrac{3}{2} = \dfrac{3}{2}.$

Therefore, the slope of the tangent line to the graph of the function f at the point (-2,2) is 3/2. To find the equation of the tangent line to the curve at the point (-2,2), we use the point–slope form of the equation of a line obtaining

$$y-2 = \tfrac{3}{2}[x-(-2)] \quad \text{or} \quad y = \tfrac{3}{2}x+5.$$

31. a. f is continuous at $x = a$ because the three conditions for continuity are satisfied at $x = a$; that is,

    i. $f(x)$ is defined   ii. $\lim\limits_{x \to a} f(x)$ exists   iii. $\lim\limits_{x \to a} f(x) = f(a)$

b. f is not differentiable at $x = a$ because the graph of f has a kink at $x = a$.

33. a. The line passes through (0, 2.4) and (5, 7.4) and has slope $m = \dfrac{7.4-2.4}{5-0} = 1.$

Letting y denote the sales, we see that an equation of the line is
$$y-2.4 = 1(t-0), \text{ or } y = t+2.4.$$
We can also write this in the form $S(t) = t+2.4.$
b. The sales in 2003 were $S(3) = 3+2.4 = 5.4$, or \$5.4 million.

35. Substituting the first equation into the second yields

$$3x - 2(\tfrac{3}{4}x + 6) + 3 = 0 \quad \text{or} \quad \tfrac{3}{2}x - 12 + 3 = 0$$

or $x = 6$. Substituting this value of x into the first equation then gives $y = 21/2$, so the point of intersection is $(6, \tfrac{21}{2})$.

37. We solve the system
$$3x + p - 40 = 0$$
$$2x - p + 10 = 0.$$
Adding these two equations, we obtain $5x - 30 = 0$, or $x = 6$. So,
$$p = 2x + 10 = 12 + 10 = 22.$$
Therefore, the equilibrium quantity is 6000 and the equilibrium price is \$22.

39. When 1000 units are produced,
$$R(1000) = -0.1(1000)^2 + 500(1000) = 400,000, \quad \text{or } \$400,000.$$

41. $N(0) = 200(4 + 0)^{1/2} = 400$, and so there are 400 members initially.
$N(12) = 200(4 + 12)^{1/2} = 800$, and so there are 800 members a year later.

43. $T = f(n) = 4n\sqrt{n - 4}$.
$$f(4) = 0, \ f(5) = 20\sqrt{1} = 20, \ f(6) = 24\sqrt{2} \approx 33.9, \ f(7) = 28\sqrt{3} \approx 48.5,$$
$$f(8) = 32\sqrt{4} = 64, \ f(9) = 36\sqrt{5} \approx 80.5, \ f(10) = 40\sqrt{6} \approx 98,$$
$$f(11) = 44\sqrt{7} \approx 116 \quad \text{and} \quad f(12) = 48\sqrt{8} \approx 135.8.$$

The graph of f follows:

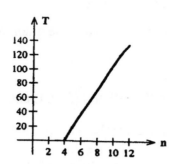

45. We solve
$$-1.1x^2 + 1.5x + 40 = 0.1x^2 + 0.5x + 15$$
$$1.2x^2 - x - 25 = 0$$

$$12x^2 - 10x - 250 = 0$$

$$6x^2 - 5x - 125 = 0; \quad (x-5)(6x+25) = 0.$$

Therefore, $x = 5$. Substituting this value of x into the second supply equation, we have $p = 0.1(5)^2 + 0.5(5) + 15 = 20$. So the equilibrium quantity is 5000 and the equilibrium price is \$20.

47.

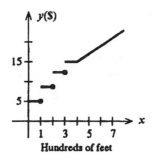

Hundreds of feet

CHAPTER 2 BEFORE MOVING ON, page 158

1. a. $f(-1) = -2(-1) + 1 = 3$ b. $f(0) = 2$ c. $f(\frac{3}{2}) = (\frac{3}{2})^2 + 2 = \frac{17}{4}$.

2. a. $(f+g)(x) = f(x) + g(x) = \dfrac{1}{x+1} + x^2 + 1$ b. $(fg)(x) = f(x)g(x) = \dfrac{x^2+1}{x+1}$

   c. $(f \pm g)(x) = f[g(x)] = \dfrac{1}{g(x)+1} = \dfrac{1}{x^2+2}$

   d. $(g \circ f)(x) = g[f(x)]^2 + 1 = \dfrac{1}{(x+1)^2} + 1$

3. $4x + \ell = 108$ so $\ell = 108 - 4x$

   Volume is $V = x^2 \ell = x^2(108 - 4x) = 108x^2 - 4x^3$

4. $\displaystyle \lim_{x \to -1} \frac{x^2 + 4x + 3}{x^2 + 3x + 2} = \lim_{x \to -1} \frac{(x+3)(x+1)}{(x+2)(x+1)} = 2$

5. a. $\displaystyle \lim_{x \to 1^-} f(x) = \lim_{x \to 1^-} (x^2 - 1) = 0$

95   *2 Functions, Limits, and the Derivative*

b. $\lim\limits_{x\to 1^+} f(x) = \lim\limits_{x\to 1^+} x^3 = 1$.

Since $\lim\limits_{x\to 1^-} f(x) \neq \lim\limits_{x\to 1^+} f(x)$, f is not continuous at 1.

6. The slope of the tangent line at any point is

$$\lim_{h\to 0}\frac{f(x+h)-f(x)}{h} = \lim_{h\to 0}\frac{(x+h)^2 - 3(x+h)+1-(x^2-3x+1)}{h}$$

$$= \lim_{h\to 0}\frac{x^2 + 2xh + h^2 - 3x - 3h - x^2 + 3x - 1}{h}$$

$$= \lim_{h\to 0}\frac{h(2x+h-3)}{h} = \lim_{h\to 0}(2x+h-3) = 2x-3$$

Therefore, the slope at 1 is $2(1) - 3 = -1$. An equation of the tangent line is

$$y - (-1) = -1(x-1)$$

$$y + 1 = -x + 1 \quad , \text{ or } \quad y = -x$$

CHAPTER 3

3.1 Problem Solving Tips

In this section, you were given four basic rules for finding the derivative of a function. As you work through the exercises that follow, first decide which rule(s) you need to find the derivative. Then write out your solution. After doing this a few times, you should have the formulas memorized. The key here is to try not to look at the formula in the text, and to work the problem just as if you were taking a test. If you train yourself to work in this manner, test-taking will be a lot easier. Also, make sure that you distinguish between the notation dy/dx and d/dx. The first notation is used for *the derivative of a function y*, where as the second notation tells us *to find the derivative of the function that follows* with respect to x.

Here are some hints for solving the problems in the exercises that follow:

1. **To find the derivative of a function involving radicals,** first rewrite the expression in exponential form. For example, if $f(x) = 2x - 5\sqrt{x}$, rewrite the function in the form $f'(x) = 2x - 5x^{1/2}$.

2. **To find the point on the graph of f where the tangent line is horizontal,** simply set $f'(x) = 0$ and solve for x. (Here we are making use of the fact that the slope of a horizontal line is zero.) This yields the x-value of the point on the graph where the

tangent line is horizontal. To find the corresponding y-value, simply evaluate the function f at this value of x.

3.1 CONCEPT QUESTIONS, page 167

1. a. The derivative of a constant is zero.
   b. The derivative of $f(x) = x^n$ is the power times x raised to the power $n - 1$.
   c. The derivative of a constant times a function is the constant times the derivative of the function.
   d. The derivative of the sum is the sum of the derivatives.

EXERCISES 3.1, page 167

1. $f'(x) = \dfrac{d}{dx}(-3) = 0.$

3. $f'(x) = \dfrac{d}{dx}(x^5) = 5x^4.$

5. $f'(x) = \dfrac{d}{dx}(x^{2.1}) = 2.1x^{1.1}.$

7. $f'(x) = \dfrac{d}{dx}(3x^2) = 6x.$

9. $f'(r) = \dfrac{d}{dr}(\pi r^2) = 2\pi r.$

11. $f'(x) = \dfrac{d}{dx}(9x^{1/3}) = \dfrac{1}{3}(9)x^{(1/3-1)} = 3x^{-2/3}.$

13. $f'(x) = \dfrac{d}{dx}(3\sqrt{x}) = \dfrac{d}{dx}(3x^{1/2}) = \dfrac{1}{2}(3)x^{-1/2} = \dfrac{3}{2}x^{-1/2} = \dfrac{3}{2\sqrt{x}}.$

15. $f'(x) = \dfrac{d}{dx}(7x^{-12}) = (-12)(7)x^{(-12-1)} = -84x^{-13}.$

17. $f'(x) = \dfrac{d}{dx}(5x^2 - 3x + 7) = 10x - 3.$

19. $f'(x) = \dfrac{d}{dx}(-x^3 + 2x^2 - 6) = -3x^2 + 4x.$

21. $f'(x) = \dfrac{d}{dx}\left(0.03x^2 - 0.4x + 10\right) = 0.06x - 0.4.$

23. If $f(x) = \dfrac{x^3 - 4x^2 + 3}{x} = x^2 - 4x + \dfrac{3}{x}$,

   then $f'(x) = \dfrac{d}{dx}\left(x^2 - 4x + 3x^{-1}\right) = 2x - 4 - \dfrac{3}{x^2}.$

25. $f'(x) = \dfrac{d}{dx}\left(4x^4 - 3x^{5/2} + 2\right) = 16x^3 - \tfrac{15}{2}x^{3/2}.$

27. $f'(x) = \dfrac{d}{dx}\left(3x^{-1} + 4x^{-2}\right) = -3x^{-2} - 8x^{-3}.$

29. $f'(t) = \dfrac{d}{dt}\left(4t^{-4} - 3t^{-3} + 2t^{-1}\right) = -16t^{-5} + 9t^{-4} - 2t^{-2}.$

31. $f'(x) = \dfrac{d}{dx}\left(2x - 5x^{1/2}\right) = 2 - \dfrac{5}{2}x^{-1/2} = 2 - \dfrac{5}{2\sqrt{x}}.$

33. $f'(x) = \dfrac{d}{dx}\left(2x^{-2} - 3x^{-1/3}\right) = -4x^{-3} + x^{-4/3} = -\dfrac{4}{x^3} + \dfrac{1}{x^{4/3}}.$

35. a. $f'(x) = \dfrac{d}{dx}\left(2x^3 - 4x\right) = 6x^2 - 4.$ $\quad f'(-2) = 6(-2)^2 - 4 = 20.$

   b. $f'(0) = 6(0) - 4 = -4.$ $\qquad\qquad$ c. $f'(2) = 6(2)^2 - 4 = 20.$

37. The given limit is $f'(1)$ where $f(x) = x^3$. Since $f'(x) = 3x^2$, we have

$$\lim_{h \to 0} \dfrac{(1+h)^3 - 1}{h} = f'(1) = 3.$$

39. Let $f(x) = 3x^2 - x$. Then $\displaystyle\lim_{h \to 0} \dfrac{3(2+h)^2 - (2+h) - 10}{h} = \lim_{h \to 0} \dfrac{f(2+h) - f(2)}{h}$

because $f(2 + h) - f(2) = 3(2 + h)^2 - (2 + h) - [3(4) - 2]$
$$= 3(2 + h)^2 - (2 + h) - 10.$$
But the last limit is $f'(2)$. Since $f'(x) = 6x - 1$, we have $f'(2) = 11$.

Therefore, $\lim\limits_{h \to 0} \dfrac{3(2+h)^2 - (2+h) - 10}{h} = 11$.

41. $f(x) = 2x^2 - 3x + 4$. The slope of the tangent line at any point $(x, f(x))$ on the graph of f is $f'(x) = 4x - 3$. In particular, the slope of the tangent line at the point $(2,6)$ is $f'(2) = 4(2) - 3 = 5$. An equation of the required tangent line is
$$y - 6 = 5(x - 2) \qquad \text{or} \qquad y = 5x - 4.$$

43. $f(x) = x^4 - 3x^3 + 2x^2 - x + 1$. $f'(x) = 4x^3 - 9x^2 + 4x - 1$.
The slope is $f'(1) = 4 - 9 + 4 - 1 = -2$. An equation
of the tangent line is $y - 0 = -2(x - 1)$ or $y = -2x + 2$.

45. a. $f'(x) = 3x^2$. At a point where the tangent line is horizontal,
$f'(x) = 0$, or $3x^2 = 0$ giving $x = 0$. Therefore, the point is $(0,0)$.

b.

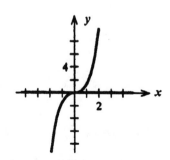

47. a. $f(x) = x^3 + 1$. The slope of the tangent line at any point $(x, f(x))$ on the graph
of f is $f'(x) = 3x^2$. At the point(s) where the slope is 12, we have
$3x^2 = 12$, or $x = \pm 2$. The required points are $(-2,-7)$ and $(2,9)$.
b. The tangent line at $(-2,-7)$ has equation
$$y - (-7) = 12[x - (-2)], \qquad \text{or} \qquad y = 12x + 17,$$
and the tangent line at $(2,9)$ has equation
$$y - 9 = 12(x - 2), \qquad \text{or} \qquad y = 12x - 15.$$

c.

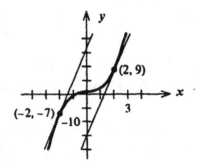

49. If $f(x) = \frac{1}{4}x^4 - \frac{1}{3}x^3 - x^2$, then $f'(x) = x^3 - x^2 - 2x$.

a. $f'(x) = x^3 - x^2 - 2x = -2x$

$\qquad x^3 - x^2 = 0$

$\qquad x^2(x - 1) = 0 \qquad$ and $\qquad x = 0$ or $x = 1$.

$\qquad f(1) = \frac{1}{4}(1)^4 - \frac{1}{3}(1)^3 - (1)^2 = -\frac{13}{12}$.

$\qquad f(0) = \frac{1}{4}(0)^4 - \frac{1}{3}(0)^3 - (0)^2 = 0$.

We conclude that the corresponding points on the graph are $(1, -\frac{13}{12})$ and $(0,0)$.

b. $\qquad f'(x) = x^3 - x^2 - 2x = 0$

$\qquad\qquad x(x^2 - x - 2) = 0$

$\qquad\qquad x(x - 2)(x + 1) = 0 \quad$ and $\quad x = 0, 2,$ or -1.

$\qquad\qquad f(0) = 0$

$$f(2) = \frac{1}{4}(2)^4 - \frac{1}{3}(2)^3 - (2)^2 = 4 - \frac{8}{3} - 4 = -\frac{8}{3}.$$

$$f(-1) = \frac{1}{4}(-1)^4 - \frac{1}{3}(-1)^3 - (-1)^2 = \frac{1}{4} + \frac{1}{3} - 1 = -\frac{5}{12}.$$

We conclude that the corresponding points are $(0,0)$, $(2, -\frac{8}{3})$ and $(-1, -\frac{5}{12})$.

c. $\qquad f'(x) = x^3 - x^2 - 2x = 10x$

$\qquad\qquad x^3 - x^2 - 12x = 0$

$\qquad\qquad x(x^2 - x - 12) = 0$

$\qquad\qquad x(x - 4)(x + 3) = 0$

and $x = 0, 4,$ or -3.

$\qquad\qquad f(0) = 0$

$\qquad\qquad f(4) = \frac{1}{4}(4)^4 - \frac{1}{3}(4)^3 - (4)^2 = 48 - \frac{64}{3} = \frac{80}{3}$.

$\qquad\qquad f(-3) = \frac{1}{4}(-3)^4 - \frac{1}{3}(-3)^3 - (-3)^2 = \frac{81}{4} + 9 - 9 = \frac{81}{4}$.

We conclude that the corresponding points are $(0,0)$, $(4, \frac{80}{3})$ and $(-3, \frac{81}{4})$.

3 Differentiation

51. $V(r) = \frac{4}{3}\pi r^3$. $V'(r) = 4\pi r^2$.

 a. $V'(\frac{2}{3}) = 4\pi(\frac{4}{9}) = \frac{16}{9}\pi$ cm³/cm. b. $V'(\frac{5}{4}) = 4\pi(\frac{25}{16}) = \frac{25}{4}\pi$ cm³/cm.

53. a. $N(1) = 16.3(1^{0.8766}) = 16.3(1) = 16.3$, or 16.3 million cameras.

 b. $N'(t) = (16.3)90.8766)t^{-0.1234}$; $N'(1) \approx 14.29$, or approximately 14.3 million cameras/yr.

 c. $N(5) = 16.3(5^{0.8766}) \approx 66.82$, or approximately 66.8 million cameras.

 d. $N'(5) = (16.3)(0.8766)(5^{-0.1234}) \approx 11.71$, or approximately 11.7 million cameras/year.

55. a.

1970 ($t = 1$)	1980($t = 2$)	1990 ($t = 3$)	2000($t = 4$)
49.6%	41.1%	36.9%	34.1%

 b. $P'(t) = (49.6)(-0.27t^{-1.27}) = -\dfrac{13.392}{t^{1.27}}$; In 1980, $P'(2) \approx -5.5$, or decreasing at 5.5%/decade. In 1990, $P'(3) \approx -3.3$, or decreasing at 3.3%/decade.

57. a. $P(9) = 24.4(9)^{0.34} = 51.5$, or 51.5%.

 b. $P(t) = 24.4t^{0.34}$; $P'(t) = \dfrac{d}{dx}(24.4t^{0.34}) = (0.34)(24.4)t^{-0.66} = 8.296t^{-0.66}$.

 $P'(9) = 8.296(9)^{-0.66} = 1.946$, or approximately 1.95%/year.

59. a. $f(t) = 120t - 15t^2$. $v = f'(t) = 120 - 30t$ b. $v(0) = 120$ ft/sec

 c. Setting $v = 0$ gives 120 - 30t = 0, or $t = 4$. Therefore, the stopping distance is $f(4) = 120(4) - 15(16)$ or 240 ft.

61. a. At the beginning of 1980, $P(0) = 5\%$. At the beginning of 1990, $P(10) = -0.0105(10^2) + 0.735(10) + 5 \approx 11.3\%$. At the beginning of 2000, $P(20) = -0.0105(20)^2 + 0.735(20) + 5 \approx 15.5\%$.

 b. $P'(t) = -0.021t + 0.735$; At the beginning of 1985, $P'(5) = -0.02(5) + 0.735 \approx 0.63\%/$ yr. At the beginning of 1990, $P'(10) = -0.021(10) + 0.735 = 0.525\%/$ yr.

63. a $f(t) = 5.303t^2 - 53.977t + 253.8$. The rate of change of the groundfish
population at any time t is given by $f'(t) = 10.606t - 53.977$. The rate of change
at the beginning of 1994 is given by $f'(5) = 10.606(5) - 53.977 = -0.947$
and so the population is decreasing at the rate of 0.9 thousand metric tons/yr. At
the beginning of 1996, the rate of change is $f'(7) = 10.606(7) - 53.977 = 20.265$
and so the population is increasing at the rate of 20.3 thousand metric tons/yr.
b. Yes.

65. $I'(t) = -0.6t^2 + 6t$.
a. In 1999, it was changing at a rate of $I'(5) = -0.6(25) + 6(5)$, or 15 points/yr. In
2001, it was $I'(7) = -0.6(49) + 6(7)$, or 12.6 pts/yr. In 2004, it was
$I'(10) = -0.6(100) + 6(10)$, or 0 pts/yr.
b. The average rate of increase of the CPI over the period from 1999 to 2004 was
$$\frac{I(10) - I(5)}{5} = \frac{[-0.2(1000) + 3(100) + 100] - [-0.2(125) + 3(25) + 100]}{5}$$
$$= \frac{200 - 150}{5} = 10, \text{ or } 10 \text{ pts/yr.}$$

67. a. $f'(x) = \frac{d}{dx}\left[0.0001x^{5/4} + 10\right] = \frac{5}{4}(0.0001x^{1/4}) = 0.000125x^{1/4}$
b. $f'(10,000) = 0.000125(10,000)^{1/4} = 0.00125$, or $0.00125/radio.

69. a. $f(t) = 20t - 40\sqrt{t} + 50$. $f'(t) = 20 - 40\left(\frac{1}{2}\right)t^{-1/2} = 20\left(1 - \frac{1}{\sqrt{t}}\right)$.
b. $f(0) = 20(0) - 40\sqrt{0} + 50 = 50$; $f(1) = 20(1) - 40\sqrt{1} + 50 = 30$
$f(2) = 20(2) - 40\sqrt{2} + 50 \approx 33.43$.
The average velocity at 6, 7, and 8 A.M. is 50 mph, 30 mph, and 33.43 mph,
respectively.
c. $f'(\frac{1}{2}) = 20 - 20(\frac{1}{2})^{-1/2} \approx -8.28$. $f'(1) = 20 - 20(1)^{-1/2} \approx 0$.
$f'(2) = 20 - 20(2)^{-1/2} \approx 5.86$.
At 6:30 A.M. the average velocity is decreasing at the rate of 8.28 mph/hr; at
7 A.M., it is unchanged, and at 8 A.M., it is increasing at the rate of 5.86 mph.

71. $N(t) = 2t^3 + 3t^2 - 4t + 1000$. $N'(t) = 6t^2 + 6t - 4$.

$N'(2) = 6(4) + 6(2) - 4 = 32$, or 32 turtles/yr.

$N'(8) = 6(64) + 6(8) - 4 = 428$, or 428 turtles/yr.

The population ten years after implementation of the conservation measures will be $N(10) = 2(10^3) + 3(10^2) - 4(10) + 1000$, or 3260 turtles.

73. a. At the beginning of 1991, $P(0) = 12\%$. At the beginning of 2004,

$P(13) = 0.0004(13^3) + 0.0036(13^2) + 0.8(13) + 12 \approx 23.9\%$.

b. $P'(t) = 0.0012t^2 + 0.0072t + 0.8$. At the beginning of 1991, $P'(0) = 0.8\%/$ yr.

At the beginning of 2004, $P'(13) = 0.0012(13^2) + 0.0072(3) + 0.8 \approx 1.1\%/$ yr.

75. True. $\dfrac{d}{dx}[2f(x) - 5g(x)] = \dfrac{d}{dx}[2f(x)] - \dfrac{d}{dx}[5g(x)] = 2f'(x) - 5g'(x)$.

77. $\dfrac{d}{dx}(x^3) = \lim_{h\to 0} \dfrac{(x+h)^3 - x^3}{h} = \lim_{h\to 0} \dfrac{x^3 + 3x^2h + 3xh^2 + h^3 - x^3}{h}$

$= \lim_{h\to 0} \dfrac{h(3x^2 + 3xh + h^2)}{h} = \lim_{h\to 0}(3x^2 + 3xh + h^2) = 3x^2$.

USING TECHNOL0GY EXERCISES 3.1, page 172

1. 1   2. 3.072   3. 0.4226   4. 0.0732   5.   0.1613   6. 3.9730

7. a.

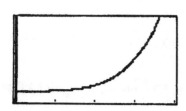

b. 3.4295 parts/million;
   105.4332 parts/million

9. a.

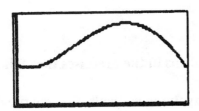

b. decreasing at the rate of 9 days/yr
   increasing at the rate of 13 days/yr

11. a.

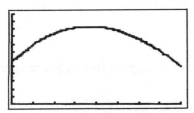

b. Increasing at the rate of 1.1557%/yr
   decreasing at the rate of 0.2116%/yr

3.2 Problem Solving Tips

The answers at the back of the book for the exercises in this section are given in both

simplified and unsimplified terms. Here, as with all of your homework, you should

make it a practice to analyze your errors. If you do not get the right answer for the

unsimplified form, it means that you are not applying the rules for differentiating

correctly. In this case you need to review the rules, making sure that you can write out

each rule. If you have the correct answer for the unsimplified form but the incorrect

answer for the simplified form, it probably means that you have made an algebraic

error. You may need to review the rules for simplifying algebraic expressions given on

page 12 of the text and then work some of the exercises given in section 1.2 to get

back into practice. In any case, you will need to simplify your answers when you work the problems on the applications of the derivative in the next chapter, so you should get in the habit of doing so now.

Here are some hints for solving the problems in the exercises that follow:

1. **To find the derivative of a function involving radicals**, first rewrite the expression in exponential form. For example, if $f(x) = 2x - 5\sqrt{x}$, rewrite the function in the form $f'(x) = 2x - 5x^{1/2}$.

2. **To find point on the graph of f where the tangent line is horizontal,** simply set $f'(x) = 0$ and solve for x. (Here we are making use of the fact that the slope of a horizontal line is zero.) This yields the x-value of the point on the graph where the tangent line is horizontal. To find the corresponding y-value, simply evaluate the function f at this value of x.

3.2 CONCEPT QUESTIONS, page 180

1. a. The derivative of the product of two functions is equal to the first function times the derivative of the second function plus the second function times the derivative of the first function.
    b. The derivative of the quotient of two functions is equal to the quotient whose numerator is given by the denominator times the derivative of the numerator minus the numerator times the derivative of the denominator and the denominator is the square of the denominator of the quotient.

EXERCISES 3.2, page 180

1. $f(x) = 2x(x^2 + 1)$.

$$f'(x) = 2x\frac{d}{dx}(x^2 + 1) + (x^2 + 1)\frac{d}{dx}(2x)$$
$$= 2x(2x) + (x^2 + 1)(2) = 6x^2 + 2.$$

3. $f(t) = (t - 1)(2t + 1)$
$$f'(t) = (t-1)\frac{d}{dt}(2t+1) + (2t+1)\frac{d}{dt}(t-1)$$
$$= (t-1)(2) + (2t+1)(1) = 4t - 1$$

5. $f(x) = (3x + 1)(x^2 - 2)$
$$f'(x) = (3x+1)\frac{d}{dx}(x^2 - 2) + (x^2 - 2)\frac{d}{dx}(3x+1)$$
$$= (3x+1)(2x) + (x^2 - 2)(3) = 9x^2 + 2x - 6.$$

7. $f(x) = (x^3 - 1)(x + 1).$
$$f'(x) = (x^3 - 1)\frac{d}{dx}(x+1) + (x+1)\frac{d}{dx}(x^3 - 1)$$
$$= (x^3 - 1)(1) + (x+1)(3x^2) = 4x^3 + 3x^2 - 1.$$

9. $f(w) = (w^3 - w^2 + w - 1)(w^2 + 2).$
$$f'(w) = (w^3 - w^2 + w - 1)\frac{d}{dw}(w^2 + 2) + (w^2 + 2)\frac{d}{dw}(w^3 - w^2 + w - 1)$$
$$= (w^3 - w^2 + w - 1)(2w) + (w^2 + 2)(3w^2 - 2w + 1)$$
$$= 2w^4 - 2w^3 + 2w^2 - 2w + 3w^4 - 2w^3 + w^2 + 6w^2 - 4w + 2$$
$$= 5w^4 - 4w^3 + 9w^2 - 6w + 2.$$

11. $f(x) = (5x^2 + 1)(2\sqrt{x} - 1)$
$$f'(x) = (5x^2 + 1)\frac{d}{dx}(2x^{1/2} - 1) + (2x^{1/2} - 1)\frac{d}{dx}(5x^2 + 1)$$

$$= (5x^2 + 1)(x^{-1/2}) + (2x^{1/2} - 1)(10x)$$

$$= 5x^{3/2} + x^{-1/2} + 20x^{3/2} - 10x \;=\; \frac{25x^2 - 10x\sqrt{x} + 1}{\sqrt{x}}.$$

13. $f(x) = (x^2 - 5x + 2)(x - \dfrac{2}{x})$

$$f'(x) = (x^2 - 5x + 2)\frac{d}{dx}(x - \frac{2}{x}) + (x - \frac{2}{x})\frac{d}{dx}(x^2 - 5x + 2)$$

$$= \frac{(x^2 - 5x + 2)(x^2 + 2)}{x^2} + \frac{(x^2 - 2)(2x - 5)}{x}$$

$$= \frac{(x^2 - 5x + 2)(x^2 + 2) + x(x^2 - 2)(2x - 5)}{x^2}$$

$$= \frac{x^4 + 2x^2 - 5x^3 - 10x + 2x^2 + 4 + 2x^4 - 5x^3 - 4x^2 + 10x}{x^2}$$

$$= \frac{3x^4 - 10x^3 + 4}{x^2}.$$

15. $f(x) = \dfrac{1}{x-2}.$ $f'(x) = \dfrac{(x-2)\dfrac{d}{dx}(1) - (1)\dfrac{d}{dx}(x-2)}{(x-2)^2} = \dfrac{0 - 1(1)}{(x-2)^2} = -\dfrac{1}{(x-2)^2}.$

17. $f(x) = \dfrac{x-1}{2x+1}.$

$$f'(x) = \frac{(2x+1)\dfrac{d}{dx}(x-1) - (x-1)\dfrac{d}{dx}(2x+1)}{(2x+1)^2}$$

$$= \frac{2x+1 - (x-1)(2)}{(2x+1)^2} = \frac{3}{(2x+1)^2}.$$

19. $f(x) = \dfrac{1}{x^2+1}.$

$$f'(x) = \frac{(x^2+1)\dfrac{d}{dx}(1) - (1)\dfrac{d}{dx}(x^2+1)}{(x^2+1)^2}$$

$$= \frac{(x^2+1)(0) - 1(2x)}{(x^2+1)^2} = -\frac{2x}{(x^2+1)^2}.$$

21. $f(s) = \dfrac{s^2-4}{s+1}.$

$$f'(s) = \frac{(s+1)\dfrac{d}{ds}(s^2-4) - (s^2-4)\dfrac{d}{ds}(s+1)}{(s+1)^2}$$

$$= \frac{(s+1)(2s) - (s^2-4)(1)}{(s+1)^2} = \frac{s^2+2s+4}{(s+1)^2}.$$

23. $f(x) = \dfrac{\sqrt{x}}{x^2+1}.$

$$f'(x) = \frac{(x^2+1)\dfrac{d}{dx}(x^{1/2}) - (x^{1/2})\dfrac{d}{dx}(x^2+1)}{(x^2+1)^2} = \frac{(x^2+1)(\frac{1}{2}x^{-1/2}) - (x^{1/2})(2x)}{(x^2+1)^2}$$

$$= \frac{(\frac{1}{2}x^{-1/2})[(x^2+1) - 4x^2]}{(x^2+1)^2} = \frac{1-3x^2}{2\sqrt{x}(x^2+1)^2}.$$

25. $f(x) = \dfrac{x^2+2}{x^2+x+1}.$

$$f'(x) = \frac{(x^2+x+1)\dfrac{d}{dx}(x^2+2) - (x^2+2)\dfrac{d}{dx}(x^2+x+1)}{(x^2+x+1)^2}$$

$$= \frac{(x^2+x+1)(2x) - (x^2+2)(2x+1)}{(x^2+x+1)^2}$$

$$= \frac{2x^3+2x^2+2x-2x^3-x^2-4x-2}{(x^2+x+1)^2} = \frac{x^2-2x-2}{(x^2+x+1)^2}.$$

27. $f(x) = \dfrac{(x+1)(x^2+1)}{x-2} = \dfrac{(x^3+x^2+x+1)}{x-2}$.

$f'(x) = \dfrac{(x-2)\dfrac{d}{dx}(x^3+x^2+x+1) - (x^3+x^2+x+1)\dfrac{d}{dx}(x-2)}{(x-2)^2}$

$= \dfrac{(x-2)(3x^2+2x+1) - (x^3+x^2+x+1)}{(x-2)^2}$

$= \dfrac{3x^3+2x^2+x-6x^2-4x-2-x^3-x^2-x-1}{(x-2)^2} = \dfrac{2x^3-5x^2-4x-3}{(x-2)^2}$.

29. $f(x) = \dfrac{x}{x^2-4} - \dfrac{x-1}{x^2+4} = \dfrac{x(x^2+4)-(x-1)(x^2-4)}{(x^2-4)(x^2+4)} = \dfrac{x^2+8x-4}{(x^2-4)(x^2+4)}$.

$f'(x) = \dfrac{(x^2-4)(x^2+4)\dfrac{d}{dx}(x^2+8x-4) - (x^2+8x-4)\dfrac{d}{dx}(x^4-16)}{(x^2-4)^2(x^2+4)^2}$

$= \dfrac{(x^2-4)(x^2+4)(2x+8) - (x^2+8x-4)(4x^3)}{(x^2-4)^2(x^2+4)^2}$

$= \dfrac{2x^5+8x^4-32x-128-4x^5-32x^4+16x^3}{(x^2-4)^2(x^2+4)^2}$

$= \dfrac{-2x^5-24x^4+16x^3-32x-128}{(x^2-4)^2(x^2+4)^2}$.

31. $h'(x) = f(x)g'(x) + f'(x)g(x)$, by the Product Rule. Therefore,
$h'(1) = f(1)g'(1) + f'(1)g(1) = (2)(3) + (-1)(-2) = 8$.

33. Using the Quotient Rule followed by the Product Rule, we have

$h'(x) = \dfrac{[x+g(x)]\dfrac{d}{dx}[xf(x)] - xf(x)\dfrac{d}{dx}[x+g(x)]}{[x+g(x)]^2}$

$= \dfrac{[x+g(x)][xf'(x)+f(x)] - xf(x)[1+g'(x)]}{[x+g(x)]^2}$

Therefore, $h'(1) = \dfrac{[1+g(1)][f'(1)+f(1)]-f(1)[1+g'(1)]}{[1+g(1)]^2}$

$$= \dfrac{(1-2)(-1+2)-2(1+3)}{(1-2)^2} = \dfrac{-1-8}{1} = -9.$$

35. $f(x) = (2x-1)(x^2+3)$

$f'(x) = (2x-1)\dfrac{d}{dx}(x^2+3) + (x^2+3)\dfrac{d}{dx}(2x-1)$

$\quad = (2x-1)(2x) + (x^2+3)(2) = 6x^2 - 2x + 6 = 2(3x^2 - x + 3).$

At $x = 1, f'(1) = 2[3(1)^2 - (1) + 3] = 2(5) = 10.$

37. $f(x) = \dfrac{x}{x^4 - 2x^2 - 1}.$

$f'(x) = \dfrac{(x^4 - 2x^2 - 1)\dfrac{d}{dx}(x) - x\dfrac{d}{dx}(x^4 - 2x^2 - 1)}{(x^4 - 2x^2 - 1)^2}$

$\quad = \dfrac{(x^4 - 2x^2 - 1)(1) - x(4x^3 - 4x)}{(x^4 - 2x^2 - 1)^2} = \dfrac{-3x^4 + 2x^2 - 1}{(x^4 - 2x^2 - 1)^2}.$

Therefore, $f'(-1) = \dfrac{-3+2-1}{(1-2-1)^2} = -\dfrac{2}{4} = -\dfrac{1}{2}.$

39. $f(x) = (x^3 + 1)(x^2 - 2).$

$f'(x) = (x^3 + 1)\dfrac{d}{dx}(x^2 - 2) + (x^2 - 2)\dfrac{d}{dx}(x^3 + 1)$

$\quad = (x^3 + 1)(2x) + (x^2 - 2)(3x^2).$

The slope of the tangent line at $(2,18)$ is $f'(2) = (8 + 1)(4) + (4 - 2)(12) = 60.$
An equation of the tangent line is $y - 18 = 60(x - 2)$, or $y = 60x - 102.$

41. $f(x) = \dfrac{x+1}{x^2 + 1}.$

$$f'(x) = \frac{(x^2+1)\dfrac{d}{dx}(x+1) - (x+1)\dfrac{d}{dx}(x^2+1)}{(x^2+1)^2}$$

$$= \frac{(x^2+1)(1) - (x+1)(2x)}{(x^2+1)^2} = \frac{-x^2-2x+1}{(x^2+1)^2}.$$

At $x=1$, $f'(1) = \dfrac{-1-2+1}{4} = -\dfrac{1}{2}$. Therefore, the slope of the tangent line at $x=1$ is -1/2. Then an equation of the tangent line is

$$y-1 = -\tfrac{1}{2}(x-1) \quad \text{or} \quad y = -\tfrac{1}{2}x + \tfrac{3}{2}.$$

43. $f(x) = (x^3+1)(3x^2-4x+2)$

$$f'(x) = (x^3+1)\frac{d}{dx}(3x^2-4x+2) + (3x^2-4x+2)\frac{d}{dx}(x^3+1)$$

$$= (x^3+1)(6x-4) + (3x^2-4x+2)(3x^2)$$

$$= 6x^4 + 6x - 4x^3 - 4 + 9x^4 - 12x^3 + 6x^2$$

$$= 15x^4 - 16x^3 + 6x^2 + 6x - 4.$$

At $x = 1$, $f'(1) = 15(1)^4 - 16(1)^3 + 6(1) + 6(1) - 4 = 7$. The slope of the tangent line at the point $x = 1$ is 7. The equation of the tangent line is

$$y - 2 = 7(x - 1), \quad \text{or} \quad y = 7x - 5.$$

45. $f(x) = (x^2+1)(2-x)$

$$f'(x) = (x^2+1)\frac{d}{dx}(2-x) + (2-x)\frac{d}{dx}(x^2+1)$$

$$= (x^2+1)(-1) + (2-x)(2x) = -3x^2 + 4x - 1.$$

At a point where the tangent line is horizontal, we have

$$f'(x) = -3x^2 + 4x - 1 = 0$$

or $3x^2 - 4x + 1 = (3x-1)(x-1) = 0$, giving $x = 1/3$ or $x = 1$.

Since $f(\tfrac{1}{3}) = (\tfrac{1}{9}+1)(2-\tfrac{1}{3}) = \tfrac{50}{27}$, and $f(1) = 2(2-1) = 2$, we see that the required points are $(\tfrac{1}{3}, \tfrac{50}{27})$ and $(1, 2)$.

47. $f(x) = (x^2+6)(x-5)$

$$f'(x) = (x^2+6)\frac{d}{dx}(x-5)+(x-5)\frac{d}{dx}(x^2+6)$$

$$= (x^2+6)(1)+(x-5)(2x) = x^2+6+2x^2-10x = 3x^2-10x+6.$$

At a point where the slope of the tangent line is -2, we have

$$f'(x) = 3x^2-10x+6 = -2.$$

This gives $3x^2 - 10x + 8 = (3x-4)(x-2) = 0$. So $x = \frac{4}{3}$ or $x = 2$.

Since $f(\frac{4}{3}) = (\frac{16}{9}+6)(\frac{4}{3}-5) = -\frac{770}{27}$ and $f(2) = (4+6)(2-5) = -30$,

the required points are $(\frac{4}{3}, -\frac{770}{27})$ and $(2,-30)$.

49. $y = \dfrac{1}{1+x^2}$. $y' = \dfrac{(1+x^2)\frac{d}{dx}(1)-(1)\frac{d}{dx}(1+x^2)}{(1+x^2)^2} = \dfrac{-2x}{(1+x^2)^2}.$

So, the slope of the tangent line at $(1,\frac{1}{2})$ is

$$y'\big|_{x=1} = \frac{-2x}{(1+x^2)^2}\bigg|_{x=1} = \frac{-2}{4} = -\frac{1}{2}$$

and the equation of the tangent line is $y-\frac{1}{2} = -\frac{1}{2}(x-1)$, or $y = -\frac{1}{2}x+1$.

Next, the slope of the required normal line is 2 and its equation is

$$y-\frac{1}{2} = 2(x-1), \quad \text{or} \quad y = 2x - \frac{3}{2}.$$

51. $C(x) = \dfrac{0.5x}{100-x}$. $C'(x) = \dfrac{(100-x)(0.5)-0.5x(-1)}{(100-x)^2} = \dfrac{50}{(100-x)^2}.$

$C'(80) = \dfrac{50}{20^2} = 0.125$; $\qquad C'(90) = \dfrac{50}{10^2} = 0.5,$

$C'(95) = \dfrac{50}{5^2} = 2;$ $\qquad\qquad C'(99) = \dfrac{50}{1} = 50.$

The rates of change of the cost in removing 80%, 90%, and 99% of the toxic waste are 0.125, 0.5, 2, and 50 million dollars per 1% more of the waste to be removed, respectively. It is too costly to remove *all* of the pollutant.

53. $N(t) = \dfrac{10,000}{1+t^2}+2000$

$$N'(t) = \frac{d}{dt}[10,000(1+t^2)^{-1}+2000] = -\frac{10,000}{(1+t^2)^2}(2t) = -\frac{20,000t}{(1+t^2)^2}.$$

The rate of change after 1 minute and after 2 minutes is

$$N'(1) = -\frac{20,000}{(1+1^2)^2} = -5000; \quad N'(2) = -\frac{20,000(2)}{(1+2^2)^2} = -1600.$$

The population of bacteria after one minute is $N(1) = \dfrac{10,000}{1+1} + 2000 = 7000$.

The population after two minutes is $N(2) = \dfrac{10,000}{1+4} + 2000 = 4000$.

55. a. $N(t) = \dfrac{60t + 180}{t+6}$.

$$N'(t) = \frac{(t+6)\dfrac{d}{dt}(60t+180) - (60t+180)\dfrac{d}{dt}(t+6)}{(t+6)^2}$$

$$= \frac{(t+6)(60) - (60t+180)(1)}{(t+6)^2} = \frac{180}{(t+6)^2}.$$

b. $N'(1) = \dfrac{180}{(1+6)^2} = 3.7, \quad N'(3) = \dfrac{180}{(3+6)^2} = 2.2, \quad N'(4) = \dfrac{180}{(4+6)^2} = 1.8,$

$N'(7) = \dfrac{180}{(7+6)^2} = 1.1$

We conclude that the rate at which the average student is increasing his or her speed one week, three weeks, four weeks, and seven weeks into the course is 3.7, 2.2, 1.8, and 1.1 words per minute, respectively.

c. Yes

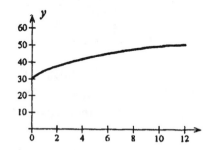

d. $N(12) = \dfrac{60(12)+180}{12+6} = 50$, or 50 words/minute.

57. $f(t) = \dfrac{0.055t + 0.26}{t + 2}$; $f'(t) = \dfrac{(t+2)(0.055) - (0.055t + 0.26)(1)}{(t+2)^2} = -\dfrac{0.15}{(t+2)^2}$.

At the beginning, the formaldehyde level is changing at the rate of

$$f'(0) = -\dfrac{0.15}{4} = -0.0375;$$

that is, it is dropping at the rate of 0.0375 parts per million per year. Next,

$$f'(3) = -\dfrac{0.15}{5^2} = -0.006,$$

and so the level is dropping at the rate of 0.006 parts per million per year at the beginning of the fourth year ($t = 3$).

59. False. Take $f(x) = x$ and $g(x) = x$. Then $f(x)g(x) = x^2$. So

$$\dfrac{d}{dx}[f(x)g(x)] = \dfrac{d}{dx}(x^2) = 2x \neq f'(x)g'(x) = 1.$$

61. False. Let $f(x) = x^3$. Then

$$\dfrac{d}{dx}\left[\dfrac{f(x)}{x^2}\right] = \dfrac{d}{dx}\left(\dfrac{x^3}{x^2}\right) = \dfrac{d}{dx}(x) = 1 \neq \dfrac{f'(x)}{2x} = \dfrac{3x^2}{2x} = \dfrac{3}{2}x.$$

63. Let $f(x) = u(x)v(x)$ and $g(x) = w(x)$. Then $h(x) = f(x)g(x)$. Therefore,
$$h'(x) = f'(x)g(x) + f(x)g'(x).$$
But $\qquad f'(x) = u(x)v'(x) + u'(x)v(x).$
Therefore, $h'(x) = [u(x)v'(x) + u'(x)v(x)]g(x) + u(x)v(x)w'(x)$
$$= u(x)v(x)w'(x) + u(x)v'(x)w(x) + u'(x)v(x)w(x).$$

USING TECHNOLOGY EXERCISES 3.2, page 184

1. 0.8750   3. 0.0774   5. -0.   7. 87,322 per year

3.3 Problem Solving Tips

Here are some hints for solving the problems in the exercises that follow:

3 Differentiation

1. It is often easier to find the derivative of a quotient when the numerator is a constant by using the General Power Rule, rather than the Quotient Rule. For example, to find the derivative of $f(x) = -\dfrac{1}{\sqrt{2x^2 - 1}}$ in self-check exercise #1 of this section, we first rewrite the function in the form $f(x) = -(2x^2 - 1)^{-1/2}$ and then use the General Power Rule to find the derivative.

2. **To simplify a function involving the powers of an expression**, factor out the lowest power of the expression. For example, to factor the expression

$$5(x+1)^{1/2} - 3(x+1)^{-1/2}$$

factor out $(x+1)^{-1/2}$ [which is the lowest power of $(x + 1)$ found in the expression] .

3.3 CONCEPT QUESTIONS, page 192

1. The derivative of $h(x) = g[f(x)]$ is equal to the derivative of g evaluated at $f(x)$ times the derivative of f.

EXERCISES 3.3, page 192

1. $f(x) = (2x - 1)^4$. $f'(x) = 4(2x - 1)^3 \dfrac{d}{dx}(2x - 1) = 4(2x - 1)^3(2) = 8(2x - 1)^3$.

3. $f(x) = (x^2 + 2)^5$. $f'(x) = 5(x^2 + 2)^4(2x) = 10x(x^2 + 2)^4$.

5. $f(x) = (2x - x^2)^3$.

 $f'(x) = 3(2x - x^2)^2 \dfrac{d}{dx}(2x - x^2) = 3(2x - x^2)^2(2 - 2x) = 6x^2(1 - x)(2 - x)^2$.

7. $f(x) = (2x+1)^{-2}$.

$$f'(x) = -2(2x+1)^{-3}\frac{d}{dx}(2x+1) = -2(2x+1)^{-3}(2) = -4(2x+1)^{-3}.$$

9. $f(x) = (x^2-4)^{3/2}$.

$$f'(x) = \tfrac{3}{2}(x^2-4)^{1/2}\frac{d}{dx}(x^2-4) = \tfrac{3}{2}(x^2-4)^{1/2}(2x) = 3x(x^2-4)^{1/2}.$$

11. $f(x) = \sqrt{3x-2} = (3x-2)^{1/2}$.

$$f'(x) = \frac{1}{2}(3x-2)^{-1/2}(3) = \frac{3}{2}(3x-2)^{-1/2} = \frac{3}{2\sqrt{3x-2}}.$$

13. $f(x) = \sqrt[3]{1-x^2}$.

$$f'(x) = \frac{d}{dx}(1-x^2)^{1/3} = \frac{1}{3}(1-x^2)^{-2/3}\frac{d}{dx}(1-x^2)$$

$$= \frac{1}{3}(1-x^2)^{-2/3}(-2x) = -\frac{2}{3}x(1-x^2)^{-2/3} = \frac{-2x}{3(1-x^2)^{2/3}}.$$

15. $f(x) = \dfrac{1}{(2x+3)^3} = (2x+3)^{-3}$.

$$f'(x) = -3(2x+3)^{-4}(2) = -6(2x+3)^{-4} = -\frac{6}{(2x+3)^4}.$$

17. $f(t) = \dfrac{1}{\sqrt{2t-3}}$.

$$f'(t) = \frac{d}{dt}(2t-3)^{-1/2} = -\frac{1}{2}(2t-3)^{-3/2}(2) = -(2t-3)^{-3/2} = -\frac{1}{(2t-3)^{3/2}}.$$

19. $y = \dfrac{1}{(4x^4+x)^{3/2}}$.

$$\frac{dy}{dx} = \frac{d}{dx}(4x^4+x)^{-3/2} = -\frac{3}{2}(4x^4+x)^{-5/2}(16x^3+1) = -\frac{3}{2}(16x^3+1)(4x^4+x)^{-5/2}.$$

21. $f(x) = (3x^2+2x+1)^{-2}$.

3 Differentiation

$$f'(x) = -2(3x^2 + 2x + 1)^{-3} \frac{d}{dx}(3x^2 + 2x + 1)$$
$$= -2(3x^2 + 2x + 1)^{-3}(6x + 2) = -4(3x + 1)(3x^2 + 2x + 1)^{-3}.$$

23. $f(x) = (x^2 + 1)^3 - (x^3 + 1)^2.$

$$f'(x) = 3(x^2 + 1)^2 \frac{d}{dx}(x^2 + 1) - 2(x^3 + 1)\frac{d}{dx}(x^3 + 1)$$
$$= 3(x^2 + 1)^2 (2x) - 2(x^3 + 1)(3x^2)$$
$$= 6x[(x^2 + 1)^2 - x(x^3 + 1)] = 6x(2x^2 - x + 1).$$

25. $f(t) = (t^{-1} - t^{-2})^3.$ $f'(t) = 3(t^{-1} - t^{-2})^2 \frac{d}{dt}(t^{-1} - t^{-2}) = 3(t^{-1} - t^{-2})^2(-t^{-2} + 2t^{-3}).$

27. $f(x) = \sqrt{x+1} + \sqrt{x-1} = (x+1)^{1/2} + (x-1)^{1/2}.$
$$f'(x) = \tfrac{1}{2}(x+1)^{-1/2}(1) + \tfrac{1}{2}(x-1)^{-1/2}(1) = \tfrac{1}{2}[(x+1)^{-1/2} + (x-1)^{-1/2}].$$

29. $f(x) = 2x^2(3 - 4x)^4.$
$$f'(x) = 2x^2(4)(3 - 4x)^3(-4) + (3 - 4x)^4(4x) = 4x(3 - 4x)^3(-8x + 3 - 4x)$$
$$= 4x(3 - 4x)^3(-12x + 3) = (-12x)(4x - 1)(3 - 4x)^3.$$

31. $f(x) = (x - 1)^2(2x + 1)^4.$
$$f'(x) = (x - 1)^2 \frac{d}{dx}(2x + 1)^4 + (2x + 1)^4 \frac{d}{dx}(x - 1)^2 \quad \text{[Product Rule]}$$
$$= (x - 1)^2(4)(2x + 1)^3 \frac{d}{dx}(2x + 1) + (2x + 1)^4(2)(x - 1)\frac{d}{dx}(x - 1)$$
$$= 8(x - 1)^2(2x + 1)^3 + 2(x - 1)(2x + 1)^4$$
$$= 2(x - 1)(2x + 1)^3(4x - 4 + 2x + 1) = 6(x - 1)(2x - 1)(2x + 1)^3.$$

33. $f(x) = \left(\dfrac{x+3}{x-2}\right)^3.$

$$f'(x) = 3\left(\frac{x+3}{x-2}\right)^2 \frac{d}{dx}\left(\frac{x-3}{x-2}\right) = 3\left(\frac{x+3}{x-2}\right)^2\left[\frac{(x-2)(1) - (x+3)(1)}{(x-2)^2}\right]$$

$$= 3\left(\frac{x+3}{x-2}\right)^2\left[-\frac{5}{(x-2)^2}\right] = -\frac{15(x+3)^2}{(x-2)^4}.$$

35. $s(t) = \left(\dfrac{t}{2t+1}\right)^{3/2}.$

$$s'(t) = \frac{3}{2}\left(\frac{t}{2t+1}\right)^{1/2}\frac{d}{dt}\left(\frac{t}{2t+1}\right) = \frac{3}{2}\left(\frac{t}{2t+1}\right)^{1/2}\left[\frac{(2t+1)(1)-t(2)}{(2t+1)^2}\right]$$

$$= \frac{3}{2}\left(\frac{t}{2t+1}\right)^{1/2}\left[\frac{1}{(2t+1)^2}\right] = \frac{3t^{1/2}}{2(2t+1)^{5/2}}.$$

37. $g(u) = \left(\dfrac{u+1}{3u+2}\right)^{1/2}.$

$$g'(u) = \frac{1}{2}\left(\frac{u+1}{3u+2}\right)^{-1/2}\frac{d}{du}\left(\frac{u+1}{3u+2}\right)$$

$$= \frac{1}{2}\left(\frac{u+1}{3u+2}\right)^{-1/2}\left[\frac{(3u+2)(1)-(u+1)(3)}{(3u+2)^2}\right] = -\frac{1}{2\sqrt{u+1}(3u+2)^{3/2}}.$$

39. $f(x) = \dfrac{x^2}{(x^2-1)^4}.$

$$f'(x) = \frac{(x^2-1)^4\dfrac{d}{dx}(x^2)-(x^2)\dfrac{d}{dx}(x^2-1)^4}{\left[(x^2-1)^4\right]^2}$$

$$= \frac{(x^2-1)^4(2x)-x^2(4)(x^2-1)^3(2x)}{(x^2-1)^8}$$

$$= \frac{(x^2-1)^3(2x)(x^2-1-4x^2)}{(x^2-1)^8} = \frac{(-2x)(3x^2+1)}{(x^2-1)^5}.$$

41. $h(x) = \dfrac{(3x^2+1)^3}{(x^2-1)^4}.$

$$h'(x) = \frac{(x^2-1)^4(3)(3x^2+1)^2(6x)-(3x^2+1)^3(4)(x^2-1)^3(2x)}{(x^2-1)^8}$$

$$= \frac{2x(x^2-1)^3(3x^2+1)^2[9(x^2-1)-4(3x^2+1)]}{(x^2-1)^8}$$

$$= -\frac{2x(3x^2+13)(3x^2+1)^2}{(x^2-1)^5}.$$

43. $f(x) = \dfrac{\sqrt{2x+1}}{x^2-1}.$

$$f'(x) = \frac{(x^2-1)(\frac{1}{2})(2x+1)^{-1/2}(2)-(2x+1)^{1/2}(2x)}{(x^2-1)^2}$$

$$= \frac{(2x+1)^{-1/2}[(x^2-1)-(2x+1)(2x)]}{(x^2-1)^2} = -\frac{3x^2+2x+1}{\sqrt{2x+1}(x^2-1)^2}.$$

45. $g(t) = \dfrac{(t+1)^{1/2}}{(t^2+1)^{1/2}}.$

$$g'(t) = \frac{(t^2+1)^{1/2}\dfrac{d}{dt}(t+1)^{1/2}-(t+1)^{1/2}\dfrac{d}{dt}(t^2+1)^{1/2}}{t^2+1}$$

$$= \frac{(t^2+1)^{1/2}(\frac{1}{2})(t+1)^{-1/2}(1)-(t+1)^{1/2}(\frac{1}{2})(t^2+1)^{-1/2}(2t)}{t^2+1}$$

$$= \frac{\frac{1}{2}(t+1)^{-1/2}(t^2+1)^{-1/2}[(t^2+1)-2t(t+1)]}{t^2+1} = -\frac{t^2+2t-1}{2\sqrt{t+1}(t^2+1)^{3/2}}.$$

47. $f(x) = (3x+1)^4(x^2-x+1)^3$

$$f'(x) = (3x+1)^4 \cdot \frac{d}{dx}(x^2-x+1)^3 + (x^2-x+1)^3\frac{d}{dx}(3x+1)^4$$

$$= (3x+1)^4 \cdot 3(x^2-x+1)^2(2x-1) + (x^2-x+1)^3 \cdot 4(3x+1)^3 \cdot 3$$

$$= 3(3x+1)^3(x^2-x+1)^2[(3x+1)(2x-1)+4(x^2-x+1)]$$

$$= 3(3x+1)^3(x^2-x+1)^2(6x^2-3x+2x-1+4x^2-4x+4)$$

$$= 3(3x+1)^3(x^2-x+1)^2(10x^2-5x+3)$$

49. $y = g(u) = u^{4/3}$ and $\dfrac{dy}{du} = \dfrac{4}{3}u^{1/3}$, $u = f(x) = 3x^2 - 1$, and $\dfrac{du}{dx} = 6x$.

So $\dfrac{dy}{dx} = \dfrac{dy}{du} \cdot \dfrac{du}{dx} = \tfrac{4}{3}u^{1/3}(6x) = \tfrac{4}{3}(3x^2-1)^{1/3}6x = 8x(3x^2-1)^{1/3}$.

51. $\dfrac{dy}{du} = -\dfrac{2}{3}u^{-5/3} = -\dfrac{2}{3u^{5/3}}$, $\dfrac{du}{dx} = 6x^2 - 1$.

$\dfrac{dy}{dx} = \dfrac{dy}{du} \cdot \dfrac{du}{dx} = -\dfrac{2(6x^2-1)}{3u^{5/3}} = -\dfrac{2(6x^2-1)}{3(2x^3-x+1)^{5/3}}$.

53. $\dfrac{dy}{du} = \tfrac{1}{2}u^{-1/2} - \tfrac{1}{2}u^{-3/2}$, $\dfrac{du}{dx} = 3x^2 - 1$.

$\dfrac{dy}{dx} = \dfrac{dy}{du} \cdot \dfrac{du}{dx} = \left[\dfrac{1}{2\sqrt{x^3-x}} - \dfrac{1}{2(x^3-x)^{3/2}} \right](3x^2-1)$

$= \dfrac{(3x^2-1)(x^3-x-1)}{2(x^3-x)^{3/2}}$.

55. $F(x) = g(f(x))$; $F'(x) = g'(f(x))f'(x)$ and $F'(2) = g'(3)(-3) = (4)(-3) = -12$

57. Let $g(x) = x^2 + 1$, then $F(x) = f(g(x))$. Next, $F'(x) = f'(g(x))g'(x)$
and $F'(1) = f'(2)(2x) = (3)(2) = 6$.

59. No. Suppose $h = g(f(x))$. Let $f(x) = x$ and $g(x) = x^2$. Then
$h = g(f(x)) = g(x) = x^2$ and $h'(x) = 2x \neq g'(f'(x)) = g'(1) = 2(1) = 2$.

61. $f(x) = (1-x)(x^2-1)^2$.

$f'(x) = (1-x)2(x^2-1)(2x) + (-1)(x^2-1)^2$

$= (x^2-1)(4x-4x^2-x^2+1) = (x^2-1)(-5x^2+4x+1)$.

Therefore, the slope of the tangent line at $(2,-9)$ is

$f'(2) = [(2)^2-1][-5(2)^2+4(2)+1] = -33$.

Then the required equation is $y+9 = -33(x-2)$, or $y = -33x + 57$.

63. $f(x) = x\sqrt{2x^2 + 7}$. $f'(x) = \sqrt{2x^2 + 7} + x(\frac{1}{2})(2x^2 + 7)^{-1/2}(4x)$.

The slope of the tangent line is $f'(3) = \sqrt{25} + (\frac{3}{2})(25)^{-1/2}(12) = \frac{43}{5}$.

An equation of the tangent line is $y - 15 = \frac{43}{5}(x - 3)$ or $y = \frac{43}{5}x - \frac{54}{5}$.

65. $N(t) = (60 + 2t)^{2/3}$. $\quad N'(t) = \frac{2}{3}(60 + 2t)^{-1/3}\frac{d}{dt}(60 + 2t) = \frac{4}{3}(60 + 2t)^{-1/3}$.

The rate of increase at the end of the second week is

$\qquad N'(2) = \frac{4}{3}(64)^{-1/3} = \frac{1}{3}$, or $\frac{1}{3}$ million/week

At the end of the 12th week, $N'(12) = \frac{4}{3}(84)^{-1/3} \approx 0.3$ million/wk. The number of viewers in the 2nd and 24th week are $N(2) = (60 + 4)^{2/3} = 16$ million and $N(24) = (60 + 48)^{2/3} = 22.7$ million, respectively.

67. $P(t) = 33.55(t + 5)^{0.205}$. $\quad P'(t) = 33.55(0.205)(t + 5)^{-0.795}(1) = 6.87775(t + 5)^{-0.795}$

The rate of change at the beginning of 2000 is

$\qquad P'(20) = 6.87775(25)^{-0.795} \approx 0.5322$ or 0.53%/yr.

The percent of these mothers was $P(20) = 33.55(25)^{0.205} \approx 64.90$, or 64.9%.

69. a. $f(t) = 23.7(0.2t + 1)^{1.32}$. The rate of change at any time t is given by

$\qquad f'(t) = (23.7)(1.32)(0.2t + 1)^{0.32}(0.2) = 6.2568(0.2t + 1)^{0.32}$

At the beginning of 2000, the rate of change is

$\qquad f'(9) = 6.2568[0.2(9) + 1]^{0.32} \approx 8.6985$, or approximately \$8.7 billion/yr.

b. The assets were $f(9) = 92.256$, or \$92.3 billion.

71. $C(t) = 0.01(0.2t^2 + 4t + 64)^{2/3}$.

a. $C'(t) = 0.01(\frac{2}{3})(0.2t^2 + 4t + 64)^{-1/3}\dfrac{d}{dt}(0.2t^2 + 4t + 64)$

$\qquad = (0.01)(0.667)(0.4t + 4)(0.2t^2 + 4t + 4)^{-1/3}$

$\qquad = 0.027(0.1t + 1)(0.2t^2 + 4t + 64)^{-1/3}$.

b. $C'(5) = 0.007[0.4(5) + 4][0.2(25) + 4(5) + 64]^{-1/3} \approx 0.009$,

or 0.009 parts per million per year.

73. a. $A(t) = 0.03t^3(t - 7)^4 + 60.2$

$\qquad A'(t) = 0.03[3t^2(t - 7)^4 + t^3(4)(t - 7)^3] = 0.03t^2(t - 7)^3[3(t - 7) + 4t]$

$$= 0.21t^2(t-3)(t-7)^3.$$

b. $A'(1) = 0.21(-2)(-6)^3 = 90.72$; $A'(3) = 0$. $A'(4) = 0.21(16)(1)(-3)^3 = -90.72$.
The amount of pollutant is increasing at the rate of 90.72 units/hr at 8 A.M. Its rate of change is 0 units/hr at 10 A.M.; its rate of change is -90.72 units/hr at 11 A.M.

75. $$P(t) = \frac{300\sqrt{\frac{1}{2}t^2 + 2t + 25}}{t+25} = \frac{300(\frac{1}{2}t^2 + 2t + 25)^{1/2}}{t+25}.$$

$$P'(t) = 300\left[\frac{(t+25)\frac{1}{2}(\frac{1}{2}t^2 + 2t + 25)^{-1/2}(t+2) - (\frac{1}{2}t^2 + 2t + 25)^{1/2}(1)}{(t+25)^2}\right]$$

$$= 300\left[\frac{(\frac{1}{2}t^2 + 2t + 25)^{-1/2}[(t+25)(t+2) - 2(\frac{1}{2}t^2 + 2t + 25)]}{(t+25)^2}\right]$$

$$= \frac{3450t}{(t+25)^2\sqrt{\frac{1}{2}t^2 + 2t + 25}}.$$

Ten seconds into the run, the athlete's pulse rate is increasing at
$$P'(10) = \frac{3450(10)}{(35)^2\sqrt{50 + 20 + 25}} \approx 2.9, \text{ or approximately 2.9 beats per minute per}$$
minute. Sixty seconds into the run, it is increasing at
$$P'(60) = \frac{3450(60)}{(85)^2\sqrt{1800 + 120 + 25}} \approx 0.65, \text{ or approximately 0.7 beats per minute per}$$
minute. Two minutes into the run, it is increasing at
$$P'(120) = \frac{3450(120)}{(145)^2\sqrt{7200 + 240 + 25}} \approx 0.23, \text{ or approximately 0.2 beats per minute}$$
per minute. The pulse rate two minutes into the run is given by
$$P(120) = \frac{300\sqrt{7200 + 240 + 25}}{120 + 25} \approx 178.8, \text{ or approximately 179 beats per minute.}$$

77. The area is given by $A = \pi r^2$. The rate at which the area is increasing is given by
dA/dt, that is, $\dfrac{dA}{dt} = \dfrac{d}{dt}(\pi r^2) = \dfrac{d}{dt}(\pi r^2)\dfrac{dr}{dt} = 2\pi r\dfrac{dr}{dt}$.

If $r = 40$ and $dr/dt = 2$, then $\dfrac{dA}{dt} = 2\pi(40)(2) = 160\pi$, that is, it is increasing at the
rate of 160π, or approximately 503, sq ft/sec.

79. $f(t) = 6.25t^2 + 19.75t + 74.75.$ $g(x) = -0.00075x^2 + 67.5.$

$$\frac{dS}{dt} = g'(x)f'(t) = (-0.0015x)(12.5t + 19.75).$$

When $t = 4$, we have $x = f(4) = 6.25(16) + 19.75(4) + 74.75 = 253.75$

and $\left.\dfrac{dS}{dt}\right|_{t=4} = (-0.0015)(253.75)[12.5(4) + 19.75] \approx -26.55;$

that is, the average speed will be dropping at the rate of approximately 27 mph per decade. The average speed of traffic flow at that time will be

$S = g(f(4)) = -0.00075(253.75^2) + 67.5 = 19.2$, or approximately 19 mph.

81. $N(x) = 1.42x$ and $x(t) = \dfrac{7t^2 + 140t + 700}{3t^2 + 80t + 550}$. The number of construction jobs as a function of time is $n(t) = N[x(t)]$. Using the Chain Rule,

$$n'(t) = \frac{dN}{dx} \cdot \frac{dx}{dt} = 1.42\frac{dx}{dt}$$

$$= (1.42)\left[\frac{(3t^2 + 80t + 550)(14t + 140) - (7t^2 + 140t + 700)(6t + 80)}{(3t^2 + 80t + 550)^2}\right]$$

$$= \frac{1.42(140t^2 + 3500t + 21000)}{(3t^2 + 80t + 550)^2}.$$

$n'(1) = \dfrac{1.42(140 + 3500 + 21000)}{(3 + 80 + 550)^2} \approx 0.0873216$, or approximately 87,322 jobs/year.

83. $x = f(p) = 10\sqrt{\dfrac{50 - p}{p}}$;

$$\frac{dx}{dp} = \frac{d}{dp}\left[10\left(\frac{50 - p}{p}\right)^{1/2}\right] = (10)(\tfrac{1}{2})\left(\frac{50 - p}{p}\right)^{-1/2}\frac{d}{dp}\left(\frac{50 - p}{p}\right)$$

$$= 5\left(\frac{50 - p}{p}\right)^{-1/2} \cdot \frac{d}{dp}\left(\frac{50}{p} - 1\right)$$

$$= 5\left(\frac{50-p}{p}\right)^{-1/2}\left(-\frac{50}{p^2}\right) = -\frac{250}{p^2\left(\frac{50-p}{p}\right)^{1/2}}$$

$$\left.\frac{dx}{dp}\right|_{p=25} = -\frac{250}{p^2\left(\frac{50-p}{p}\right)^{1/2}} = -\frac{250}{(625)\left(\frac{25}{25}\right)^{1/2}} = -0.4$$

So the quantity demanded is falling at the rate of 0.4(1000) or 400 wristwatches per dollar increase in price.

85. True. This is just the statement of the Chain Rule.

87. True. $\dfrac{d}{dx}\sqrt{f(x)} = \dfrac{d}{dx}[f(x)]^{1/2} = \dfrac{1}{2}[f(x)]^{-1/2}f'(x) = \dfrac{f'(x)}{2\sqrt{f(x)}}.$

89. Let $f(x) = x^{1/n}$ so that $[f(x)]^n = x$.
Differentiating both sides with respect to x, we get
$$n[f(x)]^{n-1}f'(x) = 1$$
$$f'(x) = \frac{1}{n[f(x)]^{n-1}} = \frac{1}{n[x^{1/n}]^{n-1}} = \frac{1}{nx^{1-(1/n)}} = \frac{1}{n}x^{(1/n)-1}.$$
as was to be shown.

USING TECHNOLOGY EXERCISES 3.3 page 197

1. 0.5774   3. 0.9390   5. –4.9498

7. a. 10,146,200/decade   b. 7,810,520/decade

3.4 Problem Solving Tips

Here are some hints for solving the problems in the exercises that follow:

1. The *marginal cost function* is the derivative of the cost function. Similarly, the

marginal profit function and the marginal *revenue function* are the derivatives of the profit function and the revenue function, respectively. The key word here is "marginal" as it indicates that we are dealing with the derivative of the function that follows.

2. The *average cost function* is given by $\bar{C}(x) = C(x)/x$ and the *marginal average cost function* is given by $\bar{C}'(x)$.

3. Remember that the revenue is *increasing* on an interval where the demand is *inelastic, decreasing* on an interval where the demand is *elastic*, and *stationary* at the point where the demand is *unitary*.

3.4 CONCEPT QUESTIONS, page 208

1. a. The marginal cost function is the derivative of the cost function.
   b. The average cost function is equal to the total cost function divided by the total number of the commodity produced.
   c. The marginal average cost function is the derivative of the average cost function.
   d. The marginal revenue function is the derivative of the revenue function.
   e. The marginal profit function is the derivative of the profit function.

EXERCISES 3.4, page 208

1. a. $C(x)$ is always increasing because as x, the number of units produced, increases, the greater the amount of money that must be spent on production.
    b. This occurs at $x = 4$, or a production level of 4000. You can see this by looking at the slopes of the tangent lines for x less than, equal to, and a little larger then $x = 4$.

3. a. The actual cost incurred in the production of the 1001st record is given by

$$C(1001) - C(1000) = [2000 + 2(1001) - 0.0001(1001)^2]$$
$$-[2000 + 2(1000) - 0.0001(1000)^2]$$
$$= 3901.7999 - 3900 = 1.7999,$$

or $1.80. The actual cost incurred in the production of the 2001st record is given by $C(2001) - C(2000) = [2000 + 2(2001) - 0.0001(2001)^2]$
$$-[2000 + 2(2000) - 0.0001(2000)^2]$$
$$= 5601.5999 - 5600 = 1.5999, \text{ or } \$1.60.$$

b. The marginal cost is $C'(x) = 2 - 0.0002x$. In particular
$$C'(1000) = 2 - 0.0002(1000) = 1.80$$
and $\quad C'(2000) = 2 - 0.0002(2000) = 1.60.$

5. a. $\overline{C}(x) = \dfrac{C(x)}{x} = \dfrac{100x + 200,000}{x} = 100 + \dfrac{200,000}{x}.$

b. $\overline{C}'(x) = \dfrac{d}{dx}(100) + \dfrac{d}{dx}(200,000x^{-1}) = -200,000x^{-2} = -\dfrac{200,000}{x^2}.$

c. $\lim\limits_{x \to \infty} \overline{C}(x) = \lim\limits_{x \to \infty}\left[100 + \dfrac{200,000}{x}\right] = 100$

and this says that the average cost approaches $100 per unit if the production level is very high.

7. $\overline{C}(x) = \dfrac{C(x)}{x} = \dfrac{2000 + 2x - 0.0001x^2}{x} = \dfrac{2000}{x} + 2 - 0.0001x.$

$\overline{C}'(x) = -\dfrac{2000}{x^2} + 0 - 0.0001 = -\dfrac{2000}{x^2} - 0.0001.$

9. a. $R'(x) = \dfrac{d}{dx}(8000x - 100x^2) = 8000 - 200x.$

b. $R'(39) = 8000 - 200(39) = 200.$ $R'(40) = 8000 - 200(40) = 0$
$R'(41) = 8000 - 200(41) = -200$

c. This suggests the total revenue is maximized if the price charged/ passenger is $40.

11. a. $P(x) = R(x) - C(x) = (-0.04x^2 + 800x) - (200x + 300,000)$
$$= -0.04x^2 + 600x - 300,000.$$

b. $P'(x) = -0.08x + 600$

c. $P'(5000) = -0.08(5000) + 600 = 200$ $P'(8000) = -0.08(8000) + 600 = -40.$

d.

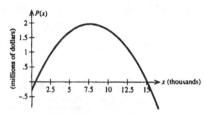

The profit realized by the company increases as production increases, peaking at a level of production of 7500 units. Beyond this level, the profit begins to fall.

13. a. The revenue function is $R(x) = px = (600 - 0.05x)x = 600x - 0.05x^2$
and the profit function is
$$P(x) = R(x) - C(x)$$
$$= (600x - 0.05x^2) - (0.000002x^3 - 0.03x^2 + 400x + 80,000)$$
$$= -0.000002x^3 - 0.02x^2 + 200x - 80,000.$$

b. $C'(x) = \dfrac{d}{dx}(0.000002x^3 - 0.03x^2 + 400x + 80,000) = 0.000006x^2 - 0.06x + 400.$

$R'(x) = \dfrac{d}{dx}(600x - 0.05x^2) = 600 - 0.1x.$

$P'(x) = \dfrac{d}{dx}(-0.000002x^3 - 0.02x^2 + 200x - 80,000) = -0.000006x^2 - 0.04x + 200.$

c. $C'(2000) = 0.000006(2000)^2 - 0.06(2000) + 400 = 304$, and this says that at a level of production of 2000 units, the cost for producing the 2001st unit is $304. $R'(2000) = 600 - 0.1(2000) = 400$ and this says that the revenue realized in selling the 2001st unit is $400. $P'(2000) = R'(2000) - C'(2000) = 400 - 304 = 96$, and this says that the revenue realized in selling the 2001st unit is $96.

d.

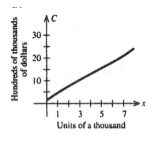

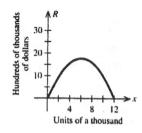

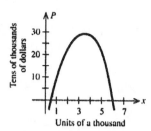

15. $\overline{C}(x) = \dfrac{C(x)}{x} = \dfrac{0.000002x^3 - 0.03x^2 + 400x + 80{,}000}{x}$

$\qquad = 0.000002x^2 - 0.03x + 400 + \dfrac{80{,}000}{x}.$

a. $\overline{C}'(x) = 0.000004x - 0.03 - \dfrac{80{,}000}{x^2}.$

b. $\overline{C}'(5000) = 0.000004(5000) - 0.03 - \dfrac{80{,}000}{5000^2} \approx -0.0132,$

and this says that, at a level of production of 5000 units, the average cost of production is dropping at the rate of approximately a penny per unit.

$\qquad \overline{C}'(10{,}000) = 0.000004(10000) - 0.03 - \dfrac{80{,}000}{10{,}000^2} \approx 0.0092,$

and this says that, at a level of production of 10,000 units, the average cost of production is increasing at the rate of approximately a penny per unit.

c.

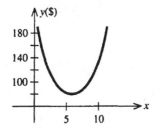

17. a. $R(x) = px = \dfrac{50x}{0.01x^2 + 1}.$ b. $R'(x) = \dfrac{(0.01x^2 + 1)50 - 50x(0.02x)}{(0.01x^2 + 1)^2} = \dfrac{50 - 0.5x^2}{(0.01x^2 + 1)^2}$

c. $R'(2) = \dfrac{50 - 0.5(4)}{[0.01(4) + 1]^2} \approx 44.379.$

This result says that at a level of sale of 2000 units, the revenue increases at the rate of approximately $44,379 per sales of 1000 units.

19. $C(x) = 0.873x^{1.1} + 20.34; \quad C'(x) = 0.873(1.1)x^{0.1}$
$C'(10) = 0.873(1.1)(10)^{0.1} = 1.21,$ or \$1.21 billion per billion dollars.

21. The consumption function is given by $C(x) = 0.712x + 95.05.$ The marginal

propensity to consume is given by $\dfrac{dC}{dx} = 0.712$. The marginal propensity to save is given by $\dfrac{dS}{dx} = 1 - \dfrac{dC}{dx} = 1 - 0.712 = 0.288$, or $0.288 billion per billion dollars.

23. Here $x = f(p) = -\frac{5}{4}p + 20$ and so $f'(p) = -\frac{5}{4}$. Therefore,

$$E(p) = -\frac{pf'(p)}{f(p)} = -\frac{p(-\frac{5}{4})}{-\frac{5}{4}p + 20} = \frac{5p}{80 - 5p}.$$

$$E(10) = \frac{5(10)}{80 - 5(10)} = \frac{50}{30} = \frac{5}{3} > 1, \quad \text{and so the demand is elastic.}$$

25. $f(p) = -\frac{1}{3}p + 20; \ f'(p) = -\frac{1}{3}$. Then the elasticity of demand is given by

$$E(p) = -\frac{p(-\frac{1}{3})}{-\frac{1}{3}p + 20}, \quad \text{and} \quad E(30) = -\frac{30(-\frac{1}{3})}{-\frac{1}{3}(30) + 20} = 1,$$

and we conclude that the demand is unitary at this price.

27. $x^2 = 169 - p$ and $f(p) = (169 - p)^{1/2}$.
    Next, $f'(p) = \frac{1}{2}(169 - p)^{-1/2}(-1) = -\frac{1}{2}(169 - p)^{-1/2}$.
    Then the elasticity of demand is given by

$$E(p) = -\frac{pf'(p)}{f(p)} = -\frac{p(-\frac{1}{2})(169 - p)^{-1/2}}{(169 - p)^{1/2}} = \frac{\frac{1}{2}p}{169 - p}.$$

Therefore, when $p = 29$, $E(p) = \dfrac{\frac{1}{2}(29)}{169 - 29} = \dfrac{14.5}{140} = 0.104.$

Since $E(p) < 1$, we conclude that demand is inelastic at this price.

29. $f(p) = \frac{1}{5}(225 - p^2); \ f'(p) = \frac{1}{5}(-2p) = -\frac{2}{5}p.$
    Then the elasticity of demand is given by

$$E(p) = -\frac{pf'(p)}{f(p)} = -\frac{p(-\frac{2}{5}p)}{\frac{1}{5}(225 - p^2)} = \frac{2p^2}{225 - p^2}.$$

a. When $p = 8$, $E(8) = \dfrac{2(64)}{225 - 64} = 0.8 < 1$ and the demand is inelastic. When $p = 10$,

$$E(10) = \frac{2(100)}{225 - 100} = 1.6 > 1$$

and the demand is elastic.

b. The demand is unitary when $E = 1$. Solving $\dfrac{2p^2}{225 - p^2} = 1$ we find $2p^2 = 225 - p^2$,

$3p^2 = 225$, and $p = 8.66$. So the demand is unitary when $p = 8.66$.

c. Since demand is elastic when $p = 10$, lowering the unit price will cause the revenue to increase.

d. Since the demand is inelastic at $p = 8$, a slight increase in the unit price will cause the revenue to increase.

31. $f(p) = \frac{2}{3}(36 - p^2)^{1/2}$

$f'(p) = \frac{2}{3}(\frac{1}{2})(36 - p^2)^{-1/2}(-2p) = -\frac{2}{3}p(36 - p^2)^{-1/2}$.

Then the elasticity of demand is given by

$$E(p) = -\frac{pf'(p)}{f(p)} = -\frac{-\frac{2}{3}p(36 - p^2)^{-1/2}\,p}{\frac{2}{3}(36 - p^2)^{1/2}} = \frac{p^2}{36 - p^2}.$$

When $p = 2$, $E(2) = \dfrac{4}{36 - 4} = \dfrac{1}{8} < 1$, and we conclude that the demand is inelastic.

b. Since the demand is inelastic, the revenue will increase when the rental price is increased.

33. We first solve the demand equation for x in terms of p. Thus,

$p = \sqrt{9 - 0.02x}$

$p^2 = 9 - 0.02x$

or $x = -50p^2 + 450$. With $f(p) = -50p^2 + 450$, we find

$$E(p) = -\frac{pf'(p)}{f(p)} = -\frac{p(-100p)}{-50p^2 + 450} = \frac{2p^2}{9 - p^2}.$$

Setting $E(p) = 1$ gives $2p^2 = 9 - p^2$, so $p = \sqrt{3}$. So the demand is inelastic in $[0, \sqrt{3}\,]$, unitary when $p = \sqrt{3}$, and elastic in $(\sqrt{3}, 3)$.

35. True. $\overline{C}'(x) = \dfrac{d}{dx}\left[\dfrac{C(x)}{x}\right] = \dfrac{xC'(x) - C(x)\dfrac{d}{dx}(x)}{x^2} = \dfrac{xC'(x) - C(x)}{x^2}$.

3 *Differentiation*

3.5 Problem Solving Tips

When you work applied problems make sure that you keep track of the units of measure used. For example, if velocity is measured in ft/sec, then the units of acceleration will be ft/sec^2. If you are working an applied problem, and the units in your answer are not correct, it may indicate that you have made an error in your calculations or in the formulation of the problem.

Here are some hints for solving the problems in the exercises that follow:

1. Make sure that you simplify an expression before differentiating it to find the next order derivative.

2. The velocity of an object moving in a straight path is given by the derivative of the position function for that object. The acceleration of the object is given by the derivative of the velocity function.

3.5 CONCEPT QUESTIONS, page 216

1. a. The second derivative of f is the derivative of f'.
   b. To find the second derivative of f we differentiate f'.
3. The relative rate of change is $I'(c)/I(c)$.

EXERCISES 3.5, page 216

1. $f(x) = 4x^2 - 2x + 1$; $f'(x) = 8x - 2$; $f''(x) = 8$.

3. $f(x) = 2x^3 - 3x^2 + 1;\ f'(x) = 6x^2 - 6x;\ f''(x) = 12x - 6 = 6(2x - 1).$

5. $h(t) = t^4 - 2t^3 + 6t^2 - 3t + 10;\ h'(t) = 4t^3 - 6t^2 + 12t - 3$
   $h''(t) = 12t^2 - 12t + 12 = 12(t^2 - t + 1).$

7. $f(x) = (x^2 + 2)^5;\ f'(x) = 5(x^2 + 2)^4(2x) = 10x(x^2 + 2)^4$ and
   $f''(x) = 10(x^2 + 2)^4 + 10x(x^2 + 2)^3(2x)$
   $= 10(x^2 + 2)^3[(x^2 + 2) + 8x^2] = 10(9x^2 + 2)(x^2 + 2)^3.$

9. $g(t) = (2t^2 - 1)^2(3t^2);$
   $g'(t) = 2(2t^2 - 1)(4t)(3t^2) - (2t^2 - 1)^2(6t)$
   $\qquad = 6t(2t^2 - 1)[4t^2 + (2t^2 - 1)] = 6t(2t^2 - 1)(6t^2 - 1)$
   $\qquad = 6t(12t^4 - 8t^2 + 1) = 72t^5 - 48t^3 + 6t.$
   $g''(t) = 360t^4 - 144t^2 + 6 = 6(60t^4 - 24t^2 + 1)$

11. $f(x) = (2x^2 + 2)^{7/2};\ f'(x) = \frac{7}{2}(2x^2 + 2)^{5/2}(4x) = 14x(2x^2 + 2)^{5/2};$
   $f''(x) = 14(2x^2 + 2)^{5/2} + 14x(\frac{5}{2})(2x^2 + 2)^{3/2}(4x)$
   $\qquad = 14(2x^2 + 2)^{3/2}[(2x^2 + 2) + 10x^2] = 28(6x^2 + 1)(2x^2 + 2)^{3/2}.$

13. $f(x) = x(x^2 + 1)^2;$
   $f'(x) = (x^2 + 1)^2 + x(2)(x^2 + 1)(2x)$
   $\qquad = (x^2 + 1)[(x^2 + 1) + 4x^2] = (x^2 + 1)(5x^2 + 1);$
   $f''(x) = 2x(5x^2 + 1) + (x^2 + 1)(10x) = 2x(5x^2 + 1 + 5x^2 + 5) = 4x(5x^2 + 3).$

15. $f(x) = \dfrac{x}{2x+1};\ f'(x) = \dfrac{(2x+1)(1) - x(2)}{(2x+1)^2} = \dfrac{1}{(2x+1)^2};$
   $f''(x) = \dfrac{d}{dx}(2x+1)^{-2} = -2(2x+1)^{-3}(2) = -\dfrac{4}{(2x+1)^3}.$

17. $f(s) = \dfrac{s-1}{s+1};\ f'(s) = \dfrac{(s+1)(1) - (s-1)(1)}{(s+1)^2} = \dfrac{2}{(s+1)^2}.$

3 Differentiation

$$f''(s) = 2\frac{d}{ds}(s+1)^{-2} = -4(s+1)^{-3} = -\frac{4}{(s+1)^3}.$$

19. $f(u) = \sqrt{4-3u} = (4-3u)^{1/2}$. $f'(u) = \frac{1}{2}(4-3u)^{-1/2}(-3) = -\frac{3}{2\sqrt{4-3u}}$.

$$f''(u) = -\frac{3}{2}\cdot\frac{d}{du}(4-3u)^{-1/2} = -\frac{3}{2}\left(-\frac{1}{2}\right)(4-3u)^{-3/2}(-3) = -\frac{9}{4(4-3u)^{3/2}}.$$

21. $f(x) = 3x^4 - 4x^3$; $f'(x) = 12x^3 - 12x^2$; $f''(x) = 36x^2 - 24x$; $f'''(x) = 72x - 24$.

23. $f(x) = \frac{1}{x}$; $f'(x) = \frac{d}{dx}(x^{-1}) = -x^{-2}$; $f''(x) = 2x^{-3}$; $f'''(x) = -6x^{-4} = -\frac{6}{x^4}$.

25. $g(s) = (3s-2)^{1/2}$; $g'(s) = \frac{1}{2}(3s-2)^{-1/2}(3) = \frac{3}{2(3s-2)^{1/2}}$;

$$g''(s) = \frac{3}{2}\left(-\frac{1}{2}\right)(3s-2)^{-3/2}(3) = -\frac{9}{4}(3s-2)^{-3/2} = -\frac{9}{4(3s-2)^{3/2}};$$

$$g'''(s) = \frac{27}{8}(3s-2)^{-5/2}(3) = \frac{81}{8}(3s-2)^{-5/2} = \frac{81}{8(3s-2)^{5/2}}.$$

27. $f(x) = (2x-3)^4$; $f'(x) = 4(2x-3)^3(2) = 8(2x-3)^3$

$f''(x) = 24(2x-3)^2(2) = 48(2x-3)^2$; $f'''(x) = 96(2x-3)(2) = 192(2x-3)$.

29. Its velocity at any time t is $v(t) = \frac{d}{dt}(16t^2) = 32t$. The hammer strikes the ground when $16t^2 = 256$ or $t = 4$ (we reject the negative root). Therefore, its velocity at the instant it strikes the ground is $v(4) = 32(4) = 128$ ft/sec. Its acceleration at time t is $a(t) = \frac{d}{dt}(32t) = 32$. In particular, its acceleration at $t = 4$ is 32 ft/sec^2.

31. $N(t) = -0.1t^3 + 1.5t^2 + 100$.
a. $N'(t) = -0.3t^2 + 3t = 0.3t(10 - t)$. Since $N'(t) > 0$ for $t = 0, 1, 2, ..., 7$, it is evident that $N(t)$ (and therefore the crime rate) was increasing from 1988 through 1995.

b. $N''(t) = -0.6t + 3 = 0.6(5 - t)$. Now $N''(4) = 0.6 > 0$, $N''(5) = 0$, $N''(6) = -0.6 < 0$ and $N''(7) = -1.2 < 0$. This shows that the rate of the rate of change was decreasing beyond $t = 5$ (1990). This shows that the program was working.

33. $N(t) = 0.00037t^3 - 0.0242t^2 + 0.52t + 5.3 \qquad (0 \le t \le 10)$

$N'(t) = 0.00111t^2 - 0.0484t + 0.52$

$N''(t) = 0.00222t - 0.0484$

So $\quad N(8) = 0.00037(8)^3 - 0.0242(8)^2 + 5.3 = 8.1$

$\qquad N'(8) = 0.00111(8)^2 - 0.0484(8) + 0.52 \approx 0.204.$

$\qquad N''(8) = 0.00222(8) - 0.0484 = -0.031.$

We conclude that at the beginning of 1998, there were 8.1 million persons receiving disability benefits, the number is increasing at the rate of 0.2 million/yr, and the rate of the rate of change of the number of persons is decreasing at the rate of 0.03 million persons/yr^2.

35. a. $h(t) = \frac{1}{16}t^4 - t^3 + 4t^2$. $h'(t) = \frac{1}{4}t^3 - 3t^2 + 8t$

b. $h'(0) = 0$ or zero feet per second.

$h'(4) = \frac{1}{4}(64) - 3(16) + 8(4) = 0$, or zero feet per second.

$h'(8) = \frac{1}{4}(8)^3 - 3(64) + 8(8) = 0$, or zero feet per second.

c. $h''(t) = \frac{3}{4}t^2 - 6t + 8$

d. $h''(0) = 8$ ft/sec^2; $\quad h''(4) = \frac{3}{4}(16) - 6(4) + 8 = -4$ ft/sec^2.

$h''(8) = \frac{3}{4}(64) - 6(8) + 8 = 8$ ft/sec^2.

e. $h(0) = 0$ feet; $h(4) = \frac{1}{16}(4)^4 - (4)^3 + 4(4)^2 = 16$ feet.

$h(8) = \frac{1}{16}(8)^4 - (8)^3 + 4(8)^2 = 0$ feet.

37. $f(t) = 10.72(0.9t + 10)^{0.3}$.

$f'(t) = 10.72(0.3)(0.9t + 10)^{-0.7}(0.9) = 2.8944(0.9t + 10)^{-0.7}$

$f''(t) = 2.8944(-0.7)(0.9t + 10)^{-1.7}(0.9) = -1.823472(0.9t + 10)^{-1.7}$

So $\quad f''(10) = -1.823472(19)^{-1.7} \approx -0.01222$. And this says that the rate of the rate of change of the population is decreasing at the rate of 0.01%/ yr^2.

39. False. If f has derivatives of order two at $x = a$, then $f''(a) = [f'(a)]^2$.

41. True. If $f(x)$ is a polynomial function of degree n, then $f^{(n+1)}(x) = 0$.

43. True. Using the chain rule, $h'(x) = f'(2x) \cdot \dfrac{d}{dx}(2x) = f'(x) \cdot 2 = 2f'(2x)$

Using the chain rule again, $h''(x) = 2f''(2x) \cdot 2 = 4f''(2x)$.

45. Consider the function $f(x) = x^{(2n+1)/2} = x^{n+(1/2)}$.

Then
$$f'(x) = (n + \tfrac{1}{2})x^{n-(1/2)}$$
$$f''(x) = (n + \tfrac{1}{2})(n - \tfrac{1}{2})x^{n-(3/2)}$$
$$\cdots\cdots\cdots\cdots\cdots\cdots\cdots\cdots$$
$$f^{(n)}(x) = (n + \tfrac{1}{2})(n - \tfrac{1}{2}) \cdots \tfrac{3}{2}x^{1/2}$$
$$f^{(n+1)}(x) = (n + \tfrac{1}{2})(n - \tfrac{1}{2}) \cdots \tfrac{1}{2}x^{-1/2}.$$

The first n derivatives exist at $x = 0$, but the $(n + 1)$st derivative fails to be defined there.

USING TECHNOLOGY EXERCISES 3.5, page 220

1. -18 3. 15.2762

5. -0.6255 7. 0.1973

9. $f''(6) = -68.46214$ and it tells us that at the beginning of 1988, the rate of the rate of the rate at which banks were failing was 68 banks per year per year per year.

3.6 Problem Solving Tips

Here are some hints for solving the problems in the exercises that follow:

1. If an equation expresses y implicitly as a function of x, then we can use implicit differentiation to find its derivative. We apply the Chain rule to find the derivative of

any term involving y. (Note that the derivative of any term involving y will include the factor dy/dx.) The terms involving only x are differentiated in the usual manner.

2. Step 3 of the guidelines for solving related rates problems, page 227 in the text, asks you to find an equation giving the relationship between the variables in the related rates problem.. Make sure that you differentiate this equation implicitly with respect to t before you substitute the values of the variables back into the equation (Step 5).

3.6 CONCEPT QUESTIONS, page 229

1. a. We differentiate both sides of $F(x, y) = 0$ with respect to x. Then solve for dy / dx.
    b. The chain rule is used to differentiate any expression involving the dependent variable y.
3. Suppose x and y are two variables that are related by an equation. Furthermore, suppose x and y are both functions of a third variable t. (Normally, t represents time). Then a related rates problem involves finding dx / dt or dy / dt.

EXERCISES 3.6, page 229

1. a. Solving for y in terms of x, we have $y = -\frac{1}{2}x + \frac{5}{2}$. Therefore, $y' = -\frac{1}{2}$.
    b. Next, differentiating $x + 2y = 5$ implicitly, we have $1 + 2y' = 0$, or $y' = -\frac{1}{2}$.

3. a. $xy = 1, y = \dfrac{1}{x}$, and $\dfrac{dy}{dx} = -\dfrac{1}{x^2}$.

    b. $$x\frac{dy}{dx} + y = 0$$
    $$x\frac{dy}{dx} = -y$$
    $$\frac{dy}{dx} = -\frac{y}{x} = \frac{-\frac{1}{x}}{x} = -\frac{1}{x^2}.$$

5. $x^3 - x^2 - xy = 4.$
   a. $-xy = 4 - x^3 + x^2$

   $$y = -\frac{4}{x} + x^2 - x \quad \text{and} \quad y' = \frac{4}{x^2} + 2x - 1.$$

   b. $x^3 - x^2 - xy = 4$

   $$-x\frac{dy}{dx} = -3x^2 + 2x + y$$

   $$\frac{dy}{dx} = 3x - 2 - \frac{y}{x}$$

   $$= 3x - 2 - \frac{1}{x}(-\frac{4}{x} + x^2 - x) = 3x - 2 + \frac{4}{x^2} - x + 1$$

   $$= \frac{4}{x^2} + 2x - 1.$$

7. a. $\dfrac{x}{y} - x^2 = 1$ is equivalent to $\dfrac{x}{y} = x^2 + 1$, or $y = \dfrac{x}{x^2 + 1}$. Therefore,

   $$y' = \frac{(x^2 + 1) - x(2x)}{(x^2 + 1)^2} = \frac{1 - x^2}{(x^2 + 1)^2}.$$

   b. Next, differentiating the equation $x - x^2y = y$ implicitly, we obtain

   $$1 - 2xy - x^2y' = y', \ y'(1 + x^2) = 1 - 2xy, \text{ or } \ y' = \frac{1 - 2xy}{(1 + x^2)}.$$

   (This may also be written in the form $-2y^2 + \dfrac{y}{x}$.) To show that this is equivalent to

   the results obtained earlier, use the value of y obtained before, to get

   $$y' = \frac{1 - 2x\left(\dfrac{x}{x^2 + 1}\right)}{1 + x^2} = \frac{x^2 + 1 - 2x^2}{(1 + x^2)^2} = \frac{1 - x^2}{(1 + x^2)^2}.$$

9. $x^2 + y^2 = 16.$ Differentiating both sides of the equation implicitly, we obtain

   $$2x + 2yy' = 0 \quad \text{and so} \quad y' = -\frac{x}{y}.$$

11. $x^2 - 2y^2 = 16.$ Differentiating implicitly with respect to x, we have

$$2x - 4y\frac{dy}{dx} = 0 \text{ and } \frac{dy}{dx} = \frac{x}{2y}.$$

13. $x^2 - 2xy = 6$. Differentiating both sides of the equation implicitly, we obtain

$2x - 2y - 2xy' = 0$ and so $y' = \dfrac{x-y}{x} = 1 - \dfrac{y}{x}$.

15. $x^2y^2 - xy = 8$. Differentiating both sides of the equation implicitly, we obtain
$$2xy^2 + 2x^2yy' - y - xy' = 0, \ 2xy^2 - y + y'(2x^2y - x) = 0$$

and so
$$y' = \frac{y(1-2xy)}{x(2xy-1)} = -\frac{y}{x}.$$

17. $x^{1/2} + y^{1/2} = 1$. Differentiating implicitly with respect to x, we have

$$\tfrac{1}{2}x^{-1/2} + \tfrac{1}{2}y^{-1/2}\frac{dy}{dx} = 0. \text{ Therefore, } \frac{dy}{dx} = -\frac{x^{-1/2}}{y^{-1/2}} = -\frac{\sqrt{y}}{\sqrt{x}}.$$

19. $\sqrt{x+y} = x$. Differentiating both sides of the equation implicitly, we obtain
$$\tfrac{1}{2}(x+y)^{-1/2}(1+y') = 1, \ 1 + y' = 2(x+y)^{1/2},$$

or
$$y' = 2\sqrt{x+y} - 1.$$

21. $\dfrac{1}{x^2} + \dfrac{1}{y^2} = 1$. Differentiating both sides of the equation implicitly, we obtain

$$-\frac{2}{x^3} - \frac{2}{y^3}y' = 0, \text{ or } y' = -\frac{y^3}{x^3}.$$

23. $\sqrt{xy} = x + y$. Differentiating both sides of the equation implicitly, we obtain
$$\tfrac{1}{2}(xy)^{-1/2}(xy' + y) = 1 + y'$$
$$xy' + y = 2\sqrt{xy}(1 + y')$$
$$y'(x - 2\sqrt{xy}) = 2\sqrt{xy} - y$$

or
$$y' = -\frac{(2\sqrt{xy} - y)}{(2\sqrt{xy} - x)} = \frac{2\sqrt{xy} - y}{x - 2\sqrt{xy}}.$$

3 Differentiation

25. $\dfrac{x+y}{x-y} = 3x$, or $x+y = 3x^2 - 3xy$. Differentiating both sides of the equation

implicitly, we obtain $\quad 1 + y' = 6x - 3xy' - 3y$ or $y' = \dfrac{6x - 3y - 1}{3x + 1}$.

27. $xy^{3/2} = x^2 + y^2$. Differentiating implicitly with respect to x, we obtain

$$y^{3/2} + x\left(\tfrac{3}{2}\right)y^{1/2}\dfrac{dy}{dx} = 2x + 2y\dfrac{dy}{dx}$$

$$2y^{3/2} + 3xy^{1/2}\dfrac{dy}{dx} = 4x + 4y\dfrac{dy}{dx} \qquad \text{(Multiplying by 2.)}$$

$$(3xy^{1/2} - 4y)\dfrac{dy}{dx} = 4x - 2y^{3/2}$$

$$\dfrac{dy}{dx} = \dfrac{2(2x - y^{3/2})}{3xy^{1/2} - 4y}.$$

29. $(x+y)^3 + x^3 + y^3 = 0$. Differentiating implicitly with respect to x, we obtain

$$3(x+y)^2\left(1 + \dfrac{dy}{dx}\right) + 3x^2 + 3y^2\dfrac{dy}{dx} = 0$$

$$(x+y)^2 + (x+y)^2\dfrac{dy}{dx} + x^2 + y^2\dfrac{dy}{dx} = 0$$

$$[(x+y)^2 + y^2]\dfrac{dy}{dx} = -[(x+y)^2 + x^2]$$

$$\dfrac{dy}{dx} = -\dfrac{2x^2 + 2xy + y^2}{x^2 + 2xy + 2y^2}.$$

31. $4x^2 + 9y^2 = 36$. Differentiating the equation implicitly, we obtain
$$8x + 18yy' = 0.$$
At the point (0,2), we have $0 + 36y' = 0$ and the slope of the tangent line is 0.
Therefore, an equation of the tangent line is $y = 2$.

33. $x^2y^3 - y^2 + xy - 1 = 0$. Differentiating implicitly with respect to x, we have
$$2xy^3 + 3x^2y^2\dfrac{dy}{dx} - 2y\dfrac{dy}{dx} + y + x\dfrac{dy}{dx} = 0.$$

At $(1,1)$, $2+3\dfrac{dy}{dx}-2\dfrac{dy}{dx}+1+\dfrac{dy}{dx}=0$, and

$$2\dfrac{dy}{dx}=-3 \quad\text{and}\quad \dfrac{dy}{dx}=-\dfrac{3}{2}.$$

Using the point-slope form of an equation of a line, we have
$y-1=-\frac{3}{2}(x-1)$, and the equation of the tangent line to the graph of the function
f at $(1,1)$ is $y=-\frac{3}{2}x+\frac{5}{2}$.

35. $xy=1$. Differentiating implicitly, we have $xy'+y=0$, or $y'=-\dfrac{y}{x}$.

Differentiating implicitly once again, we have $xy''+y'+y'=0$.

Therefore, $\quad y''=-\dfrac{2y'}{x}=\dfrac{2\left(\dfrac{y}{x}\right)}{x}=\dfrac{2y}{x^2}.$

37. $y^2-xy=8$. Differentiating implicitly we have $2yy'-y-xy'=0$

and $y'=\dfrac{y}{2y-x}$. Differentiating implicitly again, we have

$$2(y')^2+2yy''-y'-y'-xy''=0, \quad\text{or}\quad y''=\dfrac{2y'-2(y')^2}{2y-x}.$$

Then $\quad y''=\dfrac{2\left(\dfrac{y}{2y-x}\right)\left(1-\dfrac{y}{2y-x}\right)}{2y-x}=\dfrac{2y(2y-x-y)}{(2y-x)^3}=\dfrac{2y(y-x)}{(2y-x)^3}.$

39. a. Differentiating the given equation with respect to t, we obtain

$$\dfrac{dV}{dt}=\pi r^2\dfrac{dh}{dt}+2\pi rh\dfrac{dr}{dt}=\pi r\left(r\dfrac{dh}{dt}+2h\dfrac{dr}{dt}\right).$$

b. Substituting $r=2$, $h=6$, $\dfrac{dr}{dt}=0.1$ and $\dfrac{dh}{dt}=0.3$ into the expression for $\dfrac{dV}{dt}$

we obtain $\dfrac{dV}{dt}=\pi(2)[2(0.3)+2(6)(0.1)]=3.6\pi$, and so the volume is increasing at
the rate of 3.6π cu in/sec.

41. We are given $\dfrac{dp}{dt} = 2$ and are required to find $\dfrac{dx}{dt}$ when $x = 9$ and $p = 63$.

Differentiating the equation $p + x^2 = 144$ with respect to t, we obtain

$$\dfrac{dp}{dt} + 2x\dfrac{dx}{dt} = 0.$$

When $x = 9$, $p = 63$, and $\dfrac{dp}{dt} = 2$,

$$2 + 2(9)\dfrac{dx}{dt} = 0$$

and
$$\dfrac{dx}{dt} = -\dfrac{1}{9} \approx -0.111,$$

or the quantity demanded is decreasing at the rate of 111 tires per week.

43. $100x^2 + 9p^2 = 3600$. Differentiating the given equation implicitly with respect to t, we have $200x\dfrac{dx}{dt} + 18p\dfrac{dp}{dt} = 0$. Next, when $p = 14$, the given equation yields

$$100x^2 + 9(14)^2 = 3600$$
$$100x^2 = 1836,$$

or $x = 4.2849$. When $p = 14$, $\dfrac{dp}{dt} = -0.15$, and $x = 4.2849$, we have

$$200(4.2849)\dfrac{dx}{dt} + 18(14)(-0.15) = 0$$

$$\dfrac{dx}{dt} = 0.0441.$$

So the quantity demanded is increasing at the rate of 44 ten–packs per week.

45. $625p^2 - x^2 = 100$. Differentiating the given equation implicitly with respect to t, we have $1250p\dfrac{dp}{dt} - 2x\dfrac{dx}{dt} = 0$.

When $p = 1.0770$, $x = 25$, and $\dfrac{dx}{dt} = -1$, we find that

$$1250(1.077)\dfrac{dp}{dt} - 2(25)(-1) = 0,$$

and
$$\frac{dp}{dt} = -\frac{50}{1250(1.077)} = -0.037.$$

We conclude that the price is decreasing at the rate of 3.7 cents per carton.

47. $p = -0.01x^2 - 0.2x + 8$. Differentiating the given equation implicitly with respect to p, we have
$$1 = -0.02x\frac{dx}{dp} - 0.2\frac{dx}{dp} = [0.02x + 0.2]\frac{dx}{dp}$$

or
$$\frac{dx}{dp} = -\frac{1}{0.02x + 0.2}.$$

When $x = 15$, $p = -0.01(15)^2 - 0.2(15) + 8 = 2.75$

and
$$\frac{dx}{dp} = -\frac{1}{0.02(15) + 0.2} = -2.$$

Therefore, $E(p) = -\frac{pf'(p)}{f(p)} = -\frac{(2.75)(-2)}{15} = 0.37 < 1$,

and the demand is inelastic.

49. $A = \pi r^2$. Differentiating with respect to t, we obtain
$$\frac{dA}{dt} = 2\pi r\frac{dr}{dt}.$$

When the radius of the circle is 40 ft and increasing at the rate of 2 ft/sec,
$$\frac{dA}{dt} = 2\pi(40)(2) = 160\pi, \text{ or } 160\pi \text{ ft}^2/\text{sec}.$$

51. Let D denote the distance between the two cars, x the distance traveled by the car heading east, and y the distance traveled by the car heading north as shown in the diagram at the right. Then $D^2 = x^2 + y^2$.
Differentiating with respect to t, we have

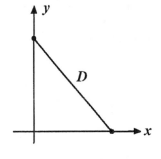

$$2D\frac{dD}{dt} = 2x\frac{dx}{dt} + 2y\frac{dy}{dt},$$

or
$$\frac{dD}{dt} = \frac{x\frac{dx}{dt} + y\frac{dy}{dt}}{D}$$

3 Differentiation

When $t = 5$, $x = 30$, $y = 40$, $\dfrac{dx}{dt} = 2(5)+1 = 11$, and $\dfrac{dy}{dt} = 2(5)+3 = 13$.

Therefore, $\dfrac{dD}{dt} = \dfrac{(30)(11)+(40)(13)}{\sqrt{900+1600}} = 17$, or 17 ft/sec.

53. Referring to the diagram at the right, we see that
$$D^2 = 120^2 + x^2.$$
Differentiating this last equation with
respect to t, we have

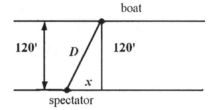

$$2D\dfrac{dD}{dt} = 2x\dfrac{dx}{dt} \quad \text{and} \quad \dfrac{dD}{dt} = \dfrac{x\dfrac{dx}{dt}}{D}.$$

When $x = 50$, $D = \sqrt{120^2 + 50^2} = 130$ and
$$\dfrac{dD}{dt} = \dfrac{(20)(50)}{130} \approx 7.69, \text{ or } 7.69 \text{ ft/sec.}$$

55. Let V and S denote its volume and surface area. Then we are given that
$\dfrac{dV}{dt} = -kS$, where k is the constant of proportionality. But from $V = \left(\dfrac{4}{3}\right)\pi r^3$,
we find, upon differentiating both sides with respect to t, that
$$\dfrac{dV}{dt} = \left(\dfrac{4}{3}\right)\pi(3\pi r^2)\dfrac{dr}{dt} = 4\pi^2 r^2\dfrac{dr}{dt}$$
and using the fact stated earlier,
$$\dfrac{dV}{dt} = 4\pi^2 r^2\dfrac{dr}{dt} = -kS = -k(4\pi r^2).$$

Therefore, $\dfrac{dr}{dt} = -\dfrac{k(4\pi r^2)}{4\pi^2 r^2} = -\dfrac{k}{\pi}$ and this proves that the radius is decreasing at
the constant rate of (k/π) units/unit time.

57. Refer to the figure at the right.

We are given that $\dfrac{dx}{dt} = 264$. Using the

Pythagorean Theorem,
$$s^2 = x^2 + 1000^2 = x^2 + 1000000.$$

We want to find $\dfrac{ds}{dt}$ when $s = 1500$.

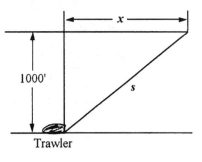

Trawler

Differentiating both sides of the equation with respect to t, we have

$$2s\frac{ds}{dt} = 2x\frac{dx}{dt} \quad \text{and so} \quad \frac{ds}{dt} = \frac{x\dfrac{dx}{dt}}{s}.$$

Now, when $s = 1500$, we have
$$1500^2 = x^2 + 10000 \quad \text{or} \quad x = \sqrt{1250000}.$$

Therefore, $\qquad \dfrac{ds}{dt} = \dfrac{\sqrt{1250000}\cdot(264)}{1500} \approx 196.8,$

that is, the aircraft is receding from the trawler at the speed of approximately 196.8 ft/sec.

59. Refer to the diagram at the right.
$$\frac{y}{6} = \frac{y+x}{18}, \quad 18y = 6(y+x)$$
$$3y = y+x, \quad 2y = x, \quad y = \tfrac{1}{2}x.$$
Then $D = y + x = \tfrac{3}{2}x$. Differentiating

implicitly, we have $\qquad \dfrac{dD}{dt} = \dfrac{3}{2}\cdot\dfrac{dx}{dt}$

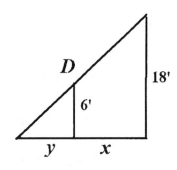

and when $\quad \dfrac{dx}{dt} = 6$,

$$\frac{dD}{dt} = \frac{3}{2}(6) = 9, \quad \text{or } 9\,\text{ft}/\text{sec}.$$

3 Differentiation

61. Differentiating $x^2 + y^2 = 13^2 = 169$ with respect to t gives
$$2x\frac{dx}{dt} + 2y\frac{dy}{dt} = 0.$$
When $x = 12$, we have
$$144 + y^2 = 169 \quad \text{or} \quad y = 5.$$

Therefore, with $x = 12$, $y = 5$, and $\dfrac{dx}{dt} = 8$, we

find $\quad 2(12)(8) + 2(5)\dfrac{dy}{dt} = 0$, or

$\dfrac{dy}{dt} = -19.2$, that is, the top of the ladder is sliding down the wall at 19.2 ft/sec.

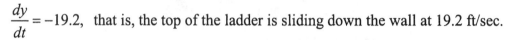

63. True. Differentiating both sides of the equation with respect to x, we have
$$\frac{d}{dx}[f(x)g(y)] = \frac{d}{dx}(0)$$
$$f(x)g'(y)\frac{dy}{dx} + f'(x)g(y) = 0$$
$$\frac{dy}{dx} = -\frac{f'(x)g(y)}{f(x)g'(y)}$$
provided $f(x) \neq 0$ and $g'(y) \neq 0$.

3.7 Problem Solving Tips

1. Δy is the *actual change* in y due to a small change Δx in x. It is approximated by dy, the differential of y, where $dy = f'(x)\Delta x = f'(x)dx$. Note that $dx = \Delta x$, *but* $dy \approx \Delta y$ for a small change Δx in x.

3.7 CONCEPT QUESTIONS, page 237

1. The differential of x is dx. The differential of y is $dy = f'(x)\,dx$.

1. $f(x) = 2x^2$ and $dy = 4x \, dx$.

3. $f(x) = x^3 - x$ and $dy = (3x^2 - 1) \, dx$.

5. $f(x) = \sqrt{x+1} = (x+1)^{1/2}$ and $dy = \dfrac{1}{2}(x+1)^{-1/2} \, dx = \dfrac{dx}{2\sqrt{x+1}}$.

7. $f(x) = 2x^{3/2} + x^{1/2}$ and $dy = (3x^{1/2} + \frac{1}{2}x^{-1/2}) \, dx = \frac{1}{2}x^{-1/2}(6x+1)dx = \dfrac{6x+1}{2\sqrt{x}} \, dx$.

9. $f(x) = x + \dfrac{2}{x}$ and $dy = \left(1 - \dfrac{2}{x^2}\right) dx = \dfrac{x^2 - 2}{x^2} \, dx$.

11. $f(x) = \dfrac{x-1}{x^2+1}$ and $dy = \dfrac{x^2 + 1 - (x-1)2x}{(x^2+1)^2} \, dx = \dfrac{-x^2 + 2x + 1}{(x^2+1)^2} \, dx$.

13. $f(x) = \sqrt{3x^2 - x} = (3x^2 - x)^{1/2}$ and

$$dy = \frac{1}{2}(3x^2 - x)^{-1/2}(6x-1)dx = \frac{6x-1}{2\sqrt{3x^2 - x}} \, dx.$$

15. $f(x) = x^2 - 1$.
    a. $dy = 2x \, dx$.   b. $dy \approx 2(1)(0.02) = 0.04$.
    c. $\Delta y = [(1.02)^2 - 1] - [1 - 1] = 0.0404$.

17. $f(x) = \dfrac{1}{x}$.

    a. $dy = -\dfrac{dx}{x^2}$.   b. $dy \approx -0.05$ c. $\Delta y = \dfrac{1}{-0.95} - \dfrac{1}{-1} = -0.05263$.

19. $y = \sqrt{x}$ and $dy = \dfrac{dx}{2\sqrt{x}}$. Therefore, $\sqrt{10} \approx 3 + \dfrac{1}{2 \cdot \sqrt{9}} = 3.167$.

21. $y = \sqrt{x}$ and $dy = \dfrac{dx}{2\sqrt{x}}$. Therefore, $\sqrt{49.5} \approx 7 + \dfrac{0.5}{2 \cdot 7} = 7.0357$.

23. $y = x^{1/3}$ and $dy = \frac{1}{3}x^{-2/3}\ dx$. Therefore, $\sqrt[3]{7.8} \approx 2 - \dfrac{0.2}{3\cdot 4} = 1.983$.

25. $y = \sqrt{x}$ and $dy = \dfrac{dx}{2\sqrt{x}}$. Therefore, $\sqrt{0.089} \approx \frac{1}{10}\sqrt{8.9} = \frac{1}{10}\left[3 - \dfrac{0.1}{2\cdot 3}\right] \approx 0.298$.

27. $y = f(x) = \sqrt{x} + \dfrac{1}{\sqrt{x}} = x^{1/2} + x^{-1/2}$. Therefore,

$$\frac{dy}{dx} = \frac{1}{2}x^{-1/2} - \frac{1}{2}x^{-3/2}$$

$$dy = \left(\frac{1}{2x^{1/2}} - \frac{1}{2x^{3/2}}\right)dx.$$

Letting $x = 4$ and $dx = 0.02$, we find

$$\sqrt{4.02} + \frac{1}{\sqrt{4.02}} - f(4) = f(4.02) - f(4) = \Delta y \approx dy$$

$$\sqrt{4.02} + \frac{1}{\sqrt{4.02}} \approx f(4) + dy\Big|_{\substack{x=4 \\ dx=0.02}}$$

$$\approx 2 + \frac{1}{2} + \left(\frac{1}{2\cdot 2} - \frac{1}{2\cdot 2\sqrt{2}}\right)(0.02) \approx 2.50146.$$

29. The volume of the cube is given by $V = x^3$. Then $dV = 3x^2\ dx$ and when $x = 12$ and $dx = 0.02$, $dV = 3(144)(\pm 0.02) = \pm 8.64$, and the possible error that might occur in calculating the volume is ± 8.64 cm^3.

31. The volume of the hemisphere is given by $V = \frac{2}{3}\pi r^3$. The amount of rust-proofer needed is

$$\Delta V = \frac{2}{3}\pi(r + \Delta r)^3 - \frac{2}{3}\pi r^3$$

$$\approx dV = \left(\frac{2}{3}\right)(3\pi r^2)dr.$$

So, with $r = 60$, and $dr = \dfrac{1}{12}(0.01)$, we have

$$\Delta V \approx 2\pi(60^2)\left(\frac{1}{12}\right)(0.01) \approx 18.85.$$

So we need approximately 18.85 ft^3 of rust-proofer.

33. $dR = \dfrac{d}{dr}(k\ell r^{-4})dr = -4k\ell r^{-5}\,dr$. With $\dfrac{dr}{r} = 0.1$, we find

$$\frac{dR}{R} = -\frac{4k\ell r^{-5}}{k\ell r^{-4}}\,dr = -4\frac{dr}{r} = -4(0.1) = -0.4.$$

In other words, the resistance will drop by 40%.

35. $f(n) = 4n\sqrt{n-4} = 4n(n-4)^{1/2}$.

Then $df = 4[(n-4)^{1/2} + \frac{1}{2}n(n-4)^{-1/2}]dn$

When $n = 85$ and $dn = 5$, $df = 4[9 + \frac{85}{2.9}]5 \approx 274$ seconds.

37. $N(r) = \dfrac{7}{1+0.02r^2}$ and $dN = -\dfrac{0.28r}{(1+0.02r^2)^2}\,dr$. To estimate the decrease in the number of housing starts when the mortgage rate is increased from 12 to 12.5 percent, we compute

$$dN = -\frac{(0.28)(12)(0.5)}{(3.88)^2} \approx -0.111595 \quad (r = 12,\ dr = 0.5)$$

or 111,595 fewer housing starts.

39. $p = \dfrac{30}{0.02x^2 + 1}$ and $dp = -\dfrac{(1.2x)}{(0.02x^2+1)^2}\,dx$. To estimate the change in the price p when the quantity demanded changed from 5000 to 5500 units ($x = 5$ to $x = 5.5$) per week, we compute $dp = \dfrac{(-1.2)(5)(0.5)}{[0.02(25)+1]^2} \approx -1.33$, or a decrease of $1.33.

41. $P(x) = -0.000032x^3 + 6x - 100$ and $dP = (-0.000096x^2 + 6)\,dx$. To determine the error in the estimate of Trappee's profits corresponding to a maximum error in the forecast of 15 percent [$dx = \pm0.15(200)$], we compute

$$dP = [(-0.000096)(200)^2 + 6]\,(\pm30) = (2.16)(30) = \pm64.80$$

or $\pm$ \$64,800.

43. $N(x) = \dfrac{500(400+20x)^{1/2}}{(5+0.2x)^2}$ and

$N'(x) = \dfrac{(5+0.2x)^2 250(400+20x)^{-1/2}(20) - 500(400+20x)^{1/2}(2)(5+0.2x)(0.2)}{(5+0.2x)^4}\,dx.$

To estimate the change in the number of crimes if the level of reinvestment changes from 20 cents per dollars deposited to 22 cents per dollar deposited, we compute

$dN = \dfrac{(5+4)^2(250)(800)^{-1/2}(20) - 500(400+400)^{1/2}(2)(9)(0.2)}{(5+4)^4}(2) \quad (2)$

$= \dfrac{(14318.91 - 50911.69)}{9^4}(2) \approx -11$

or a decrease of approximately 11 crimes per year.

45. $A = 10,000\left(1 + \dfrac{r}{12}\right)^{120}$.

a. $dA = 10,000(120)\left(1 + \dfrac{r}{12}\right)^{119}\left(\dfrac{1}{12}\right)dr = 100,000\left(1 + \dfrac{r}{12}\right)^{119} dr.$

b. At 8.1%, it will be worth $100,000\left(1 + \dfrac{0.08}{12}\right)^{119}(0.001)$, or $220.49 more.

At 8.2%, it will be worth $100,000\left(1 + \dfrac{0.08}{12}\right)^{119}(0.002)$, or $440.99 more.

At 8.3%, it will be worth $100,000\left(1 + \dfrac{0.08}{12}\right)^{119}(0.003)$, or $661.48 more.

47. True. $dy = f'(x)\,dx = \dfrac{d}{dx}(ax+b)\,dx = a\,dx$. On the other hand,

$\Delta y = f(x+\Delta x) - f(x) = [a(x+\Delta x)+b] - (ax+b) = a\Delta x = a\,dx.$

USING TECHNOLOGY EXERCISES 3.7, page 242

1. $dy = f'(3)\,dx = 757.87(0.01) \approx 7.5787.$

3. $dy = f'(1) dx = 1.04067285926(0.03) \approx 0.031220185778$.

5. $dy = f'(4)(0.1) = -0.198761598(0.1) = -0.0198761598$.

7. If the interest rate changes from 10% to 10.3% per year, the monthly payment will increase by
$$dP = f'(0.1)(0.003) \approx 26.60279,$$
or approximately $26.60 per month. If the interest rate changes from 10% to 10.4% per year, it will be $35.47 per month. If the interest rate changes from 10% to 10.5% per year, it will be $44.34 per month.

9. $dx = f'(40)(2) \approx -0.625$. That is, the quantity demanded will decrease by 625 watches per week.

CHAPTER 3, CONCEPT REVIEW QUESTIONS, page 243

1. a. 0   b. nx^{n-1} c. $cf'(x)$ d. $f'(x) \pm g'(x)$
3. a. $g'[f(x)]f'(x)$ b. $n[f(x)]^{n-1} f'(x)$

5. a. $-\dfrac{pf'(p)}{f(p)}$ b. Elastic; unitary; inelastic
7. $y;\ dy/dt;\ a$
9. a. $x_2 - x_1$ b. $f(x + \Delta x) - f(x)$

CHAPTER 3 REVIEW, page 243

1. $f'(x) = \dfrac{d}{dx}(3x^5 - 2x^4 + 3x^2 - 2x + 1) = 15x^4 - 8x^3 + 6x - 2.$

3. $g'(x) = \dfrac{d}{dx}(-2x^{-3} + 3x^{-1} + 2) = 6x^{-4} - 3x^{-2}.$

5. $g'(t) = \dfrac{d}{dt}(2t^{-1/2} + 4t^{-3/2} + 2) = -t^{-3/2} - 6t^{-5/2}.$

7. $f'(t) = \dfrac{d}{dt}(t + 2t^{-1} + 3t^{-2}) = 1 - 2t^{-2} - 6t^{-3} = 1 - \dfrac{2}{t^2} - \dfrac{6}{t^3}.$

9. $h'(x) = \dfrac{d}{dx}(x^2 - 2x^{-3/2}) = 2x + 3x^{-5/2} = 2x + \dfrac{3}{x^{5/2}}.$

11. $g(t) = \dfrac{t^2}{2t^2 + 1}.$

$$g'(t) = \frac{(2t^2 + 1)\dfrac{d}{dt}(t^2) - t^2 \dfrac{d}{dt}(2t^2 + 1)}{(2t^2 + 1)^2}$$

$$= \frac{(2t^2 + 1)(2t) - t^2(4t)}{(2t^2 + 1)^2} = \frac{2t}{(2t^2 + 1)^2}.$$

13. $f(x) = \dfrac{\sqrt{x} - 1}{\sqrt{x} + 1} = \dfrac{x^{1/2} - 1}{x^{1/2} + 1}.$

$$f'(x) = \frac{(x^{1/2} + 1)(\frac{1}{2}x^{-1/2}) - (x^{1/2} - 1)(\frac{1}{2}x^{-1/2})}{(x^{1/2} + 1)^2}$$

$$= \frac{\frac{1}{2} + \frac{1}{2}x^{-1/2} - \frac{1}{2} + \frac{1}{2}x^{-1/2}}{(x^{1/2} + 1)^2} = \frac{x^{-1/2}}{(x^{1/2} + 1)^2} = \frac{1}{\sqrt{x}(\sqrt{x} + 1)^2}.$$

15. $f(x) = \dfrac{x^2(x^2 + 1)}{x^2 - 1}.$

$$f'(x) = \frac{(x^2 - 1)\dfrac{d}{dx}(x^4 + x^2) - (x^4 + x^2)\dfrac{d}{dx}(x^2 - 1)}{(x^2 - 1)^2}$$

$$= \frac{(x^2 - 1)(4x^3 + 2x) - (x^4 + x^2)(2x)}{(x^2 - 1)^2}$$

$$= \frac{4x^5 + 2x^3 - 4x^3 - 2x - 2x^5 - 2x^3}{(x^2 - 1)^2}$$

$$= \frac{2x^5 - 4x^3 - 2x}{(x^2 - 1)^2} = \frac{2x(x^4 - 2x^2 - 1)}{(x^2 - 1)^2}.$$

17. $f(x) = (3x^3 - 2)^8;\ f'(x) = 8(3x^3 - 2)^7(9x^2) = 72x^2(3x^3 - 2)^7.$

19. $f'(t) = \dfrac{d}{dt}(2t^2 + 1)^{1/2} = \dfrac{1}{2}(2t^2 + 1)^{-1/2}\dfrac{d}{dt}(2t^2 + 1)$

$$= \frac{1}{2}(2t^2 + 1)^{-1/2}(4t) = \frac{2t}{\sqrt{2t^2 + 1}}.$$

21. $s(t) = (3t^2 - 2t + 5)^{-2}$

$$s'(t) = -2(3t^2 - 2t + 5)^{-3}(6t - 2) = -4(3t^2 - 2t + 5)^{-3}(3t - 1)$$

$$= -\frac{4(3t - 1)}{(3t^2 - 2t + 5)^3}.$$

23. $h(x) = \left(x + \dfrac{1}{x}\right)^2 = (x + x^{-1})^2.$

$$h'(x) = 2(x + x^{-1})(1 - x^{-2}) = 2\left(x + \frac{1}{x}\right)\left(1 - \frac{1}{x^2}\right)$$

$$= 2\left(\frac{x^2 + 1}{x}\right)\left(\frac{x^2 - 1}{x^2}\right) = \frac{2(x^2 + 1)(x^2 - 1)}{x^3}.$$

25. $h'(t) = (t^2 + t)^4 \dfrac{d}{dt}(2t^2) + 2t^2 \dfrac{d}{dt}(t^2 + t)^4$

$$= (t^2 + t)^4(4t) + 2t^2 \cdot 4(t^2 + t)^3(2t + 1)$$

$$= 4t(t^2 + t)^3[(t^2 + t) + 4t^2 + 2t] = 4t^2(5t + 3)(t^2 + t)^3.$$

27. $g(x) = x^{1/2}(x^2 - 1)^3.$

$$g'(x) = \frac{d}{dx}[x^{1/2}(x^2 - 1)^3] = x^{1/2} \cdot 3(x^2 - 1)^2(2x) + (x^2 - 1)^3 \cdot \tfrac{1}{2}x^{-1/2}$$

$$= \tfrac{1}{2}x^{-1/2}(x^2 - 1)^2[12x^2 + (x^2 - 1)]$$

$$= \frac{(13x^2 - 1)(x^2 - 1)^2}{2\sqrt{x}}.$$

29. $h(x) = \dfrac{(3x + 2)^{1/2}}{4x - 3}.$

$$h'(x) = \frac{(4x - 3)\tfrac{1}{2}(3x + 2)^{-1/2}(3) - (3x + 2)^{1/2}(4)}{(4x - 3)^2}$$

$$= \frac{\frac{1}{2}(3x+2)^{-1/2}[3(4x-3)-8(3x+2)]}{(4x-3)^2} = -\frac{12x+25}{2\sqrt{3x+2}(4x-3)^2}.$$

31. $f(x) = 2x^4 - 3x^3 + 2x^2 + x + 4.$

$$f'(x) = \frac{d}{dx}(2x^4 - 3x^3 + 2x^2 + x + 4) = 8x^3 - 9x^2 + 4x + 1.$$

$$f''(x) = \frac{d}{dx}(8x^3 - 9x^2 + 4x + 1) = 24x^2 - 18x + 4 = 2(12x^2 - 9x + 2).$$

33. $h(t) = \dfrac{t}{t^2+4}.$ $h'(t) = \dfrac{(t^2+4)(1)-t(2t)}{(t^2+4)^2} = \dfrac{4-t^2}{(t^2+4)^2}.$

$$h''(t) = \frac{(t^2+4)^2(-2t)-(4-t^2)2(t^2+4)(2t)}{(t^2+4)^4}$$

$$= \frac{-2t(t^2+4)[(t^2+4)+2(4-t^2)]}{(t^2+4)^4} = \frac{2t(t^2-12)}{(t^2+4)^3}.$$

35. $f'(x) = \dfrac{d}{dx}(2x^2+1)^{1/2} = \dfrac{1}{2}(2x^2+1)^{-1/2}(4x) = 2x(2x^2+1)^{-1/2}.$

$$f''(x) = 2(2x^2+1)^{-1/2} + 2x \cdot (-\tfrac{1}{2})(2x^2+1)^{-3/2}(4x)$$

$$= 2(2x^2+1)^{-3/2}[(2x^2+1)-2x^2] = \frac{2}{(2x^2+1)^{3/2}}.$$

37. $6x^2 - 3y^2 = 9$ so $12x - 6y\dfrac{dy}{dx} = 0$ and $-6y\dfrac{dy}{dx} = -12x.$

Therefore, $\dfrac{dy}{dx} = \dfrac{-12x}{-6y} = \dfrac{2x}{y}.$

39. $y^3 + 3x^2 = 3y,$ so $3y^2 y' + 6x = 3y',$ $3y^2 y' - 3y' = -6x,$

and $y'(3y^2 - 3) = -6x.$ Therefore, $y' = -\dfrac{6x}{3(y^2-1)} = -\dfrac{2x}{y^2-1}.$

41. $x^2 - 4xy - y^2 = 12$ so $2x - 4xy' - 4y - 2yy' = 0$ and $y'(-4x - 2y) = -2x + 4y$.

So $y' = \dfrac{-2(x - 2y)}{-2(2x + y)} = \dfrac{x - 2y}{2x + y}$.

43. $df = f'(x)dx = (2x - 2x^{-3})dx = \left(2x - \dfrac{2}{x^3}\right)dx = \dfrac{2(x^4 - 1)}{x^3}dx$

45. a. $df = f'(x)dx = \dfrac{d}{dx}(2x^2 + 4)^{1/2}\,dx = \dfrac{1}{2}(2x^2 + 4)^{-1/2}(4x) = \dfrac{2x}{\sqrt{2x^2 + 4}}\,dx$

b. $\Delta f \approx df\big|_{\substack{x=4 \\ dx=0.1}} = \dfrac{2(4)(0.1)}{\sqrt{2(16) + 4}} = \dfrac{0.8}{6} = \dfrac{8}{60} = \dfrac{2}{15}.$

c. $\Delta f = f(4.1) - f(4) = \sqrt{2(4.1)^2 + 4} - \sqrt{2(16) + 4} = 0.1335$

From (b), $\Delta f \approx \dfrac{2}{15} \approx 0.1333.$

47. $f(x) = 2x^3 - 3x^2 - 16x + 3$ and $f'(x) = 6x^2 - 6x - 16$.

a. To find the point(s) on the graph of f where the slope of the tangent line is equal to -4, we solve
$$6x^2 - 6x - 16 = -4,\ 6x^2 - 6x - 12 = 0,\ 6(x^2 - x - 2) = 0$$
$$6(x - 2)(x + 1) = 0$$
and $x = 2$ or $x = -1$. Then $f(2) = 2(2)^3 - 3(2)^2 - 16(2) + 3 = -25$ and $f(-1) = 2(-1)^3 - 3(-1)^2 - 16(-1) + 3 = 14$ and the points are $(2,-25)$ and $(-1,14)$.
b. Using the point-slope form of the equation of a line, we find
that $\quad y - (-25) = -4(x - 2),\ y + 25 = -4x + 8,$ or $y = -4x - 17$
and $\quad y - 14 = -4(x + 1),$ or $y = -4x + 10$
are the equations of the tangent lines at $(2,-25)$ and $(-1,14)$.

49. $y = (4 - x^2)^{1/2}$. $y' = \tfrac{1}{2}(4 - x^2)^{-1/2}(-2x) = -\dfrac{x}{\sqrt{4 - x^2}}.$

The slope of the tangent line is obtained by letting $x = 1$, giving
$$m = -\dfrac{1}{\sqrt{3}} = -\dfrac{\sqrt{3}}{3}.$$
Therefore, an equation of the tangent line is

$$y - \sqrt{3} = -\frac{\sqrt{3}}{3}(x-1), \quad \text{or} \quad y = -\frac{\sqrt{3}}{3}x + \frac{4\sqrt{3}}{3}.$$

51. $f(x) = (2x-1)^{-1}$; $f'(x) = -2(2x-1)^{-2}$, $f''(x) = 8(2x-1)^{-3} = \dfrac{8}{(2x-1)^3}$.

$f'''(x) = -48(2x-1)^{4} = -\dfrac{48}{(2x-1)^{4}}$.

Since $(2x-1)^{4} = 0$ when $x = 1/2$, we see that the domain of f''' is $(-\infty, \frac{1}{2}) \cup (\frac{1}{2}, \infty)$.

53. $x = \dfrac{25}{\sqrt{p}} - 1$; $\quad f'(p) = -\dfrac{25}{2p^{3/2}}$; $\quad E(p) = -\dfrac{p\left(-\frac{25}{2p^{3/2}}\right)}{\frac{25}{p^{1/2}} - 1} = \dfrac{\frac{25}{2p^{1/2}}}{\frac{25 - p^{1/2}}{p^{1/2}}} = \dfrac{25}{2(25 - p^{1/2})}$

Since $\quad E(p) = 1,$
$$2(25 - p^{1/2}) = 25,$$
$$25 - p^{1/2} = \tfrac{25}{2}, \quad p^{1/2} = \tfrac{25}{2}, \text{ and } p = \tfrac{625}{4}.$$

$E(p) > 1$ and demand is elastic if $p > 156.25$; $E(p) = 1$ and demand is unitary if $p = 156.25$; and $E(p) < 1$ and demand is inelastic, if $p < 156.25$.

55. $p = 9\sqrt[3]{1000 - x}$; $\quad \sqrt[3]{1000 - x} = \dfrac{p}{9}$; $\quad 1000 - x = \dfrac{p^3}{729}$; $\quad x = 1000 - \dfrac{p^3}{729}$

Therefore, $\quad x = f(p) = \dfrac{729,000 - p^3}{729}$ and $\quad f'(x) = -\dfrac{3p^2}{729} = -\dfrac{p^2}{243}$.

Then $\quad E(p) = -\dfrac{p\left(-\frac{p^2}{243}\right)}{\frac{729,000 - p^3}{729}} = \dfrac{3p^3}{729,000 - p^3}$.

So $\quad E(60) = \dfrac{3(60)^3}{729,000 - 60^3} = \dfrac{648,000}{513,000} = \dfrac{648}{513} > 1$, and so demand is elastic.

Therefore, raising the price slightly will cause the revenue to decrease.

57. $N(x) = 1000(1 + 2x)^{1/2}$. $\quad N'(x) = 1000(\tfrac{1}{2})(1 + 2x)^{-1/2}(2) = \dfrac{1000}{\sqrt{1 + 2x}}$.

The rate of increase at the end of the twelfth week is $N'(12) = \dfrac{1000}{\sqrt{25}} = 200$, or 200 subscribers/week.

59. He can expect to live $f(100) = 46.9[1+1.09(100)]^{0.1} \approx 75.0433$, or approximately 75.04 years. $f'(t) = 46.9(0.1)(1+1.09t)^{-0.9}(1.09) = 5.1121(1+1.09t)^{-0.9}$
So the required rate of change is $f'(100) = 5.1121(1+1.09)^{-0.9} = 0.074$, or approximately 0.07 yr/yr.

61. a. $R(x) = px = (-0.02x + 600)x = -0.02x^2 + 600x$
b. $R'(x) = -0.04x + 600$
c. $R'(10,000) = -0.04(10,000) + 600 = 200$ and this says that the sale of the 10,001st phone will bring a revenue of $200.

CHAPTER 3, BEFORE MOVING ON, page 245

1. $f'(x) = 2(3x^2) - 3(\frac{1}{3}x^{-2/3}) + 5(-\frac{2}{3}x^{-5/3}) = 6x^2 - x^{-2/3} - \frac{10}{3}x^{-5/3}$

2. $g'(x) = \dfrac{d}{dx}[x(2x^2-1)^{1/2}] = (2x^2-1)^{1/2} + x(\frac{1}{2})(2x^2-1)^{-1/2}\dfrac{d}{dx}(2x^2-1)$

$= (2x^2-1)^{1/2} + \frac{1}{2}x(2x^2-1)^{-1/2}(4x) = (2x^2-1)^{-1/2}[(2x^2-1)+2x^2]$

$= \dfrac{4x^2-1}{\sqrt{2x^2-1}}$

3. $\dfrac{dy}{dx} = \dfrac{(x^2+x+1)(2) - (2x+1)(2x+1)}{(x^2+x+1)^2} = \dfrac{2x^2+2x+2-(4x^2+4x+1)}{(x^2+x+1)^2}$

$= -\dfrac{2x^2+2x-1}{(x^2+x+1)^2}.$

4. $f'(x) = \dfrac{d}{dx}(x+1)^{-1/2} = -\dfrac{1}{2}(x+1)^{-3/2} = -\dfrac{1}{2(x+1)^{3/2}}$

$f''(x) = -\dfrac{1}{2}\left(-\dfrac{3}{2}\right)(x+1)^{-5/2} = \dfrac{3}{4}(x+1)^{-5/2} = \dfrac{3}{4(x+1)^{5/2}}$

$f'''(x) = \dfrac{3}{4}\left(-\dfrac{5}{2}\right)(x+1)^{-7/2} = -\dfrac{15}{8}(x+1)^{-7/2} = -\dfrac{15}{8(x+1)^{7/2}}$

5. Differentiating both sides of the equation with respect to x gives

$$y^2 + x(2yy') - 2xy - x^2 y' + 3x^2 = 0$$

$$(2xy - x^2)y' + (y^2 - 2xy + 3x^2) = 0$$

$$y' = \frac{-y^2 + 2xy - 3x^2}{2xy - x^2} = \frac{-y^2 + 2xy - 3x^2}{x(2y - x)}$$

6. a. $dy = \dfrac{d}{dx}[x(x^2 + 5)^{1/2}]dx = [(x^2 + 5)^{1/2}(2x)]dx$

$$= (x^2 + 5)^{-1/2}[(x^2 + 5) + x^2]dx = \frac{2x^2 + 5}{\sqrt{x^2 + 5}}dx$$

Here $dx = \Delta x = 2.01 - 2 = 0.01$. Therefore,

$$\Delta y \approx dy\Big|_{\substack{x=2 \\ dx=0.01}} = \frac{2(4) + 5}{\sqrt{4 + 5}}(0.01) = \frac{0.13}{3} \approx 0.043$$

CHAPTER 4

4.1 Problem Solving Tips

1. The critical number of a function f is any number x in the domain of f such that $f'(x) = 0$, or $f'(x)$ does not exist. Note that the definition requires that x be in the domain of f. Consider the function $f(x) = x + 1/x$ in Example 8 on page 257 in the text. Even though f' is discontinuous at $x = 0$, $x = 0$ did not qualify as a critical number because it does not lie in the domain of f.

2. Note that when you use test values to find the sign of a derivative over an interval, you don't need to evaluate the derivative function at a test value. You only need to find the *sign* of the derivative function at that test value. For example, to find the sign of $f'(x) = \dfrac{(x+1)(x-1)}{x^2}$ in the interval $(0,1)$ using the test value $x = \frac{1}{2}$, we simply note that sign of the numerator is $(+) \times (-) = (-)$. Since x^2 is always positive, the denominator is always positive. Therefore, the sign of f' over the interval $(0,1)$ is negative.

4.1 CONCEPT QUESTIONS, page 258

1. a. f is increasing on I if whenever x_1 and x_2 are in I with $x_1 < x_2$, then $f(x_1) < f(x_2)$.
   b. f is decreasing on I if whenever x_1 and x_2 are in I with $x_1 < x_2$, then $f(x_1) > f(x_2)$.

3. a. f has a relative maximum at $x = a$ if there is an open interval I containing a such that $f(x) \le f(a)$ for all x in I.

b. f has a relative minimum at $x = a$ if there is an open interval I containing a such that $f(x) \le f(a)$ for all x in I.

5. See text page 255.

EXERCISES 4.1, page 258

1. f is decreasing on $(-\infty, 0)$ and increasing on $(0, \infty)$.

3. f is increasing on $(-\infty, -1) \cup (1, \infty)$, and decreasing on $(-1, 1)$.

5. f is increasing on $(0, 2)$ and decreasing on $(-\infty, 0) \cup (2, \infty)$.

7. f is decreasing on $(-\infty, -1) \cup (1, \infty)$ and increasing on $(-1, 1)$.

9. Increasing on $(20.2, 20.6) \cup (21.7, 21.8)$, constant on $(19.6, 20.2) \cup (20.6, 21.1)$, and decreasing on $(21.1, 21.7) \cup (21.8, 22.7)$,

11. $f(x) = 3x + 5; f'(x) = 3 > 0$ for all x and so f is increasing on $(-\infty, \infty)$.

13. $f(x) = x^2 - 3x.$ $f'(x) = 2x - 3$ is continuous everywhere and is equal to zero when $x = 3/2$. From the following sign diagram

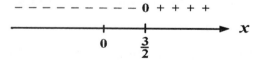

we see that f is decreasing on $(-\infty, \frac{3}{2})$ and increasing on $(\frac{3}{2}, \infty)$.

15. $g(x) = x - x^3.$ $g'(x) = 1 - 3x^2$ is continuous everywhere and is equal to zero when $1 - 3x^2 = 0$, or $x = \pm\frac{\sqrt{3}}{3}$. From the following sign diagram

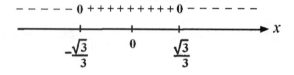

we see that f is decreasing on $(-\infty, -\frac{\sqrt{3}}{3}) \cup (\frac{\sqrt{3}}{3}, \infty)$ and increasing on $(-\frac{\sqrt{3}}{3}, \frac{\sqrt{3}}{3})$.

17. $g(x) = x^3 + 3x^2 + 1$; $g'(x) = 3x^2 + 6x = 3x(x+2)$.
From the following sign diagram

$+ + + + + + + + 0 - - 0 + + + + + +$

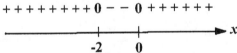

$\quad\quad\quad\quad\quad\quad -2 \quad\quad 0$

we see that g is increasing on $(-\infty,-2) \cup (0,\infty)$ and decreasing on $(-2,0)$.

19. $f(x) = \frac{1}{3}x^3 - 3x^2 + 9x + 20$; $f'(x) = x^2 - 6x + 9 = (x-3)^2 > 0$ for all x except
$x = 3$, at which point $f'(3) = 0$. Therefore, f is increasing on $(-\infty,3) \cup (3,\infty)$.

21. $h(x) = x^4 - 4x^3 + 10$; $h'(x) = 4x^3 - 12x^2 = 4x^2(x-3)$ if $x = 0$ or 3. From the sign
diagram of h',

$- - - - - - 0 - - - - - 0 + + + + + +$

$\quad\quad\quad\quad\quad 0 \quad\quad\quad\quad 3$

we see that h is increasing on $(3,\infty)$ and decreasing on $(-\infty,0) \cup (0,3)$.

23. $f(x) = \frac{1}{x-2} = (x-2)^{-1}$. $f'(x) = -1(x-2)^{-2}(1) = -\frac{1}{(x-2)^2}$ is discontinuous at
$x = 2$ and is continuous everywhere else. From the sign diagram

f' not defined here
$\downarrow$

$- -$

$\quad\quad\quad\quad\quad 0 \quad\quad\quad 2$

we see that f is decreasing on $(-\infty,2) \cup (2,\infty)$.

25. $h(t) = \frac{t}{t-1}$. $h'(t) = \frac{(t-1)(1) - t(1)}{(t-1)^2} = -\frac{1}{(t-1)^2}$.
From the following sign diagram,

h' not defined here
$\downarrow$

$- - - - - - - - - - - - - - - - -$

$\quad\quad\quad\quad\quad 0 \quad\quad 1$

we see that $h'(t) < 0$ whenever it is defined. We conclude that h is decreasing on $(-\infty,1) \cup (1,\infty)$.

27. $f(x) = x^{3/5}$. $f'(x) = \dfrac{3}{5}x^{-2/5} = \dfrac{3}{5x^{2/5}}$. Observe that $f'(x)$ is not defined at $x = 0$,

but is positive everywhere else and therefore increasing on $(-\infty,0) \cup (0,\infty)$.

29. $f(x) = \sqrt{x+1}$. $f'(x) = \dfrac{d}{dx}(x+1)^{1/2} = \dfrac{1}{2}(x+1)^{-1/2} = \dfrac{1}{2\sqrt{x+1}}$ and we see that

$f'(x) > 0$ if $x > -1$. Therefore, f is increasing on $(-1, \infty)$.

31. $f(x) = \sqrt{16 - x^2} = (16 - x^2)^{1/2}$. $f'(x) = \dfrac{1}{2}(16 - x^2)^{-1/2}(-2x) = -\dfrac{x}{\sqrt{16-x^2}}$.

Since the domain of f is [-4,4], we consider the sign diagram for f' on this interval. Thus,

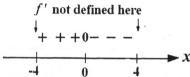

and we see that f is increasing on (-4,0) and decreasing on (0,4).

33. $f'(x) = \dfrac{d}{dx}(x - x^{-1}) = 1 + \dfrac{1}{x^2} = \dfrac{x^2 + 1}{x^2}$ and so $f'(x) > 0$ for all $x \neq 0$.

Therefore, f is increasing on $(-\infty,0) \cup (0,\infty)$.

35. f has a relative maximum of $f(0) = 1$ and relative minima of $f(-1) = 0$ and $f(1) = 0$.

37. f has a relative maximum of $f(-1) = 2$ and a relative minimum of $f(1) = -2$.

39. f has a relative maximum of $f(1) = 3$ and a relative minimum of $f(2) = 2$.

41. f has a relative minimum at (0,2).

43. a   45. d

47. $f(x) = x^2 - 4x$. $f'(x) = 2x - 4 = 2(x - 2)$ has a critical point at $x = 2$. From the

4 Applications of the Derivative   **162**

following sign diagram

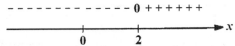

we see that $f(2) = -4$ is a relative minimum by the First Derivative Test.

49. $h(t) = -t^2 + 6t + 6$; $h'(t) = -2t + 6 = -2(t - 3) = 0$ if $t = 3$, a critical point. The sign

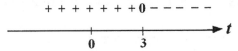

diagram and the First Derivative Test imply that h has a relative maximum at 3 with value $f(3) = -9 + 18 + 6 = 15$.

51. $f(x) = x^{5/3}$. $f'(x) = \frac{5}{3}x^{2/3}$ giving $x = 0$ as the critical point of f.
From the sign diagram

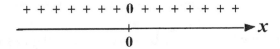

we see that f' does not change sign as we move across $x = 0$, and conclude that f has no relative extremum.

53. $g(x) = x^3 - 3x^2 + 4$. $g'(x) = 3x^2 - 6x = 3x(x - 2) = 0$ if $x = 0$ or 2. From the sign diagram, we see that the critical point $x = 0$ gives a relative maximum, whereas,

$$+ + + + + 0 - - - 0 + + + +$$
$$0 \qquad 2$$

$x = 2$ gives a relative minimum. The values are $g(0) = 4$ and $g(2) = 8 - 12 + 4 = 0$.

55. $f(x) = \frac{1}{2}x^4 - x^2$. $f'(x) = 2x^3 - 2x = 2x(x^2 - 1) = 2x(x + 1)(x - 1)$ is continuous everywhere and has zeros as $x = -1$, $x = 0$, and $x = 1$, the critical points of f. Using the First Derivative Test and the following sign diagram of f'

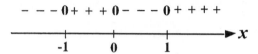

we see that $f(-1) = -1/2$ and $f(1) = -1/2$ are relative minima of f and $f(0) = 0$ is a relative maximum of f.

57. $F(x) = \frac{1}{3}x^3 - x^2 - 3x + 4$. Setting $F'(x) = x^2 - 2x - 3 = (x-3)(x+1) = 0$ gives $x = -1$ and $x = 3$ as critical points. From the sign diagram

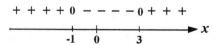

we see that $x = -1$ gives a relative maximum and $x = 3$ gives a relative minimum. The values are
$$F(-1) = -\frac{1}{3} - 1 + 3 + 4 = \frac{17}{3} \quad \text{and} \quad F(3) = 9 - 9 - 9 + 4 = -5,$$
respectively.

59. $g(x) = x^4 - 4x^3 + 8$. Setting $g'(x) = 4x^3 - 12x^2 = 4x^2(x - 3) = 0$ gives $x = 0$ and $x = 3$ as critical points. From the sign diagram

$$\begin{array}{c} - \ - \ - \ - \ -0- \ - \ - \ + \ + \ + \ + \\ \hline \quad\quad 0 \quad\quad\quad 3 \quad\quad\quad \rightarrow x \end{array}$$

we see that $x = 3$ gives a relative minimum. Its value is $g(3) = 3^4 - 4(3)^3 + 8 = -19$.

61. $g'(x) = \dfrac{d}{dx}\left(1 + \dfrac{1}{x}\right) = -\dfrac{1}{x^2}$. Observe that g' is never zero for all values of x.

Furthermore, g' is undefined at $x = 0$, but $x = 0$ is not in the domain of g. Therefore g has no critical points and so g has no relative extrema.

63. $f(x) = x + \dfrac{9}{x} + 2$. Setting $f'(x) = 1 - \dfrac{9}{x^2} = \dfrac{x^2 - 9}{x^2} = \dfrac{(x+3)(x-3)}{x^2} = 0$

gives $x = -3$ and $x = 3$ as critical points. From the sign diagram

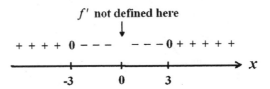

we see that (-3,-4) is a relative maximum and (3,8) is a relative minimum.

65. $f(x) = \dfrac{x}{1+x^2}$. $f'(x) = \dfrac{(1+x^2)(1)-x(2x)}{(1+x^2)^2} = \dfrac{1-x^2}{(1+x^2)^2} = \dfrac{(1-x)(1+x)}{(1+x^2)^2} = 0$ if $x = \pm 1$,

and these are critical points of f. From the sign diagram of f'

$$- - - - 0 + + + + + + + 0 - - - -$$

$$\xrightarrow{\hspace{0.5cm} \underset{-1}{|} \hspace{1cm} \underset{0}{|} \hspace{1cm} \underset{1}{|} \hspace{1cm}} x$$

we see that f has a relative minimum at $(-1, -\frac{1}{2})$ and a relative maximum at $(1, \frac{1}{2})$.

67. $f(x) = (x - 1)^{2/3}$. $f'(x) = \dfrac{2}{3}(x-1)^{-1/3} = \dfrac{2}{3(x-1)^{1/3}}$.

$f'(x)$ is discontinuous at $x = 1$. The sign diagram for f' is

f' not defined here

$$- - - - - - - - - - - - \downarrow + + + + + + +$$

$$\xrightarrow{\hspace{0.5cm} \underset{0}{|} \hspace{1cm} \underset{1}{|} \hspace{1cm}} x$$

We conclude that $f(1) = 0$ is a relative minimum.

69. $h(t) = -16t^2 + 64t + 80$. $h'(t) = -32t + 64 = -32(t - 2)$ and has sign diagram

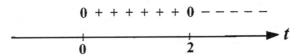

This tells us that the stone is rising on the time interval (0,2) and falling when $t > 2$. It hits the ground when $h(t) = -16t^2 + 64t + 80 = 0$
or $t^2 - 4t - 5 = (t - 5)(t + 1) = 0$ or $t = 5$ (we reject the root $t = -1$.)

165   *4 Applications of the Derivative*

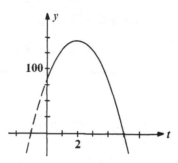

71. $P'(x) = \dfrac{d}{dx}(0.0726x^2 + 0.7902x + 4.9623) = 0.1452x + 0.7902.$

Since $P'(x) > 0$ on $(0, 25)$, we see that P is increasing on the interval in question. Our result tells us that the percent of the population afflicted with Alzheimer's disease increases with age for those that are 65 and over.

73. $h(t) = -\frac{1}{3}t^3 + 16t^2 + 33t + 10$; $h'(t) = -t^2 + 32t + 33 = -(t+1)(t-33).$

The sign diagram for h' is

$$0 + + + + + + 0 - - - - -$$

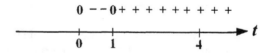

The rocket is rising on the time interval $(0,33)$ and descending on $(33,T)$ for some positive number T. The parachute is deployed 33 seconds after liftoff.

75. $f(t) = 20t - 40\sqrt{t} + 50 = 20t - 40t^{1/2} + 50.$

$$f'(t) = 20 - 40(\frac{1}{2}t^{-1/2}) = 20(1 - \frac{1}{\sqrt{t}}) = \frac{20(\sqrt{t}-1)}{\sqrt{t}}.$$

Then f' is continuous on $(0,4)$ and is equal to zero at $t = 1$. From the sign diagram

$$0 - -0 + + + + + + + + +$$

we see that f is decreasing on $(0,1)$ and increasing on $(1,4)$. We conclude that the average speed decreases from 6 A.M. to 7 A.M. and then picks up from 7 A.M. to 10 A. M.

77. a. $f'(t) = \dfrac{d}{dt}(-0.05t^3 + 0.56t^2 + 5.47t + 7.5) = -0.15t^2 + 1.12t + 5.47.$

Setting $f'(t) = 0$ gives $-0.15t^2 + 1.12t + 5.47 = 0.$ Using the quadratic formula, we find

$$t = \frac{-1.12 \pm \sqrt{(1.12)^2 - 4(-0.15)(5.47)}}{-0.3}$$

that is, $t = $ -3.37, or 10.83. Since f' is continuous, the only critical points of f are $t = -3.4$ and $t = 10.8,$ both of which lie outside the interval of interest.

Nevertheless this result can be used to tell us that f' does not change sign in the interval (-3.4, 10.8). Using $t = 0$ as the test point, we see that $f'(0) = 5.47 > 0$ and so we see that f is increasing on (-3.4, 10.8), and , in particular, in the interval (0, 6). Thus, we conclude that f is increasing on (0, 6).
b. The result of part (a) tells us that sales in the Web-hosting industry will be increasing throughout the years from 1999 through 2005.

79. a. $P'(t) = 0.00279t^2 - 0.036t - 0.51.$ Setting $P'(t) = 0$ and solving the resulting equation, we have

$$t = \frac{0.036 \pm \sqrt{(-0.036)^2 - 4(0.00279)(-0.51)}}{2(0.00279)}$$

≈ -8.53 or 21.43

The sign diagram for P' follows. [Take $t = $ -10, 0, 25 as test points.]

From the sign diagram, we see that P is decreasing on [0, 21.43] and increasing on [21.43, 30].
b. The percent of men 65 years and older in the workforce was decreasing from 1970 through about the middle of 1991, then starting increasing from then through the year 2000.

81. $S'(t) = \dfrac{d}{dt}(0.46t^3 - 2.22t^2 + 6.21t + 17.25) = 0.96t^2 - 4.44t + 6.21.$

Observe that S' is continuous everywhere. Setting $S'(t) = 0$ and solving, we find

$$t = \frac{4.44\sqrt{4.44^2 - 4(0.96)(6.21)}}{2(0.96)}$$

Now, the discriminant is $-4.1328 < 0$ which shows that the equation has no real roots. Since $S'(0) = 6.21 > 0$, we conclude that $S'(t) > 0$ for all t, in particular, for t in the interval $(0,4)$. This shows that S is increasing on $(0,4)$.

83. $S'(t) = -6.945t^2 + 68.65t + 1.32$. Setting $S'(t) = 0$ and solving the resulting equation, we obtain

$$t = \frac{-68.65 \pm \sqrt{(68.65)^2 - 4(-6.945)(1.32)}}{2(-6.945)} \approx -0.02 \text{ or } 9.90.$$

The sign diagram for S' follows.

From the sign diagram, we see that S' is increasing on the interval $[0, 5]$. We conclude that U.S. telephone company spending was projected to be increasing from 2001 through 2006

85. $C(t) = \dfrac{t^2}{2t^3 + 1};\quad C'(t) = \dfrac{(2t^3 + 1)(2t) - t^2(6t^2)}{(2t^3 + 1)^2} = \dfrac{2t - 2t^4}{(2t^3 + 1)^2} = \dfrac{2t(1 - t^3)}{(2t^3 + 1)^2}.$

From the sign diagram of C' on $(0,\infty)$,

```
        + +0  - - - - - - -
  ──────+──+──────────+────────▶ t
        0  1          4
```

We see that the drug concentration is increasing on $(0,1)$ and decreasing on $(1,4)$.

87. $A(t) = \dfrac{136}{1 + 0.25(t - 4.5)^2} + 28.$

$$A'(t) = 136\frac{d}{dt}[1+0.25(t-4.5)^2]^{-1} = -136[1+0.25(t-4.5)^2]^{-2}2(0.25)(t-4.5)$$

$$= -\frac{68(t-4.5)}{[1+0.25(t-4.5)^2]^2}.$$

Observe that $A'(t) > 0$ if $t < 4.5$ and $A'(t) < 0$ if $t > 4.5$, so the pollution is increasing from 7 A.M. to 11:30 A.M. and decreasing from 11:30 A.M. to 6 P.M.

89. a. $G(t) = (D - S)(t) = D(t) - S(t)$
$$= (0.0007t^2 + 0.0265t + 2) - (-0.0014t^2 + 0.0326t + 1.9)$$
$$= 0.0021t^2 - 0.0061t + 0.1$$
 b. $G'(t) = 0.0042t - 0.0061 = 0$ implies $t \approx 1.45$. The sign diagram of G' follows:

We see that G is decreasing on $(0, 1.5)$ and increasing on $(1.5, 15)$. This shows that the gap between the demand and supply of nurses eas increasing from 2000 through the middle of 2001 but starts widening from the middle of 2001 through 2015.
 b. The relative minimum of G occurs at $t = 1.5$ and is $f(1.45) \approx 0.0956$. This says that at its best there is a shortage of approximately 96,000.

91. True. Let $a < x_1 < x_2 < b$. Then $f(x_2) > f(x_1)$ and $g(x_2) > g(x_1)$. Therefore,
$$(f + g)(x_2) = f(x_2) + g(x_2) > f(x_1) + g(x_1) = (f + g)(x_1)$$
and so $f + g$ is increasing on (a, b).

93. True. Let $a < x_1 < x_2 < b$, then $f(x_1) < f(x_2)$ and $g(x_1) < g(x_2)$. We find
$$(fg)(x_2) - (fg)(x_1) = f(x_2)g(x_2) - f(x_1)g(x_1)$$
$$= f(x_2)g(x_2) - f(x_2)g(x_1) + f(x_2)g(x_1) - f(x_1)g(x_1)$$
$$= f(x_2)[g(x_2) - g(x_1)] + g(x_1)[f(x_2) - f(x_1)]$$
$$> 0$$
So $(fg)(x_2) > (fg)(x_1)$ and fg is increasing on (a, b).

95. False. Let $f(x) = |x|$. Then f has a relative minimum at $x = 0$, but $f'(0)$ does not

exist.

97. $f'(x) = 3x^2 + 1$ is continuous on $(-\infty, \infty)$ and is always greater than or equal to 1. So f has no critical points in $(-\infty, \infty)$. Therefore f has no relative extrema on $(-\infty, \infty)$.

99. a. $f'(x) = -2x$ if $x \neq 0$. $f'(-1) = 2$ and $f'(1) = -2$ so $f'(x)$ changes sign from positive to negative as we move across $x = 0$.

   b. f does not have a relative maximum at $x = 0$ because $f(0) = 2$ but a neighborhood of $x = 0$, for example $(-\frac{1}{2}, \frac{1}{2})$, contains points with values larger than 2. This does not contradict the First Derivative Test because f is not continuous at $x = 0$.

101. $f(x) = ax^2 + bx + c$. Setting $f'(x) = 2ax + b = 2a\left(x + \frac{b}{2a}\right) = 0$ gives $x = -\frac{b}{2a}$ as the

   only critical point of f. If $a < 0$, we have the sign diagram

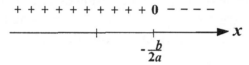

   from which we see that $x = -b/2a$ gives a relative maximum. Similarly, you can show that if $a > 0$, then $x = -b/2a$ gives a relative minimum.

103. a. $f'(x) = 3x^2 + 1$ and so $f'(x) > 1$ on the interval $(0,1)$. Therefore, f is increasing on $(0,1)$.

   b. $f(0) = -1$ and $f(1) = 1 + 1 - 1 = 1$. So the Intermediate Value Theorem guarantees that there is at least one root of $f(x) = 0$ in $(0,1)$. Since f is increasing on $(0,1)$, the graph of f can cross the x-axis at only one point in $(0,1)$. So $f(x) = 0$ has exactly one root.

USING TECHNOLOGY EXERCISES 4.1, page 226

1. a. f is decreasing on $(-\infty, -0.2934)$ and increasing on $(-0.2934, \infty)$.

   b. Relative minimum: $f(-0.2934) = -2.5435$

3. a. f is increasing on $(-\infty,-1.6144) \cup (0.2390,\infty)$ and decreasing on $(-1.6144, 0.2390)$

    b. Relative maximum: $f(-1.6144) = 26.7991$; relative minimum: $f(0.2390) = 1.6733$

5. a. f is decreasing on $(-\infty,-1) \cup (0.33,\infty)$ and increasing on $(-1,0.33)$

    b. Relative maximum: $f(0.33) = 1.11$; relative minimum: $f(-1) = -0.63$.

7. a. f is decreasing on $(-1,-0.71)$ and increasing on $(-0.71,1)$.

    b. f has a relative minimum at $(-0.71,-1.41)$.

9. a.

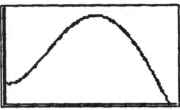

    b. f is decreasing on $(0,0.2398) \cup (6.8758,12)$ and increasing on $(0.2398,6.8758)$

    c. $(6.8758, 200.14)$; The rate at which the number of banks were failing reached a peak of 200/yr during the latter part of 1988 $(t = 6.8758)$.

11. a.

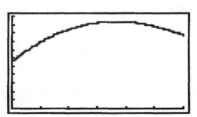

    b. increasing on $(0,3.6676)$ and decreasing on $(3.6676, 6)$.

13. f is decreasing on the interval $(0,1)$ and increasing on $(1,4)$. The relative minimum occurs at the point $(1,32)$. These results indicate that the speed of traffic flow drops between 6 A.M. and 7 A.M. reaching a low of 32 mph. Thereafter, it increases till 10 A.M.

4.2 Problem Solving Tips

1. As an aid for remembering the shape of a graph,

If $f''(x) > 0$, then the graph of f "holds

water" and we say the graph of f is

concave upward.

If $f''(x) < 0$, then the graph of f "loses

water" and we say the graph of f is concave

downward.

2. To find the inflection points of a function f, we determine the number(s) in the domain of f for which $f''(x) = 0$ or $f''(x)$ does not exist. Note that each of these numbers c provides us with a *candidate* $(c, f(c))$ for an inflection point of f.

3. Note that the second derivative test is not valid when $f''(c) = 0$ or $f''(c)$ does not exist. In these cases you need to use the first derivative test to determine the relative extrema.

4.2 CONCEPT QUESTIONS, page 276

1. a. f is concave upward on (a,b) if f' is increasing on (a,b).
   f is concave downward on (a,b) if f' is decreasing on (a,b).
   b. To determine where f is concave upward and concave downward, see page 269.

3. The second derivatve test is stated in the text on page 274. In general, if f'' is easy to compute, then use the second derivative test. However, keep in mind that (1) in order to use this test f'' must exist, (2) the test is inconclusive if $f''(c) = 0$, and (3) the test is inconvenient to use if f'' is difficult to compute.

EXERCISES 4.2, page 276

1. f is concave downward on $(-\infty,0)$ and concave upward on $(0,\infty)$. f has an inflection point at $(0,0)$.

3. f is concave downward on $(-\infty,0) \cup (0,\infty)$.

5. f is concave upward on $(-\infty,0) \cup (1,\infty)$ and concave downward on $(0,1)$. $(0,0)$ and $(1,-1)$ are inflection points of f.

7. f is concave downward on $(-\infty,-2) \cup (-2,2) \cup (2,\infty)$.

9. a   11. b

13. a. $D_1'(t) > 0$, $D_2'(t) > 0$, $D_1''(t) > 0$, and $D_2''(t) < 0$ on $(0,12)$.
    b. With or without the proposed promotional campaign, the deposits will increase, but with the promotion, the deposits will increase at an increasing rate whereas without the promotion, the deposits will increase at a decreasing rate.

15. The significance of the inflection point Q is that the restoration process is working at its peak at the time t_0 corresponding to its t-coordinate.

17. $f(x) = 4x^2 - 12x + 7$. $f'(x) = 8x - 12$ and $f''(x) = 8$. So, $f''(x) > 0$ everywhere and therefore f is concave upward everywhere.

19. $f(x) = \dfrac{1}{x^4} = x^{-4}$; $f'(x) = -\dfrac{4}{x^5}$ and $f''(x) = \dfrac{20}{x^6} > 0$ for all values of x in $(-\infty,0) \cup (0,\infty)$ and so f is concave upward everywhere.

21. $f(x) = 2x^2 - 3x + 4$; $f'(x) = 4x - 3$ and $f''(x) = 4 > 0$ for all values of x. So f is concave upward on $(-\infty,\infty)$.

23. $f(x) = x^3 - 1$. $f'(x) = 3x^2$ and $f''(x) = 6x$. The sign diagram of f'' follows.

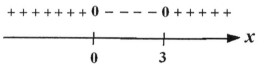

We see that f is concave downward on $(-\infty, 0)$ and concave upward on $(0, \infty)$.

25. $f(x) = x^4 - 6x^3 + 2x + 8$; $f'(x) = 4x^3 - 18x^2 + 2$ and $f''(x) = 12x^2 - 36x = 12x(x - 3)$. The sign diagram of f''

$$+++++++0----0+++++$$

$$\xrightarrow{\qquad\quad 0 \qquad\quad 3 \qquad\qquad} x$$

shows that f is concave upward on $(-\infty, 0) \cup (3, \infty)$ and concave downward on $(0, 3)$.

27. $f(x) = x^{4/7}$. $f'(x) = \dfrac{4}{7}x^{-3/7}$ and $f''(x) = -\dfrac{12}{49}x^{-10/7} = -\dfrac{12}{49x^{10/7}}$.

Observe that $f''(x) < 0$ for all x different from zero. So f is concave downward on $(-\infty, 0) \cup (0, \infty)$.

29. $f(x) = (4-x)^{1/2}$. $f'(x) = \dfrac{1}{2}(4-x)^{-1/2}(-1) = -\dfrac{1}{2}(4-x)^{-1/2}$;

$f''(x) = \dfrac{1}{4}(4-x)^{-3/2}(-1) = -\dfrac{1}{4(4-x)^{3/2}} < 0$.

whenever it is defined. So f is concave downward on $(-\infty, 4)$.

31. $f'(x) = \dfrac{d}{dx}(x-2)^{-1} = -(x-2)^{-2}$ and $f''(x) = 2(x-2)^{-3} = \dfrac{2}{(x-2)^3}$.

The sign diagram of f'' shows that f is concave downward on $(-\infty, 2)$ and concave upward on $(2, \infty)$.

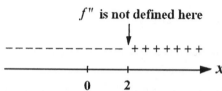

33. $f'(x) = \dfrac{d}{dx}(2+x^2)^{-1} = -(2+x^2)^{-2}(2x) = -2x(2+x^2)^{-2}$ and

$$f''(x) = -2(2+x^2)^{-2} - 2x(-2)(2+x^2)^{-3}(2x)$$

$$= 2(2+x^2)^{-3}[-(2+x^2)+4x^2] = \frac{2(3x^2-2)}{(2+x^2)^3} = 0 \text{ if } x = \pm\sqrt{2/3}.$$

From the sign diagram of f''

```
     + + + + 0  - - - - - -   0+ + + +
   ──────────┼──────┼──────┼──────────→ x
           -√2/3    0      √2/3
```

we see that f is concave upward on $(-\infty, -\sqrt{2/3}) \cup (\sqrt{2/3}, \infty)$ and concave downward on $(-\sqrt{2/3}, \sqrt{2/3})$.

35. $h(t) = \dfrac{t^2}{t-1}$; $h'(t) = \dfrac{(t-1)(2t) - t^2(1)}{(t-1)^2} = \dfrac{t^2 - 2t}{(t-1)^2}$;

$$h''(t) = \frac{(t-1)^2(2t-2) - (t^2-2t)2(t-1)}{(t-1)^4}$$

$$= \frac{(t-1)(2t^2 - 4t + 2 - 2t^2 + 4t)}{(t-1)^4} = \frac{2}{(t-1)^3}.$$

The sign diagram of h'' is

```
              h " is not defined here
                      ↓
   - - - - - -      + + + + + + + +
   ──────────┼──────────┼──────────→ t
             0          1
```

and tells us that h is concave downward on $(-\infty, 1)$ and concave upward on $(1, \infty)$.

37. $g(x) = x + \dfrac{1}{x^2}$. $g'(x) = 1 - 2x^{-3}$ and $g''(x) = 6x^{-4} = \dfrac{6}{x^4} > 0$ whenever $x \neq 0$.

Therefore, g is concave upward on $(-\infty, 0) \cup (0, \infty)$.

39. $g(t) = (2t-4)^{1/3}$. $g'(t) = = \dfrac{1}{3}(2t-4)^{-2/3}(2) = \dfrac{2}{3}(2t-4)^{-2/3}$.

$g''(t) = -\dfrac{4}{9}(2t-4)^{-5/3} = -\dfrac{4}{9(2t-4)^{5/3}}$. The sign diagram of g''

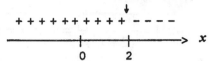

g" is not defined here

tells us that g is concave upward on $(-\infty,2)$ and concave downward on $(2,\infty)$.

41. $f(x) = x^3 - 2. f'(x) = 3x^2$ and $f''(x) = 6x. f''(x)$ is continuous everywhere and has a zero at $x = 0$. From the sign diagram of f''

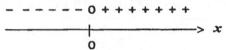

we conclude that $(0,-2)$ is an inflection point of f.

43. $f(x) = 6x^3 - 18x^2 + 12x - 15; f'(x) = 18x^2 - 36x + 12$ and
$f''(x) = 36x - 36 = 36(x - 1) = 0$ if $x = 1$. The sign diagram of f''

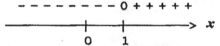

tells us that f has an inflection point at $(1,-15)$.

45. $f(x) = 3x^4 - 4x^3 + 1. f'(x) = 12x^3 - 12x^2$ and $f''(x) = 36x^2 - 24x = 12x(3x - 2) = 0$ if $x = 0$ or $2/3$. These are candidates for inflection points. The sign diagram of f''

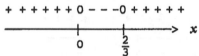

shows that $(0,1)$ and $(\frac{2}{3},\frac{11}{27})$ are inflection points of f.

47. $g(t) = t^{1/3}, g'(t) = \frac{1}{3}t^{-2/3}$ and $g''(t) = -\frac{2}{9}t^{-5/3} = -\dfrac{2}{9t^{5/3}}$. Observe that $t = 0$ is in the domain of g. Next, since $g''(t) > 0$ if $t < 0$ and $g''(t) < 0$, if $t > 0$, we see that $(0,0)$ is an inflection point of g.

49. $f(x) = (x - 1)^3 + 2. f'(x) = 3(x - 1)^2$ and $f''(x) = 6(x - 1)$. Observe that $f''(x) < 0$ if $x < 1$ and $f''(x) > 0$ if $x > 1$ and so $(1,2)$ is an inflection point of f.

51. $f(x) = \dfrac{2}{1+x^2} = 2(1+x^2)^{-1}. f'(x) = -2(1+x^2)^{-2}(2x) = -4x(1+x^2)^{-2}$.

$$f''(x) = -4(1+x^2)^{-2} - 4x(-2)(1+x^2)^{-3}(2x)$$

$$= 4(1+x^2)^{-3}[-(1+x^2)+4x^2] = \frac{4(3x^2-1)}{(1+x^2)^3},$$

is continuous everywhere and has zeros at $x = \pm\frac{\sqrt{3}}{3}$. From the sign diagram of f'' we conclude that $(-\frac{\sqrt{3}}{3}, \frac{3}{2})$ and $(\frac{\sqrt{3}}{3}, \frac{3}{2})$ are inflection points of f.

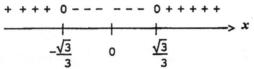

53. $f(x) = -x^2 + 2x + 4$ and $f'(x) = -2x + 2$. The critical point of f is $x = 1$. Since $f''(x) = -2$ and $f''(1) = -2 < 0$, we conclude that $f(1) = 5$ is a relative maximum of f.

55. $f(x) = 2x^3 + 1; f'(x) = 6x^2 = 0$ if $x = 0$ and this is a critical point of f. Next, $f''(x) = 12x$ and so $f''(0) = 0$. Thus, the Second Derivative Test fails. But the First Derivative Test shows that $(0,0)$ is not a relative extremum.

57. $f(x) = \frac{1}{3}x^3 - 2x^2 - 5x - 10$. $f'(x) = x^2 - 4x - 5 = (x-5)(x+1)$ and this gives $x = -1$ and $x = 5$ as critical points of f. Next, $f''(x) = 2x - 4$. Since $f''(-1) = -6 < 0$, we see that $(-1, -\frac{22}{3})$ is a relative maximum. Next, $f''(5) = 6 > 0$ and this shows that $(5, -\frac{130}{3})$ is a relative minimum.

59. $g(t) = t + \frac{9}{t}$. $g'(t) = 1 - \frac{9}{t^2} = \frac{t^2 - 9}{t^2} = \frac{(t+3)(t-3)}{t^2}$ and this shows that $t = \pm 3$ are critical points of g. Now, $g''(t) = 18t^{-3} = \frac{18}{t^3}$. Since $g''(-3) = -\frac{18}{27} < 0$ the Second Derivative Test implies that g has a relative maximum at $(-3, -6)$. Also, $g''(3) = \frac{18}{27} > 0$ and so g has a relative minimum at $(3, 6)$.

61. $f(x) = \frac{x}{1-x}$. $f'(x) = \frac{(1-x)(1) - x(-1)}{(1-x)^2} = \frac{1}{(1-x)^2}$ is never zero. So there are no critical points and f has no relative extrema.

63. $f(t) = t^2 - \dfrac{16}{t}$. $\quad f'(t) = 2t + \dfrac{16}{t^2} = \dfrac{2t^3 + 16}{t^2} = \dfrac{2(t^3 + 8)}{t^2}$. Setting

$f'(t) = 0$ gives $t = -2$ as a critical point. Next, we compute

$f''(t) = \dfrac{d}{dt}(2t + 16t^{-2}) = 2 - 32t^{-3} = 2 - \dfrac{32}{t^3}$. Since $f''(-2) = 2 - \dfrac{32}{(-8)} = 6 > 0$, we

see that $(-2, 12)$ is a relative minimum.

65. $g(s) = \dfrac{s}{1+s^2}$; $g'(s) = \dfrac{(1+s^2)(1) - s(2s)}{(1+s^2)^2} = \dfrac{1-s^2}{(1+s^2)^2} = 0$ gives $s = -1$ and $s = 1$

as critical points of g. Next, we compute

$g''(s) = \dfrac{(1+s^2)^2(-2s) - (1-s^2)2(1+s^2)(2s)}{(1+s^2)^4}$

$= \dfrac{2s(1+s^2)(-1-s^2-2+2s^2)}{(1+s^2)^4} = \dfrac{2s(s^2-3)}{(1+s^2)^3}$.

Now, $g''(-1) = \tfrac{1}{2} > 0$ and so $g(-1) = -\tfrac{1}{2}$ is a relative minimum of g. Next,

$g''(1) = -\tfrac{1}{2} < 0$ and so $g(1) = \tfrac{1}{2}$ is a relative maximum of g.

67. $f(x) = \dfrac{x^4}{x-1}$.

$f'(x) = \dfrac{(x-1)(4x^3) - x^4(1)}{(x-1)^2} = \dfrac{4x^4 - 4x^3 - x^4}{(x-1)^2} = \dfrac{3x^4 - 4x^3}{(x-1)^2} = \dfrac{x^3(3x-4)}{(x-1)^2}$

and so $x = 0$ and $x = 4/3$ are critical points of f. Next,

$f''(x) = \dfrac{(x-1)^2(12x^3 - 12x^2) - (3x^4 - 4x^3)(2)(x-1)}{(x-1)^4}$

$= \dfrac{(x-1)(12x^4 - 12x^3 - 12x^3 + 12x^2 - 6x^4 + 8x^3)}{(x-1)^4}$

$= \dfrac{6x^4 - 16x^3 + 12x^2}{(x-1)^3} = \dfrac{2x^2(3x^2 - 8x + 6)}{(x-1)^3}$.

Since $f''(\tfrac{4}{3}) > 0$, we see that $f(\tfrac{4}{3}) = \tfrac{256}{27}$ is a relative minimum. Since $f''(0) = 0$, the Second Derivative Test fails. Using the sign diagram for f',

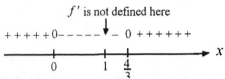

f' is not defined here

and the First Derivative Test, we see that $f(0) = 0$ is a relative maximum.

69.

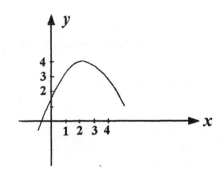

71.

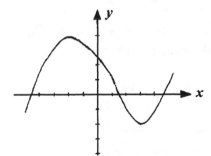

73.

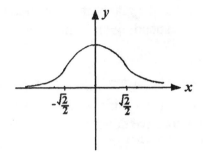

75. a. $N'(t)$ is positive because N is increasing on $(0,12)$.
   b. $N''(t) < 0$ on $(0,6)$ and $N''(t) > 0$ on $(6,12)$.
   c. The rate of growth of the number of help-wanted advertisements was decreasing over the first six months of the year and increasing over the last six months.

77. $f(t)$ increases at an increasing rate until the water level reaches the middle of the vase at which time (and this corresponds to the inflection point of f), $f(t)$ is increasing at the fastest rate. Though $f(t)$ still increases until the vase is filled, it does so at a decreasing rate.

79. a. $S'(t) = 0.39t + 0.32 > 0$ on $[0, 7]$. So sales were increasing through the years in question.
   b. $S''(t) = 0.39 > 0$ on $[0,7]$. So sales continued to accelerate through the years.

81. We wish to find the inflection point of the function $N(t) = -t^3 + 6t^2 + 15t$. Now, $N'(t) = -3t^2 + 12t + 15$ and $N''(t) = -6t + 12 = -6(t - 2)$ giving $t = 2$ as the only candidate for an inflection point of N. From the sign diagram

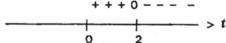

   for N'', we conclude that $t = 2$ gives an inflection point of N. Therefore, the average worker is performing at peak efficiency at 10 A.M.

83. $S'(t) = -5.64t^2 + 60.66t - 76.14$; $\quad S''(t) = -11.28t + 60.66 = 0$ if $t \approx 5.38$
   The sign diagram of S'' follows:

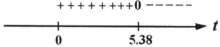

   From the sign diagram for S'', we see that the graph of S is concave upward on $(0,5)$. This says that the rate of business spending on technology is increasing from 2000 through 2005.

85. a. $R'(x) = -0.009x^2 + 2.7x + 2$; $R''(x) = -0.018x + 2.7$. Setting $R''(x) = 0$ gives $x = 150$. Since $R''(x) > 0$ if $x < 150$ and $R''(x) < 0$ if $x > 150$, we see that the graph of R is concave upward on $(0, 150)$ and concave downward on $(150,400)$. So $x = 150$ gives rise to an inflection point of R. $R(150) = 28,850$. So the inflection point is $(150, 28,850)$.

b. $R''(140) = 0.18$; $R''(160) = -0.18$ This shows that at $x = 140$, a slight increase in x (spending) would result in the revenue increasing. At $x = 160$, the opposite conclusion holds. So it would be more beneficial to increase the expenditure when it is \$140,000 than when it's at \$160,000.

87. a. $A'(t) = 0.92(0.61)(t+1)^{-0.39} = \dfrac{0.5612}{(t+1)^{0.39}} > 0$ on (0,4), so A is increasing on (0,4).

his tells us that the spending is increasing over the years in question.

b. $A''(t) = (0.5612)(-0.39)(t+1)^{-1.39} = -\dfrac{0.218868}{(t+1)^{1.39}} < 0$ on (0,4). And so A'' is

concave downward on (0,4). This tells us that the spending is increasing but at a decreasing rate.

89. $S'(t) = -5.418t^2 + 20.476t + 93.35$; $S''(t) = -10.836t + 20.476 = 0$ if $t \approx 1.9$. The

sign diagram for S'' follows.

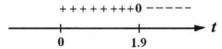

Since $S(1.9) = 784.9$, the inflection point is approximately (1.9, 784.9). The rate of annual spending slows down near the end of 2000.

91. a. $R'(t) = 74.925t^2 - 99.62t + 41.25$
   $R''(t) = 149.85t - 99.62$
   b. In solving the equation $R'(t) = 0$, we see that the discriminant is
   $(-99.62)^2 - 4(74.925)(41.25) = -2438.4806 < 0$
   and so R' has no zeros. Since $R'(0) = 41.25 > 0$, we see that $R'(t) > 0$ in (0,4). This shows that the revenue is always increasing from 1999 through 2003.
   c. $R''(t) = 0$ implies $t = 0.66$. The sign diagram of R'' follows.

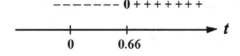

93. $A(t) = 1.0974t^3 - 0.0915t^4$. $A'(t) = 3.2922t^2 - 0.366t^3$ and $A''(t) = 6.5844t - 1.098t^2$. Setting $A'(t) = 0$, we obtain $t^2(3.2922 - 0.366t) = 0$, and this gives $t = 0$ or

$t \approx 8.995 \approx 9$. Using the Second Derivative Test, we find $A''(9) = 6.5844(9) - 1.098(81) = -29.6784 < 0$, and this tells us that $t \approx 9$ gives rise to a relative maximum of A. Our analysis tells us that on that May day, the level of ozone peaked at approximately 4 P.M. in the afternoon.

95. a. $N'(t) = \dfrac{d}{dt}(-0.9307t^3 + 74.04t^2 + 46.8667t + 3967)$

$= -2.79211t^2 + 148.08t + 46.8667$

N' is continuous everywhere and has no zeros at

$$t = \frac{-146.08 \pm \sqrt{(148.08)^2 - 4(-0.9307)(46.86667)}}{2(-2.7921)}$$

that is, at $t = -0.1053$ or 53.1406. Both these points lie outside the interval of interest. Picking $t = 0$ for a test point, we see that $N'(0) = 46.86667 > 0$ and conclude that N is increasing on $(0, 16)$. This shows that the number of participants is increasing over the years in question.

b. $N''(t) = \dfrac{d}{dt}(-2.79211t^2 + 148.08t + 46.86667) = -5.5842t + 148.08 = 0$

if $t = 26.518$. So $N''(t)$ does not change sign in the interval $(0, 16)$. Since $N''(0) = 148.08 > 0$, we see that $N'(t)$ is increasing on $(0, 16)$ and the desired conclusion follows.

97. True. If f' is increasing on (a,b), then $-f'$ is decreasing on (a,b), and so if the graph of f is concave upward on (a,b), the graph of $-f$ must be concave downward on (a,b).

99. True. The given conditions imply that $f''(0) < 0$ and the Second Derivative Test gives the desired conclusion.

101. $f(x) = ax^2 + bx + c$. $f'(x) = 2ax + b$ and $f''(x) = 2a$. So $f''(x) > 0$ if $a > 0$, and the parabola opens upward. If $a < 0$, then $f''(x) < 0$ and the parabola opens downward.

USING TECHNOLOGY EXERCISES 4.2, page 283

1. a. f is concave upward on $(-\infty, 0) \cup (1.1667, \infty)$ and concave downward on $(0, 1.1667)$.
   b. $(1.1667, 1.1153)$; $(0, 2)$

3. a. *f* is concave downward on (-∞,0) and concave upward on (0, ∞).
    b. (0,2)

5. a. *f* is concave downward on (-∞,0) and concave upward on (0, ∞).
    b. (0,0)

7. a. *f* is concave downward on (-∞,-2.4495) ∪ (0, 2.4495); *f* is concave upward on
    (-2.4495, 0) ∪ (2.4495, ∞).   b. (-2.4495, -0.3402); (2.4495, 0.3402)

9. a.

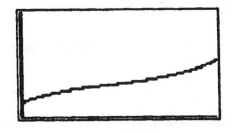

    b. (5.5318, 35.9483)   c. *t* = 5.5318

11. a.

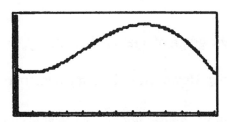

    b. (3.9024, 77.0919);
    sales of houses were increasing
    at the fastest rate in late 1988.

13. a.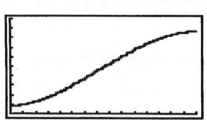

b. April 1993 ($t = 7.36$)

4.3 Problem Solving Tips

1. To find the horizontal asymptotes of a function f, find the limit of f as $x \to \infty$ and as $x \to -\infty$. If the limit is equal to a real number b, then $y = b$ is a horizontal asymptote of f.

2. To find the vertical asymptotes of a rational function $f(x) = P(x)/Q(x)$, determine the values a for which $Q(x) = 0$. If $Q(a) = 0$ but $P(a) \neq 0$, then the line $x = a$ is a vertical asymptote of f.

3. If a line $x = a$ is a vertical asymptote of the graph of a rational function f, then the denominator of $f(x)$ is equal to zero at $x = a$. However, if both the numerator and the denominator of $f(x)$ are equal to zero, then $x = a$ *need not* be a vertical asymptote.

4.3 CONCEPT QUESTIONS, page 293

1. a. See the definition on page 287 of the text.
   b. See the definition on page 287 of the text.

3. See the procedure given on page 287 of the text.

EXERCISES 4.3, page 293

1. $y = 0$ is a horizontal asymptote.

3. $y = 0$ is a horizontal asymptote and $x = 0$ is a vertical asymptote.

5. $y = 0$ is a horizontal asymptote and $x = -1$ and $x = 1$ are vertical asymptotes.

7. $y = 3$ is a horizontal asymptote and $x = 0$ is a vertical asymptote.

9. $y = 1$ and $y = -1$ are horizontal asymptotes.

11. $\lim\limits_{x \to \infty} \dfrac{1}{x} = 0$ and so $y = 0$ is a horizontal asymptote. Next, since the numerator of the rational expression is not equal to zero and the denominator is zero at $x = 0$, we see that $x = 0$ is a vertical asymptote.

13. $f(x) = -\dfrac{2}{x^2}$. $\lim\limits_{x \to \infty} -\dfrac{2}{x^2} = 0$, so $y = 0$ is a horizontal asymptote. Next, the denominator of $f(x)$ is equal to zero at $x = 0$. Since the numerator of $f(x)$ is not equal to zero at $x = 0$, we see that $x = 0$ is a vertical asymptote.

15. $\lim\limits_{x \to \infty} \dfrac{x-1}{x+1} = \lim\limits_{x \to \infty} \dfrac{1-\frac{1}{x}}{1+\frac{1}{x}} = 1$, and so $y = 1$ is a horizontal asymptote. Next, the denominator is equal to zero at $x = -1$ and the numerator is not equal to zero at this point, so $x = -1$ is a vertical asymptote.

17. $h(x) = x^3 - 3x^2 + x + 1$. $h(x)$ is a polynomial function and, therefore, it does not have any horizontal or vertical asymptotes.

19. $\lim\limits_{t \to \infty} \dfrac{t^2}{t^2 - 9} = \lim\limits_{t \to \infty} \dfrac{1}{1 - \frac{9}{t^2}} = 1$, and so $y = 1$ is a horizontal asymptote. Next, observe that the denominator of the rational expression $t^2 - 9 = (t+3)(t-3) = 0$ if $t = -3$ and $t = 3$. But the numerator is not equal to zero at these points. Therefore, $t = -3$ and $t = 3$ are vertical asymptotes.

21. $\lim\limits_{x\to\infty}\dfrac{3x}{x^2-x-6}=\lim\limits_{x\to\infty}\dfrac{\frac{3}{x}}{1-\frac{1}{x}-\frac{6}{x^2}}=0$ and so $y=0$ is a horizontal asymptote. Next,

observe that the denominator $x^2-x-6=(x-3)(x+2)=0$ if $x=-2$ or $x=3$. But the numerator $3x$ is not equal to zero at these points. Therefore, $x=-2$ and $x=3$ are vertical asymptotes.

23. $\lim\limits_{t\to\infty}\left[2+\dfrac{5}{(t-2)^2}\right]=2$, and so $y=2$ is a horizontal asymptote. Next observe that

$\lim\limits_{t\to 2^+}g(t)=\lim\limits_{t\to 2^-}\left[2+\dfrac{5}{(t-2)^2}\right]=\infty$, and so $t=2$ is a vertical asymptote.

25. $\lim\limits_{x\to\infty}\dfrac{x^2-2}{x^2-4}=\lim\limits_{x\to\infty}\dfrac{1-\frac{2}{x^2}}{1-\frac{4}{x^2}}=1$ and so $y=1$ is a horizontal asymptote. Next, observe

that the denominator $x^2-4=(x+2)(x-2)=0$ if $x=-2$ or 2. Since the numerator x^2-2 is not equal to zero at these points, the lines $x=-2$ and $x=2$ are vertical asymptotes.

27. $g(x)=\dfrac{x^3-x}{x(x+1)}$; Rewrite $g(x)$ as $g(x)=\dfrac{x^2-1}{x+1}$ $(x\neq 0)$ and note that

$\lim\limits_{x\to-\infty}g(x)=\lim\limits_{x\to-\infty}\dfrac{x-\frac{1}{x}}{1+\frac{1}{x}}=-\infty$ and $\lim\limits_{x\to\infty}g(x)=\infty$. Therefore, there are no horizontal

asymptotes. Next, note that the denominator of $g(x)$ is equal to zero at $x=0$ and $x=-1$. However, since the numerator of $g(x)$ is also equal to zero when $x=0$, we see that $x=0$ is not a vertical asymptote. Also, the numerator of $g(x)$ is equal to zero when $x=-1$, so $x=-1$ is not a vertical asymptote.

29. f is the derivative function of the function g. Observe that at a relative maximum (relative minimum) of g, $f(x)=0$.

31.

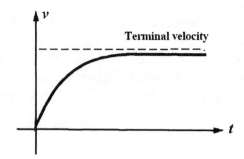

33.

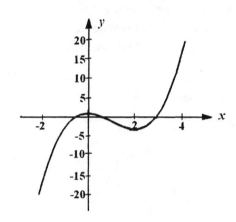

35.

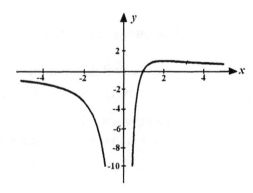

37. $g(x) = 4 - 3x - 2x^3$.

We first gather the following information on the graph of f.

1. The domain of f is $(-\infty, \infty)$.

2. Setting $x = 0$ gives $y = 4$ as the y-intercept. Setting $y = g(x) = 0$ gives a cubic equation which is not easily solved and we will not attempt to find the x-intercepts.

3. $\lim_{x \to -\infty} g(x) = \infty$ and $\lim_{x \to \infty} g(x) = -\infty$. 4. There are no asymptotes of g.

5. $g'(x) = -3 - 6x^2 = -3(2x^2 + 1) < 0$ for all values of x and so g is decreasing on $(-\infty, \infty)$.

6. The results of 5 show that g has no critical points and hence has no relative extrema.

7. $g''(x) = -12x$. Since $g''(x) > 0$ for $x < 0$ and $g''(x) < 0$ for $x > 0$, we see that g is concave upward on $(-\infty, 0)$ and concave downward on $(0, \infty)$.

8. From the results of (7), we see that $(0, 4)$ is an inflection point of g. The graph of g follows.

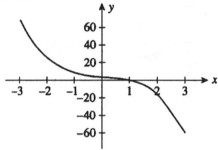

39. $h(x) = x^3 - 3x + 1$

We first gather the following information on the graph of h.

1. The domain of h is $(-\infty, \infty)$.

2. Setting $x = 0$ gives 1 as the y-intercept. We will not find the x-intercept.

3. $\lim_{x \to -\infty} (x^3 - 3x + 1) = -\infty$ and $\lim_{x \to \infty} (x^3 - 3x + 1) = \infty$

4. There are no asymptotes since $h(x)$ is a polynomial.

5. $h'(x) = 3x^2 - 3 = 3(x + 1)(x - 1)$, and we see that $x = -1$ and $x = 1$ are critical points. From the sign diagram

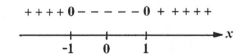

we see that h is increasing on $(-\infty,-1) \cup (1, \infty)$ and decreasing on $(-1,1)$.

6. The results of (5) shows that $(-1,3)$ is a relative maximum and $(1,-1)$ is a relative minimum.

7. $h''(x) = 6x$ and $h''(x) < 0$ if $x < 0$ and $h''(x) > 0$ if $x > 0$. So the graph of h is concave downward on $(-\infty,0)$ and concave upward on $(0, \infty)$.

8. The results of (7) show that $(0,1)$ is an inflection point of h.

The graph of h follows.

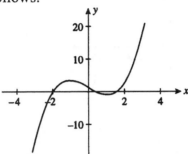

41. $f(x) = -2x^3 + 3x^2 + 12x + 2$

We first gather the following information on the graph of f.

1. The domain of f is $(-\infty, \infty)$.

2. Setting $x = 0$ gives 2 as the y-intercept.

3. $\lim_{x \to -\infty} (-2x^3 + 3x^2 + 12x + 2) = \infty$ and $\lim_{x \to \infty} (-2x^3 + 3x^2 + 12x + 2) = -\infty$

4. There are no asymptotes because $f(x)$ is a polynomial function.

5. $f'(x) = -6x^2 + 6x + 12 = -6(x^2 - x - 2) = -6(x - 2)(x + 1) = 0$ if $x = -1$ or $x = 2$, the critical points of f. From the sign diagram

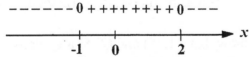

we see that f is decreasing on $(-\infty,-1) \cup (2, \infty)$ and increasing on $(-1,2)$.

6. The results of (5) show that $(-1,-5)$ is a relative minimum and $(2,22)$ is a relative maximum.

7. $f''(x) = -12x + 6 = 0$ if $x = 1/2$. The sign diagram of f''

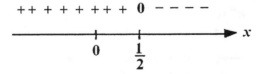

shows that the graph of f is concave upward on $(-\infty, 1/2)$ and concave downward on $(1/2, \infty)$.

8. The results of (7) show that $(\frac{1}{2}, \frac{17}{2})$ is an inflection point.

The graph of f follows.

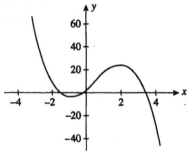

43. $h(x) = \frac{3}{2}x^4 - 2x^3 - 6x^2 + 8$

We first gather the following information on the graph of h.

1. The domain of h is $(-\infty, \infty)$.

2. Setting $x = 0$ gives 8 as the y-intercept.

3. $\lim\limits_{x \to -\infty} h(x) = \lim\limits_{x \to \infty} h(x) = \infty$

4. There are no asymptotes.

5. $h'(x) = 6x^3 - 6x^2 - 12x = 6x(x^2 - x - 2) = 6x(x-2)(x+1) = 0$ if $x = -1, 0,$ or $2,$ and these are the critical points of h. The sign diagram of h' is

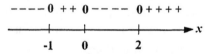

and this tells us that h is increasing on $(-1, 0) \cup (2, \infty)$ and decreasing on $(-\infty, -1) \cup (0, 2)$.

6. The results of (5) show that $(-1, \frac{11}{2})$ and $(2, -8)$ are relative minima of h and $(0, 8)$ is   a relative maximum of h.

7. $h''(x) = 18x^2 - 12x - 12 = 6(3x^2 - 2x - 2)$. The zeros of h'' are

$$x = \frac{2 \pm \sqrt{4 + 24}}{6} \approx -0.5 \text{ or } 1.2.$$

The sign diagram of h'' is

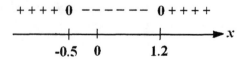

and tells us that the graph of h is concave upward on $(-\infty,-0.5) \cup (1.2, \infty)$ and is concave downward on $(0.5,1.2)$.

8. The results of (7) also show that $(-0.5,6.8)$ and $(1.2,-1)$ are inflection points. The graph of h follows.

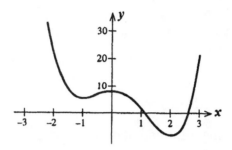

45. $f(t) = \sqrt{t^2 - 4}$.

We first gather the following information on f.

1. The domain of f is found by solving $t^2 - 4 \geq 0$ giving it as $(-\infty,-2] \cup [2,\infty)$.

2. Since $t \neq 0$, there is no y-intercept. Next, setting $y = f(t) = 0$ gives the t-intercepts as -2 and 2.

3. $\lim\limits_{t \to -\infty} f(t) = \lim\limits_{t \to \infty} f(t) = \infty$ 4. There are no asymptotes.

5. $f'(t) = \dfrac{1}{2}(t^2 - 4)^{-1/2}(2t) = t(t^2 - 4)^{-1/2} = \dfrac{t}{\sqrt{t^2 - 4}}$.

Setting $f'(t) = 0$ gives $t = 0$. But $t = 0$ is not in the domain of f and so there are no critical points. The sign diagram for f' is

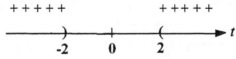

We see that f is increasing on $(2,\infty)$ and decreasing on $(-\infty,-2)$.

6. From the results of (5) we see that there are no relative extrema.

7. $f''(t) = (t^2 - 4)^{-1/2} + t(-\tfrac{1}{2})(t^2 - 4)^{-3/2}(2t) = (t^2 - 4)^{-3/2}(t^2 - 4 - t^2)$

$$= -\dfrac{4}{(t^2 - 4)^{3/2}}.$$

8. Since $f''(t) < 0$ for all t in the domain of f, we see that f is concave downward everywhere. From the results of (7), we see that there are no inflection points.

The graph of f follows.

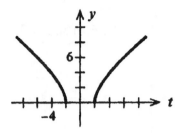

47. $g(x) = \frac{1}{2}x - \sqrt{x}$.

We first gather the following information on g.

1. The domain of g is $[0,\infty)$.

2. The y-intercept is 0. To find the x-intercept, set $y = 0$, giving
$$\frac{1}{2}x - \sqrt{x} = 0$$
$$x = 2\sqrt{x}$$
$$x^2 = 4x$$
$$x(x-4) = 0, \text{ and } x = 0 \text{ or } x = 4$$

3. $\lim\limits_{x \to \infty} (\frac{1}{2}x - \sqrt{x}) = \lim\limits_{x \to \infty} \frac{1}{2}x(1 - \frac{2}{\sqrt{x}}) = \infty$.

4. There are no asymptotes.

5. $g'(x) = \frac{1}{2} - \frac{1}{2}x^{-1/2} = \frac{1}{2}x^{-1/2}(x^{1/2} - 1) = \dfrac{\sqrt{x}-1}{2\sqrt{x}}$

which is zero when $x = 1$. From the sign diagram for g'

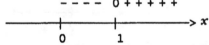

we see that g is decreasing on $(0,1)$ and increasing on $(1,\infty)$.

6. From the sign diagram of g', we see that $g(1) = -1/2$ is a relative minimum.

7. $g''(x) = (-\frac{1}{2})(-\frac{1}{2})x^{-3/2} = \dfrac{1}{4x^{3/2}} > 0$ for $x > 0$, and so g is concave upward on

$(0,\infty)$.

8. There are no inflection points.

The graph of g follows.

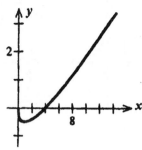

49. $g(x) = \dfrac{2}{x-1}$.   We first gather the following information on g.

1. The domain of g is $(-\infty,1) \cup (1,\infty)$.

2. Setting $x = 0$ gives -2 as the y-intercept. There are no x-intercepts since $\dfrac{2}{x-1} \neq 0$ for all values of x.

3. $\displaystyle\lim_{x\to-\infty} \dfrac{2}{x-1} = 0$ and $\displaystyle\lim_{x\to\infty} \dfrac{2}{x-1} = 0$.

4. The results of (3) show that $y = 0$ is a horizontal asymptote. Furthermore, the denominator of $g(x)$ is equal to zero at $x = 1$ but the numerator is not equal to zero there. Therefore, $x = 1$ is a vertical asymptote.

5. $g'(x) = -2(x-1)^{-2} = -\dfrac{2}{(x-1)^2} < 0$ for all $x \neq 1$ and so g is decreasing on $(-\infty,1)$ and $(1,\infty)$.

6. Since g has no critical points, there are no relative extrema.

7. $g''(x) = \dfrac{4}{(x-1)^3}$ and so $g''(x) < 0$ if $x < 1$ and $g''(x) > 0$ if $x > 1$. Therefore, the graph of g is concave downward on $(-\infty,1)$ and concave upward on $(1,\infty)$.

8. Since $g''(x) \neq 0$, there are no inflection points.
The graph of g follows.

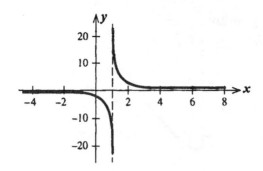

51. $h(x) = \dfrac{x+2}{x-2}$.

We first gather the following information on the graph of h.

1. The domain of h is $(-\infty,2) \cup (2,\infty)$.

2. Setting $x = 0$ gives $y = -1$ as the y-intercept. Next, setting $y = 0$ gives $x = -2$ as the x-intercept.

3. $\displaystyle\lim_{x\to\infty} h(x) = \lim_{x\to-\infty} \dfrac{1+\dfrac{2}{x}}{1-\dfrac{2}{x}} = \lim_{x\to-\infty} h(x) = 1.$

4. Setting $x - 2 = 0$ gives $x = 2$. Furthermore,

$$\lim_{x\to 2^+} \frac{x+2}{x-2} = \infty \quad \text{and} \quad \lim_{x\to 2^+} \frac{x+2}{x-2} = -\infty$$

So $x = 2$ is a vertical asymptote of h. Also, from the resultsof (3), we see that $y = 1$ is a horizontal asymptote of h.

5. $h'(x) = \dfrac{(x-2)(1)-(x+2)(1)}{(x-2)^2} = -\dfrac{4}{(x-2)^2}.$

We see that there are no critical points of h. (Note $x = 2$ does not belong to the domain of h.) The sign diagram of h' follows.

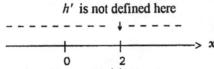

We see that h is decreasing on $(-\infty,2) \cup (2,\infty)$.

6. From the results of (5), we see that there is no relative extremum.

7. $h''(x) = \dfrac{8}{(x-2)^3}.$ Note that $x = 2$ is not a candidate for an inflection point

because $h(2)$ is not defined. Since $h''(x) < 0$ for $x < 2$ and $h''(x) > 0$ for $x > 2$, we see that h is concave downward on $(-\infty, 2)$ and concave upward on $(2, \infty)$.
8. From the results of (7), we see that there are no inflection points. The graph of h follows.

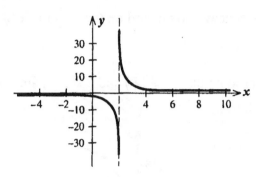

53. $f(t) = \dfrac{t^2}{1+t^2}$.

We first gather the following information on the graph of f.
1. The domain of f is $(-\infty, \infty)$.
2. Setting $t = 0$ gives the y-intercept as 0. Similarly, setting $y = 0$ gives the t-intercept   as 0.

3. $\displaystyle\lim_{t \to -\infty} \frac{t^2}{1+t^2} = \lim_{t \to \infty} \frac{t^2}{1+t^2} = 1.$

4. The results of (3) show that $y = 1$ is a horizontal asymptote. There are no vertical asymptotes since the denominator is not equal to zero.

5. $f'(t) = \dfrac{(1+t^2)(2t) - t^2(2t)}{(1+t^2)^2} = \dfrac{2t}{(1+t^2)^2} = 0$, if $t = 0$, the only critical point of f.

Since $f'(t) < 0$ if $t < 0$ and $f'(t) > 0$ if $t > 0$, we see that f is decreasing on $(-\infty, 0)$ and increasing on $(0, \infty)$.
6. The results of (5) show that $(0, 0)$ is a relative minimum.

7. $f''(t) = \dfrac{(1+t^2)^2(2) - 2t(2)(1+t^2)(2t)}{(1+t^2)^4} = \dfrac{2(1+t^2)[(1+t^2) - 4t^2]}{(1+t^2)^4}$

$= \dfrac{2(1-3t^2)}{(1+t^2)^3} = 0$ if $t = \pm\dfrac{\sqrt{3}}{3}$.

The sign diagram of f'' is

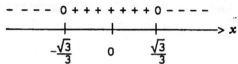

and shows that f is concave downward on $(-\infty, -\frac{\sqrt{3}}{3}) \cup (\frac{\sqrt{3}}{3}, \infty)$ and concave

upward on $(-\frac{\sqrt{3}}{3}, \frac{\sqrt{3}}{3})$.

8. The results of (7) show that $(-\frac{\sqrt{3}}{3}, \frac{1}{4})$ and $(\frac{\sqrt{3}}{3}, \frac{1}{4})$ are inflection points.
The graph of f follows.

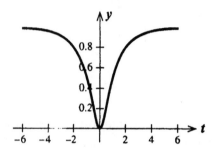

55. $g(t) = -\dfrac{t^2 - 2}{t - 1}$.

First we obtain the following information on g.
1. The domain of g is $(-\infty, 1) \cup (1, \infty)$.
2. Setting $t = 0$ gives -2 as the y-intercept.
3. $\displaystyle\lim_{t \to -\infty} -\frac{t^2 - 2}{t - 1} = \infty$ and $\displaystyle\lim_{t \to \infty} -\frac{t^2 - 2}{t - 1} = -\infty$.

4. There are no horizontal asymptotes. The denominator is equal to zero at $t = 1$ at which point the numerator is not equal to zero. Therefore $t = 1$ is a vertical asymptote.

5. $g'(t) = -\dfrac{(t-1)(2t) - (t^2 - 2)(1)}{(t-1)^2} = -\dfrac{t^2 - 2t + 2}{(t-1)^2} \neq 0$

for all values of t. The sign diagram of g'

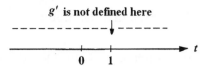

shows that *g* is decreasing on $(-\infty,1) \cup (1, \infty)$.

6. Since there are no critical points, *g* has no relative extrema.

7. $g''(t) = -\dfrac{(t-1)^2(2t-2)-(t^2-2t+2)(2)(t-1)}{(t-1)^4}$

$$= \dfrac{-2(t-1)(t^2-2t+1-t^2+2t-2)}{(t-1)^4} = \dfrac{2}{(t-1)^3}.$$

The sign diagram of *g"*

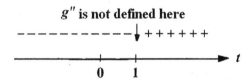

g" is not defined here

shows that the graph of *g* is concave upward on $(1,\infty)$ and concave downward on $(-\infty,1)$.

8. There are no inflection points since $g''(x) \neq 0$ for all *x*.
The graph of *g* follows.

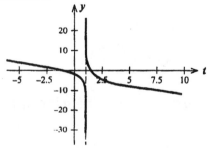

57. $g(t) = \dfrac{t^2}{t^2-1}$.

We first gather some information on the graph of *g*.

1. Since $t^2 - 1 = 0$ if $t = \pm 1$, we see that the domain of g is $(-\infty, -1) \cup (-1,1) \cup (1, \infty)$.

2. Setting $t = 0$ gives 0 as the y-intercept. Setting $y = 0$ gives 0 as the t-intercept.

3. $\lim\limits_{t \to -\infty} g(t) = \lim\limits_{t \to \infty} g(t) = 1$.

4. The results of (3) show that $y = 1$ is a horizontal asymptote. Since the denominator (but not the numerator) is zero at $t = \pm 1$, we see that $t = \pm 1$ are vertical asymptotes.

5. $g'(t) = \dfrac{(t^2 - 1)(2t) - (t^2)(2t)}{(t^2 - 1)^2} = -\dfrac{2t}{(t^2 - 1)^2} = 0$, if $t = 0$.

The sign diagram of g' is

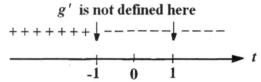

We see that g is increasing on $(-\infty,-1) \cup (-1, 0)$ and decreasing on $(0, 1) \cup (1, \infty)$.

6. From the results of (5), we see that g has a relative maximum at $t = 0$.

7. $g''(t) = \dfrac{(t^2 - 1)^2(-2) - (-2t)(2)(t^2 - 1)(2t)}{(t^2 - 1)^4}$

$= \dfrac{2(t^2 - 1)^2[-(t^2 - 1) + 4t^2]}{(t^2 - 1)^3}$

$= \dfrac{2(-t^2 + 1 + 4t^2)}{(t^2 - 1)^3} = \dfrac{2(3t^2 + 1)}{(t^2 - 1)^3}$

g'' is not defined here

$$+ + + + + + + + + + \downarrow + + + + + + + + + + + \downarrow + + + +$$

$$\begin{array}{ccc} & & \\ \hline -1 & 0 & 1 \end{array} \quad t$$

From the sign diagram we see that the graph of g is concave up on $(-\infty, -1) \cup (-1,1) \cup (1, \infty)$.

8. From (7), we see that the graph of g has no inflection points. The graph of g follows.

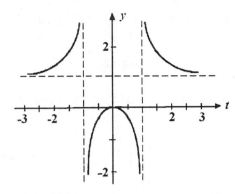

59. $h(x) = (x-1)^{2/3} + 1.$

We begin by obtaining the following information on h.

1. The domain of h is $(-\infty, \infty)$.

2. Setting $x = 0$ gives 2 as the y-intercept; since $h(x) \neq 0$ there is no x-intercept.

3. $\lim_{x \to \infty} [(x-1)^{2/3} + 1] = \infty$. Similarly, $\lim_{x \to -\infty} [(x-1)^{2/3} + 1] = \infty$.

4. There are no asymptotes.

5. $h'(x) = \frac{2}{3}(x-1)^{-1/3}$ and is positive if $x > 1$ and negative if $x < 1$. So h is increasing on $(1,\infty)$, and decreasing on $(-\infty,1)$.

6. From (5), we see that h has a relative minimum at $(1,1)$.

7. $h''(x) = \frac{2}{3}(-\frac{1}{3})(x-1)^{-4/3} = -\frac{2}{9}(x-1)^{-4/3} = -\frac{2}{(x-1)^{4/3}}$. Since $h''(x) < 0$ on

$(-\infty,1) \cup (1,\infty)$, we see that h is concave downward on $(-\infty,1) \cup (1,\infty)$. Note that $h''(x)$ is not defined at $x = 1$.

8. From the results of (7), we see h has no inflection points.

The graph of h follows.

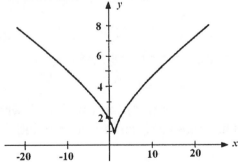

4 Applications of the Derivative

61. a. The denominator of $C(x)$ is equal to zero if $x = 100$. Also,

$$\lim_{x \to 100^-} \frac{0.5x}{100-x} = \infty \quad \text{and} \quad \lim_{x \to 100^+} \frac{0.5x}{100-x} = -\infty$$

Therefore, $x = 100$ is a vertical asymptote of C.

b. No, because the denominator will be equal to zero in that case.

63. a. Since $\displaystyle\lim_{t \to \infty} C(t) = \lim_{t \to \infty} \frac{0.2t}{t^2+1} = \lim_{t \to \infty} \left[\frac{0.2}{t + \frac{1}{t^2}} \right] = 0$, $y = 0$ is a horizontal asymptote.

b. Our results reveal that as time passes, the concentration of the drug decreases and approaches zero.

65. $G(t) = -0.2t^3 + 2.4t^2 + 60$.

We first gather the following information on the graph of G.

1. The domain of G is $(0, \infty)$.

2. Setting $t = 0$ gives 60 as the y-intercept.

Note that Step 3 is not necessary in this case because of the restricted domain.

4. There are no asymptotes since G is a polynomial function.

5. $G'(t) = -0.6t^2 + 4.8t = -0.6t(t - 8) = 0$, if $t = 0$ or $t = 8$. But these points do not lie in the interval $(0,8)$, so they are not critical points. The sign diagram of G'

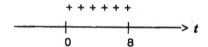

shows that G is increasing on $(0,8)$.

6. The results of (5) tell us that there are no relative extrema.

7. $G''(t) = -1.2t + 4.8 = -1.2(t - 4)$. The sign diagram of G'' is

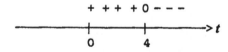

and shows that G is concave upward on $(0,4)$ and concave downward on $(4,8)$.

6. The results of (7) shows that $(4, 85.6)$ is an inflection point.

The graph of G follows.

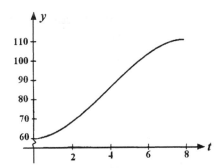

67. $C(t) = \dfrac{0.2t}{t^2 + 1}$.

We first gather the following information on the function C.
1. The domain of C is $[0,\infty)$.
2. If $t = 0$, then $y = 0$. Also, if $y = 0$, then $t = 0$.
3. $\lim\limits_{t \to \infty} \dfrac{0.2t}{t^2 + 1} = 0$.
4. The results of (3) imply that $y = 0$ is a horizontal asymptote.
5. $C'(t) = \dfrac{(t^2 + 1)(0.2) - 0.2t(2t)}{(t^2 + 1)^2} = \dfrac{0.2(t^2 + 1 - 2t^2)}{(t^2 + 1)^2} = \dfrac{0.2(1 - t^2)}{(t^2 + 1)^2}$

and this is equal to zero at $t = \pm 1$, so $t = 1$ is a critical point of C. The sign diagram of C' is

$$+\ +\ +\ 0\ -\ -\ -\ -$$
$$\xrightarrow{\underset{0}{|}\underset{1}{|}} t$$

and tells us that C is decreasing on $(1,\infty)$ and increasing on $(0,1)$.
6. The results of (5) tell us that $(1, 0.1)$ is a relative maximum.

7. $C''(t) = 0.2\left[\dfrac{(t^2 + 1)^2(-2t) - (1 - t^2)2(t^2 + 1)(2t)}{(t^2 + 1)^4}\right]$

$= \dfrac{0.2(t^2 + 1)(2t)(-t^2 - 1 - 2 + 2t^2)}{(t^2 + 1)^4} = \dfrac{0.4t(t^2 - 3)}{(t^2 + 1)^3}$.

The sign diagram of C'' is

$$0\ -\ -\ 0\ +\ +\ +$$
$$\xrightarrow{\underset{-\sqrt{3}}{|}\underset{0}{|}\underset{\sqrt{3}}{|}} t$$

and so the graph of C is concave downward on $(0, \sqrt{3}\,)$ and concave upward on $(\sqrt{3}, \infty)$.

4 Applications of the Derivative

8. The results of (7) show that $(\sqrt{3}, 0.05\sqrt{3})$ is an inflection point. The graph of C follows.

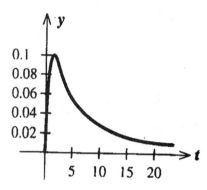

69. $T(x) = \dfrac{120x^2}{x^2 + 4}$.

We first gather the following information on the function T.
1. The domain of T is $[0,\infty)$.
2. Setting $x = 0$ gives 0 as the y-intercept.

3. $\displaystyle\lim_{x \to \infty} \frac{120x^2}{x^2 + 4} = 120$.

4. The results of (3) show that $y = 120$ is a horizontal asymptote.

5. $T'(x) = 120\left[\dfrac{(x^2 + 4)2x - x^2(2x)}{(x^2 + 4)^2}\right] = \dfrac{960x}{(x^2 + 4)^2}$. Since $T'(x) > 0$

if $x > 0$, we see that T is increasing on $(0,\infty)$.
6. There are no relative extrema in $(0,\infty)$.

7. $T''(x) = 960\left[\dfrac{(x^2 + 4)^2 - x(2)(x^2 + 4)(2x)}{(x^2 + 4)^4}\right]$

$= \dfrac{960(x^2 + 4)[(x^2 + 4) - 4x^2]}{(x^2 + 4)^4} = \dfrac{960(4 - 3x^2)}{(x^2 + 4)^3}$.

The sign diagram for T'' is

$$+ + + \; 0 \; - - -$$

$$0 \qquad \frac{2\sqrt{3}}{3}$$

$\longrightarrow x$

We see that T is concave downward on $(\frac{2\sqrt{3}}{3},\infty)$ and concave upward on $(0, \frac{2\sqrt{3}}{3})$.

8. We see from the results of (7) that $(\frac{2\sqrt{3}}{3}, 30)$ is an inflection point.

The graph of T follows.

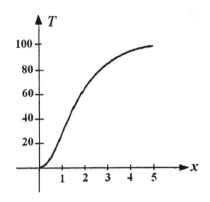

USING TECHNOLOGY EXERCISES 4.3, page 299

1.
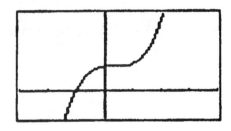

3.
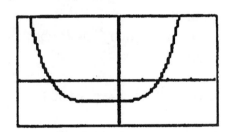

5.   -0.9733; 2.3165, 4.6569

7   -1.1301; 2.9267

9.   1.5142

4.4 Problem Solving Tips

1. To determine the absolute maximum and absolute minimum of a continuous function f on a closed interval $[a,b]$, find the critical numbers of f that lie in (a,b). Then compute the value of f at each critical number of f and compute $f(a)$ and $f(b)$. The largest and smallest of these values will be the absolute maximum value and absolute minimum value of f, respectively.

2. Note that the procedure in (1) only holds for a continuous function f over a closed interval $[a,b]$.

4.4 . CONCEPT QUESTIONS, page 307

1. a. A function f has an absolute maximum at a if $f(x) \le f(a)$ for all x in the domain of f.
   b. A function f has an absolute minimum at a if $f(x) \ge f(a)$ for all x in the domain of f.

EXERCISES 4.4, page 308

1. f has no absolute extrema.

3. f has an absolute minimum at $(0,0)$.

5. f has an absolute minimum at $(0,-2)$ and an absolute maximum at $(1,3)$.

7. f has an absolute minimum at $(\frac{3}{2}, -\frac{27}{16})$ and an absolute maximum at $(-1,3)$.

9. The graph of $f(x) = 2x^2 + 3x - 4$ is a parabola that opens upward. Therefore, the vertex of the parabola is the absolute minimum of f. To find the vertex, we solve the equation $f'(x) = 4x + 3 = 0$ giving $x = -3/4$. We conclude that the absolute

4 Applications of the Derivative   **204**

minimum value is $f(-\frac{3}{4}) = -\frac{41}{8}$.

11. Since $\lim\limits_{x \to -\infty} x^{1/3} = -\infty$ and $\lim\limits_{x \to \infty} x^{1/3} = \infty$, we see that h is unbounded. Therefore it has no absolute extrema.

13. $f(x) = \dfrac{1}{1+x^2}$.

Using the techniques of graphing, we sketch the graph of f (see Figure 40, page 272, in the text). The absolute maximum of f is $f(0) = 1$. Alternatively, observe that $1 + x^2 \geq 1$ for all real values of x. Therefore, $f(x) \leq 1$ for all x, and we see that the absolute maximum is attained when $x = 0$.

15. $f(x) = x^2 - 2x - 3$ and $f'(x) = 2x - 2 = 0$, so $x = 1$ is a critical point. From the table,

x	-2	1	3
$f(x)$	5	-4	0

we conclude that the absolute maximum value is $f(-2) = 5$ and the absolute minimum value is $f(1) = -4$.

17. $f(x) = -x^2 + 4x + 6$; The function f is continuous and defined on the closed interval $[0,5]$. $f'(x) = -2x + 4$ and $x = 2$ is a critical point. From the table

x	0	2	5
$f(x)$	6	10	1

we conclude that $f(2) = 10$ is the absolute maximum value and $f(5) = 1$ is the absolute minimum value.

19. The function $f(x) = x^3 + 3x^2 - 1$ is continuous and defined on the closed interval $[-3,2]$ and differentiable in $(-3,2)$. The critical points of f are found by solving
$$f'(x) = 3x^2 + 6x = 3x(x + 2)$$
giving $x = -2$ and $x = 0$. Next, we compute the values of f given in the following

table.

x	-3	-2	0	2
$f(x)$	-1	3	-1	19

From the table, we see that the absolute maximum value of f is $f(2) = 19$ and the absolute minimum value is $f(-3) = -1$ and $f(0) = -1$.

21. The function $g(x) = 3x^4 + 4x^3$ is continuous and differentiable on the closed interval $[-2,1]$ and differentiable in $(-2,1)$. The critical points of g are found by solving
$$g'(x) = 12x^3 + 12x^2 = 12x^2(x + 1)$$
giving $x = 0$ and $x = -1$. We next compute the values of g shown in the following table.

x	-2	-1	0	1
$g(x)$	16	-1	0	7

From the table we see that $g(-2) = 16$ is the absolute maximum value of g and $g(-1) = -1$ is the absolute minimum value of g.

23. $f(x) = \dfrac{x+1}{x-1}$ on $[2,4]$. Next, we compute,
$$f'(x) = \frac{(x-1)(1)-(x+1)(1)}{(x-1)^2} = -\frac{2}{(x-1)^2}.$$
Since there are no critical points, ($x = 1$ is not in the domain of f), we need only test the endpoints. From the table

x	2	4
$g(x)$	3	5/3

we conclude that $f(4) = 5/3$ is the absolute minimum value and $f(2) = 3$ is the absolute maximum value.

25. $f(x) = 4x + \dfrac{1}{x}$ is continuous on $[1,3]$ and differentiable in $(1,3)$. To find the

critical points of f, we solve $f'(x) = 4 - \frac{1}{x^2} = 0$, obtaining $x = \pm\frac{1}{2}$. Since these critical points lie outside the interval $[1,3]$, they are not candidates for the absolute extrema of f. Evaluating f at the endpoints of the interval $[1,3]$, we find that the absolute maximum value of f is $f(3) = \frac{37}{3}$, and the absolute minimum value of f is $f(1) = 5$.

27. $f(x) = \frac{1}{2}x^2 - 2\sqrt{x} = \frac{1}{2}x^2 - 2x^{1/2}$. To find the critical points of f, we solve
$$f'(x) = x - x^{-1/2} = 0, \quad \text{or} \quad x^{3/2} - 1 = 0,$$
obtaining $x = 1$. From the table

x	0	1	3
$f(x)$	0	$-\frac{3}{2}$	$\frac{9}{2} - 2\sqrt{3} \approx 1.04$

we conclude that $f(3) \approx 1.04$ is the absolute maximum value and $f(1) = -3/2$ is the absolute minimum value.

29. The graph of $f(x) = 1/x$ over the interval $(0,\infty)$ follows.

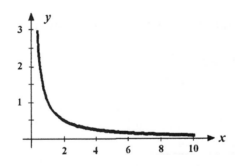

From the graph of f, we conclude that f has no absolute extrema.

31. $f(x) = 3x^{2/3} - 2x$. The function f is continuous on $[0,3]$ and differentiable on $(0,3)$. To find the critical points of f, we solve
$$f'(x) = 2x^{-1/3} - 2 = 0$$
obtaining $x = 1$ as the critical point. From the table,

4 Applications of the Derivative

x	0	1	3
$f(x)$	0	1	$3^{5/3} - 6 \approx 0.24$

we conclude that the absolute maximum value is $f(1) = 1$ and the absolute minimum value is $f(0) = 0$.

33. $f(x) = x^{2/3}(x^2 - 4)$.

$f'(x) = x^{2/3}(2x) + \frac{2}{3}x^{-1/3}(x^2 - 4) = \frac{2}{3}x^{-1/3}[3x^2 + (x^2 - 4)]$

$$= \frac{8(x^2 - 1)}{3x^{1/3}} = 0.$$

Observe that f' is not defined at $x = 0$. Furthermore, $f'(x) = 0$ at $x \pm 1$. So the critical points of f are -1, 0, 1. From the following table,

x	-1	0	1	2
$f(x)$	-3	0	-3	0

we see that f has an absolute minimum at $(-1,-3)$ and $(1,-3)$ and absolute maxima at $(0,0)$ and $(2,0)$.

35. $f(x) = \dfrac{x}{x^2 + 2}$. To find the critical points of f, we solve

$$f'(x) = \frac{(x^2 + 2) - x(2x)}{(x^2 + 2)^2} = \frac{2 - x^2}{(x^2 + 2)^2} = 0$$

obtaining $x = \pm\sqrt{2}$. Since $x = -\sqrt{2}$ lies outside $[-1,2]$, $x = \sqrt{2}$ is the only critical point in the given interval. From the table

x	-1	$\sqrt{2}$	2
$f(x)$	$-\frac{1}{3}$	$\sqrt{2}/4 \approx 0.35$	$\frac{1}{3}$

we conclude that $f(\sqrt{2})) = \sqrt{2}/4 \approx 0.35$ is the absolute maximum value and $f(-1) = -1/3$ is the absolute minimum value.

37. The function $f(x) = \dfrac{x}{\sqrt{x^2 + 1}} = \dfrac{x}{(x^2 + 1)^{1/2}}$ is continuous and defined on the closed interval $[-1,1]$ and differentiable on $(-1,1)$. To find the critical points of f, we first compute

$$f'(x) = \frac{(x^2 + 1)^{1/2}(1) - x(\tfrac{1}{2})(x^2 + 1)^{-1/2}(2x)}{[(x^2 + 1)^{1/2}]^2}$$

$$= \frac{(x^2 + 1)^{-1/2}[x^2 + 1 - x^2]}{x^2 + 1} = \frac{1}{(x^2 + 1)^{3/2}}$$

which is never equal to zero. Next, we compute the values of f shown in the following table.

x	-1	1
$f(x)$	$-\sqrt{2}/2$	$\sqrt{2}/2$

We conclude that $f(-1) = -\sqrt{2}/2$ is the absolute minimum value and $f(1) = \sqrt{2}/2$ is the absolute maximum value.

39. $h(t) = -16t^2 + 64t + 80$. To find the maximum value of h, we solve
$$h'(t) = -32t + 64 = -32(t - 2) = 0$$
giving $t = 2$ as the critical point of h. Furthermore, this value of t gives rise to the absolute maximum value of h since the graph of h is a parabola that opens downward. The maximum height is given by
$$h(2) = -16(4) + 64(2) + 80 = 144, \text{ or } 144 \text{ feet.}$$

41. $P'(t) = 0.027t - 1.126 = 0$ implies $t \approx 41.7$, a critical number for P. $P''(t) = 0.027$ and $P''(41.7) = 0.027 > 0$. Therefore, $t = 41.7$ gives a minimum of P. This is around September of 1991. The percent is approximately $P(41.7) \approx 17.7$.

43. $N(t) = 0.81t - 1.14\sqrt{t} + 1.53$. $N'(t) = 0.81 - 1.14(\tfrac{1}{2}t^{-1/2}) = 0.81 - \dfrac{0.57}{t^{1/2}}$. Setting

$N'(t) = 0$ gives $t^{1/2} = \dfrac{0.57}{0.81}$, or $t = 0.4952$ as a critical point of N. Evaluating $N(t)$

at the endpoints $t = 0$ and $t = 6$ as well as at the critical point, we have

t	0	0.4952	6
$N(t)$	1.53	1.13	3.60

From the table, we see that the absolute maximum of N occurs at $t = 6$ and the absolute minimum occurs at $t \approx 0.5$. Our results tell us that the number of nonfarm full-time self-employed women over the time interval from 1963 to 1993 was the highest in 1993 and stood at approximately 3.6 million.

45. $P(x) = -0.000002x^3 + 6x - 400$. $P'(x) = -0.000006x^2 + 6 = 0$ if $x = \pm 1000$. We reject the negative root. Next, we compute $P''(x) = -0.000012x$. Since $P''(1000) = -0.012 < 0$, the Second Derivative Test shows that $x = 1000$ affords a relative maximum of f. From physical considerations, or from a sketch of the graph of f, we see that the maximum profit is realized if 1000 cases are produced per day. The profit is $P(1000) = -0.000002(1000)^3 + 6(1000) - 400$, or \$3600/day.

47. The revenue is $R(x) = px = -0.0004x^2 + 10x$, and the profit is
$$P(x) = R(x) - C(x) = -0.0004x^2 + 10x - (400 + 4x + 0.0001x^2)$$
$$= -0.0005x^2 + 6x - 400.$$
$$P'(x) = -0.001x + 6 = 0$$
if $x = 6000$, a critical point. Since $P''(x) = -0.001 < 0$ for all x, we see that the graph of P is a parabola that opens downward. Therefore, a level of production of 6000 rackets/day will yield a maximum profit.

49. The total cost function is given by
$$C(x) = V(x) + 20,000$$
$$= 0.000001x^3 - 0.01x^2 + 50x + 20,000$$
The profit function is
$$P(x) = R(x) - C(x)$$
$$= -0.02x^2 + 150x - 0.000001x^3 + 0.01x^2 - 50x + 20,000$$
$$= -0.000001x^3 - 0.01x^2 + 100x - 20,000$$
We want to maximize P on [0, 7000].
$$P'(x) = -0.000003x^2 - 0.02x + 100$$
Setting $P'(x) = 0$ gives $3x^2 + 20,000x - 100,000,000 = 0$
or $x = \dfrac{-20,000 \pm \sqrt{20,000^2 + 1,200,000,000}}{6} = -10,000$ or 3,333.3.

So $x = 3,333.30$ is a critical point in the interval $[0, 7000]$.

x	0	3,333	7,000
$P(x)$	-20,000	165,185	-519,700

From the table, we see that a level of production of 3,333 pagers per week will yield a maximum profit of $165,185 per week.

51. a. $\overline{C}(x) = \dfrac{C(x)}{x} = 0.0025x + 80 + \dfrac{10,000}{x}$.

b. $\overline{C}'(x) = 0.0025 - \dfrac{10,000}{x^2} = 0$ if $0.0025x^2 = 10,000$, or $x = 2000$.

Since $\overline{C}''(x) = \dfrac{20,000}{x^3}$, we see that $\overline{C}''(x) > 0$ for $x > 0$ and so $\overline{C}$ is concave upward on $(0,\infty)$. Therefore, $x = 2000$ yields a minimum.

c. We solve $\overline{C}(x) = C'(x)$. $0.0025x + 80 + \dfrac{10,000}{x} = 0.005x + 80$,

$0.0025x^2 = 10,000$, or $x = 2000$.

d. It appears that we can solve the problem in two ways.
NOTE This can be proved.

53. The demand equation is $p = \sqrt{800 - x} = (800 - x)^{1/2}$. The revenue function is $R(x) = xp = x(800 - x)^{1/2}$. To find the maximum of R, we compute
$$R'(x) = \tfrac{1}{2}(800 - x)^{-1/2}(-1)(x) + (800 - x)^{1/2}$$
$$= \tfrac{1}{2}(800 - x)^{-1/2}[-x + 2(800 - x)]$$
$$= \tfrac{1}{2}(800 - x)^{-1/2}(1600 - 3x).$$
Next, $R'(x) = 0$ implies $x = 800$ or $x = 1600/3$ are critical points of R. Next, we compute the values of R given in the following table.

x	0	800	1600/3
$R(x)$	0	0	8709

We conclude that $R(\tfrac{1600}{3}) = 8709$ is the absolute maximum value. Therefore, the revenue is maximized by producing $1600/3 \approx 533$ dresses.

4 Applications of the Derivative

55. $f(t) = 100\left[\dfrac{t^2 - 4t + 4}{t^2 + 4}\right].$

a. $f'(t) = 100\left[\dfrac{(t^2 + 4)(2t - 4) - (t^2 - 4t + 4)(2t)}{(t^2 + 4)^2}\right] = \dfrac{400(t^2 - 4)}{(t^2 + 4)^2}$

$= \dfrac{400(t - 2)(t + 2)}{(t^2 + 4)^2}.$

From the sign diagram for f'

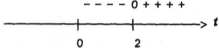

we see that $t = 2$ gives a relative minimum, and we conclude that the oxygen content is the lowest 2 days after the organic waste has been dumped into the pond.

b.

$f''(t) = 400\left[\dfrac{(t^2 + 4)^2(2t) - (t^2 - 4)2(t^2 + 4)(2t)}{(t + 4)^4}\right] = 400\left[\dfrac{(2t)(t^2 + 4)(t^2 + 4 - 2t^2 + 8)}{(t^2 + 4)^4}\right]$

$= -\dfrac{800t(t^2 - 12)}{(t^2 + 4)^3}$

and $f''(t) = 0$ when $t = 0$ and $t = \pm 2\sqrt{3}$. We reject $t = 0$ and $t = -2\sqrt{3}$. From the sign diagram for f'',

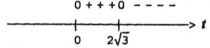

we see that $f'(2\sqrt{3})$ gives an inflection point of f and we conclude that this is an absolute maximum. Therefore, the rate of oxygen regeneration is greatest 3.5 days after the organic waste has been dumped into the pond.

57. We compute $\overline{R}'(x) = \dfrac{xR'(x) - R(x)}{x^2}$. Setting $\overline{R}'(x) = 0$ gives $xR'(x) - R(x) = 0$

or $R'(x) = \dfrac{R(x)}{x} = \overline{R}(x),$ so a critical point of $\overline{R}$ occurs when $\overline{R}(x) = R'(x).$

Next, we compute

$$\bar{R}''(x) = \frac{x^2[R'(x)+xR''(x)-R'(x)]-[xR'(x)-R(x)](2x)}{x^4} = \frac{R''(x)}{x} < 0.$$

So, by the Second Derivative Test, the critical point does give a maximum revenue.

59. The growth rate is $G'(t) = -0.6t^2 + 4.8t$. To find the maximum growth rate, we compute $G''(t) = -1.2t + 4.8$. Setting $G''(t) = 0$ gives $t = 4$ as a critical point.

t	0	4	8
$G'(t)$	0	9.6	0

From the table, we see that G is maximal at $t = 4$; that is, the growth rate is greatest in 2001.

61. $P'(t) = 0.13089t^2 - 0.534t - 1.59 = 0$ gives

$$t = \frac{0.534 \pm \sqrt{(0.534)^2 - 4(0.13089)(-1.59)}}{2(0.13089)} \approx -2 \text{ or } 6.08$$

We reject the negative root. Since
$$P''(t) = 0.26178t - 0.534 \quad \text{and} \quad P''(6.08) \approx 1.06 > 0$$
we conclude that t = 6.08 gives a minimum of P, and this number corresponds to approximately early 1970.

63. $R'(t) = -2.133t^2 + 7.52t + 0.2 = 0$ implies

$$t = \frac{-7.52 \pm \sqrt{7.52^2 - 4(-2.133)(0.2)}}{2(-2.133)} \approx -0.026, \text{ or } 3.55.$$

The root -0.026 lies outside the interval $[0, 5]$, and is rejected.
$$R''(t) = -4.266t + 7.52 \text{ and } R''(3.55) \approx -7.62 > 0$$
and so $t = 3.55$ gives a relative maximum. This is around the middle of 2000. The highest office space rent was given by $R(3.55) \approx 52.79$, or approximately $52.79/sq ft.

65. a. On $[0,3]$: $f(t) = 0.6t^2 + 2.4t + 7.6$; $f'(t) = 1.2t + 2.4 = 0$ implies $t = -2$ which lies outside the interval $[0, 3]$.

t	0	3
$f(t)$	7.6	20.2

On [3, 5]: $f(t) = 3t^2 + 18.8t - 63.2$, $f'(t) = 6t + 18.8 = 0$ implies $t = -3.13$ which lies outside the interval [3, 5].

t	3	5
$f(t)$	20.2	105.8

On [5, 8]: $f(t) = -3.3167t^3 + 80.1t^2 - 642.583t + 1730.8025$

$$f'(t) = -9.9501t^2 + 160.2t - 642.583$$

Solving the equation $f'(t) = 0$, we find

$$t = \frac{-160.2 \pm \sqrt{160.2^2 - 4(-9.9501)(642.583)}}{2(-9.9501)} \approx 7.58, \text{ or } 8.52.$$

Only the critical number t = 7.58 lies inside the interval [5, 8].

t	5	7.58	8
$f(t)$	105.8	17.8	18.4

From the tables, we see that the investment peaked when $t = 5$, that is, in 2000. The amount was \$105.8 billion.

b. The investment (\$7.6 billion) was lowest when $t = 0$.

67. $R = D^2 \left(\dfrac{k}{2} - \dfrac{D}{3} \right) = \dfrac{kD^2}{2} - \dfrac{D^3}{3}$. $\dfrac{dR}{dD} = \dfrac{2kD}{2} - \dfrac{3D^2}{3} = kD - D^2 = D(k - D)$

Setting $\dfrac{dR}{dD} = 0$, we have $D = 0$ or $k = D$. We only consider $k = D$

(since $D > 0$). If $k > 0$, $\dfrac{dR}{dD} > 0$ and if $k < 0$, $\dfrac{dR}{dD} < 0$. Therefore $k = D$ provides a relative maximum. The nature of the problem suggests that $k = D$ gives the absolute maximum of R. We can also verify this by graphing R.

69. False. Let $f(x) = \begin{cases} |x| & \text{if } x \neq 0 \\ 1 & \text{if } x = 0 \end{cases}$ on [-1, 1].

71. False. Let $f(x) = \begin{cases} -x & \text{if } -1 \le x < 0 \\ \dfrac{1}{2} & \text{if } 0 \le x < 1 \end{cases}$. Then f is discontinuous at $x = 0$. But f

   has an absolute maximum value of 1 attained at $x = -1$.

73. Since $f(x) = c$ for all x, the function f satisfies $f(x) \le c$ for all x and so f has an absolute maximum at all points of x. Similarly, f has an absolute minimum at all points of x.

75. a. f is not continuous at $x = 0$ because $\lim\limits_{x \to 0} f(x)$ does not exist.

   b. $\lim\limits_{x \to 0} f(x) = \lim\limits_{x \to 0^-} \dfrac{1}{x} = -\infty$ and $\lim\limits_{x \to 0^+} f(x) = \lim\limits_{x \to 0^+} \dfrac{1}{x} = \infty$.

   c.

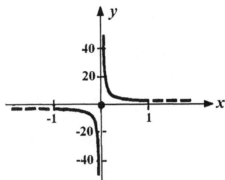

USING TECHNOLOGY EXERCISES 4.4, page 314

1. Absolute maximum value: 145.8985; absolute minimum value: -4.3834

3. Absolute maximum value: 16; absolute minimum value: -0.1257

5. Absolute maximum value: 2.8889; absolute minimum value: 0

7. a.

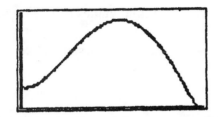

   b. 200.1410 banks/yr

9. a.

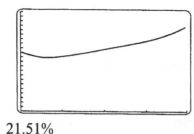

b. 21.51%

11. b. 1145

13. a.

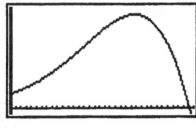

b. 1994

4.5 Problem Solving Tips

1. Follow the guidelines given on page 315 of the text to solve these optimization

problems. Remember, Theorem 3 in Section 4.4 provides us with a method of

computing the absolute extrema of a continuous function over a closed interval $[a,b]$. If

the problem involves a function that is to be optimized over an interval that is not

closed, then use the graphical method to find the optimal values of f. You might review

Example 4 on page 319 in the text to make sure that you understand how to use the

graphical method.

4.5 CONCEPT QUESTIONS, page 322

1. We could solve the problem by sketching the graph of f and checking to see if there is (are) an absolute extrema.

EXERCISES 4.5, page 322

1. Refer to the following figure.

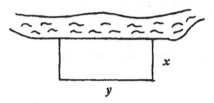

We have $2x + y = 3000$ and we want to maximize the function
$$A = f(x) = xy = x(3000 - 2x) = 3000x - 2x^2$$
on the interval $[0,1500]$. The critical point of A is obtained by solving
$f'(x) = 3000 - 4x = 0$, giving $x = 750$. From the table of values

x	0	750	1500
$f(x)$	0	1,125,000	0

we conclude that $x = 750$ yields the absolute maximum value of A. Thus, the required dimensions are 750×1500 yards. The maximum area is $1,125,000$ sq yd.

3. Let x denote the length of the side made of wood and y the length of the side made of steel. The cost of construction will be $C = 6(2x) + 3y$. But $xy = 800$. So

$y = 800/x$ and therefore $C = f(x) = 12x + 3\left(\dfrac{800}{x}\right) = 12x + \dfrac{2400}{x}$. To minimize C,

we compute
$$f'(x) = 12 - \frac{2400}{x^2} = \frac{12x^2 - 2400}{x^2} = \frac{12(x^2 - 200)}{x^2}.$$
Setting $f'(x) = 0$ gives $x = \pm\sqrt{200}$ as critical points of f. The sign diagram of f'

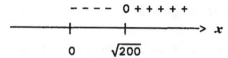

shows that $x = \pm\sqrt{200}$ gives a relative minimum of f. $f''(x) = \dfrac{4800}{x^3} > 0$

if $x > 0$ and so f is concave upward for $x > 0$. Therefore $x = \sqrt{200} = 10\sqrt{2}$ actually yields the absolute minimum. So the dimensions of the enclosure should be

$$10\sqrt{2} \text{ ft } \times \frac{800}{10\sqrt{2}} \text{ ft, or } 14.1 \text{ ft } x \text{ } 56.6 \text{ ft.}$$

5. Let the dimensions of each square that is cut out be $x'' \times x''$. Refer to the following diagram.

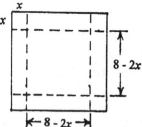 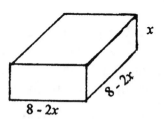

Then the dimensions of the box will be $(8 - 2x)''$ by $(8 - 2x)''$ by x''. Its volume will be $V = f(x) = x(8 - 2x)^2$. We want to maximize f on $[0,4]$.

$$\begin{aligned} f'(x) &= (8 - 2x)^2 + x(2)(8 - 2x)(-2) \qquad \text{[Using the Product Rule.]} \\ &= (8 - 2x)[(8 - 2x) - 4x] = (8 - 2x)(8 - 6x) = 0 \end{aligned}$$

if $x = 4$ or $4/3$. The latter is a critical point in $(0,4)$.

x	0	4/3	4
$f(x)$	0	1024/27	0

We see that $x = 4/3$ yields an absolute maximum for f. So the dimensions of the box should be $\frac{16}{3}'' \times \frac{16}{3}'' \times \frac{4}{3}''$.

7. Let x denote the length of the sides of the box and y denote its height. Referring to the following figure, we see that the volume of the box is given by $x^2y = 128$. The

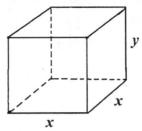

amount of material used is given by

$$S = f(x) = 2x^2 + 4xy$$

$$= 2x^2 + 4x\left(\frac{128}{x^2}\right)$$

$$= 2x^2 + \frac{512}{x} \text{ square inches.}$$

We want to minimize f subject to the condition that $x > 0$. Now

$$f'(x) = 4x - \frac{512}{x^2} = \frac{4x^3 - 512}{x^2} = \frac{4(x^3 - 128)}{x^2}.$$

Setting $f'(x) = 0$ yields $x = 5.04$, a critical point of f. Next,

$$f''(x) = 4 + \frac{1024}{x^3} > 0$$

for all $x > 0$. Thus, the graph of f is concave upward and so $x = 5.04$ yields an absolute minimum of f. Thus, the required dimensions are 5.04" × 5.04" × 5.04".

9. The length plus the girth of the box is $4x + h = 108$ and $h = 108 - 4x$. Then
$$V = x^2h = x^2(108 - 4x) = 108x^2 - 4x^3$$
and $V' = 216x - 12x^2$. We want to maximize V on the interval $[0,27]$. Setting

$V'(x) = 0$ and solving for x, we obtain $x = 18$ and $x = 0$. Evaluating $V(x)$ at $x = 0$, $x = 18$, and $x = 27$, we obtain
$$V(0) = 0, \ V(18) = 11{,}664, \text{ and } V(27) = 0$$
Thus, the dimensions of the box are 18" × 18" × 36" and its maximum volume is approximately 11,664 cu in.

11. We take $2\pi r + \ell = 108$. We want to maximize

$$V = \pi r^2 \ell = \pi r^2(-2\pi r + 108) = -2\pi^2 r^3 + 108\pi r^2$$

subject to the condition that $0 \le r \le \frac{54}{\pi}$. Now

$$V'(r) = -6\pi^2 r^2 + 216\pi r = -6\pi r(\pi r - 36).$$

Since $V' = 0$, we find $r = 0$ or $r = 36/\pi$, the critical points of V. From the table

r	0	$36/\pi$	$54/\pi$
V	0	46,656/π	0

we conclude that the maximum volume occurs when $r = 36/\pi \approx 11.5$ inches and $\ell = 108 - 2\pi\left(\frac{36}{\pi}\right) = 36$ inches and its volume is 46,656/π cu in .

13. Let y denote the height and x the width of the cabinet. Then $y = (3/2)x$. Since the volume is to be 2.4 cu ft, we have $xyd = 2.4$, where d is the depth of the cabinet.

We have $\quad x\left(\frac{3}{2}x\right)d = 2.4 \quad$ or $\quad d = \frac{2.4(2)}{3x^2} = \frac{1.6}{x^2}$.

The cost for constructing the cabinet is

$$C = 40(2xd + 2yd) + 20(2xy) = 80\left[\frac{1.6}{x} + \left(\frac{3}{2}x\right)\left(\frac{1.6}{x^2}\right)\right] + 40x\left(\frac{3}{2}x\right)$$

$$= \frac{320}{x} + 60x^2.$$

$$C'(x) = -\frac{320}{x^2} + 120x = \frac{120x^3 - 320}{x^2} = 0 \quad \text{if } x = \sqrt[3]{\frac{8}{3}} = \frac{2}{\sqrt[3]{3}} = \frac{2}{3}\sqrt[3]{9}$$

Therefore, $x = \frac{2}{3}\sqrt[3]{9}$ is a critical point of C. The sign diagram

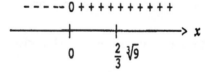

shows that $x = \frac{2}{3}\sqrt[3]{9}$ gives a relative minimum. Next, $C''(x) = \frac{640}{x^3} + 120 > 0$

for all $x > 0$ tells us that the graph of C is concave upward. So $x = \frac{2}{3}\sqrt[3]{9}$ yields an absolute minimum. The required dimensions are $\frac{2}{3}\sqrt[3]{9}' \times \sqrt[3]{9}' \times \frac{2}{5}\sqrt[3]{9}'$.

15. We want to maximize the function
$$R(x) = (200 + x)(300 - x) = -x^2 + 100x + 60000.$$
Then $\quad R'(x) = -2x + 100 = 0$
gives $x = 50$ and this is a critical point of R. Since $R''(x) = -2 < 0$, we see that $x = 50$ gives an absolute maximum of R. Therefore, the number of passengers should be 250. The fare will then be \$250/passenger and the revenue will be \$62,500.

17. Let x denote the number of people beyond 20 who sign up for the cruise. Then the revenue is $R(x) = (20 + x)(600 - 4x) = -4x^2 + 520x + 12,000$. We want to maximize R on the closed bounded interval $[0, 70]$.
$$R'(x) = -8x + 520 = 0 \text{ implies } x = 65,$$
a critical point of R. Evaluating R at this critical point and the endpoints, we have

x	0	65	70
$R(x)$	12,000	28,900	28,800

From this table, we see that R is maximized if $x = 65$. Therefore, 85 passengers will result in a maximum revenue of \$28,900. The fare would be \$340/passenger.

19. We want to maximize $S = kh^2w$. But $h^2 + w^2 = 24^2$ or $h^2 = 576 - w^2$. So $S = f(w) = kw(576 - w^2) = k(576w - w^3)$. Now, setting
$$f'(w) = k(576 - 3w^2) = 0$$
gives $w = \pm\sqrt{192} \approx \pm13.86$. Only the positive root is a critical point of interest. Next, we find $f''(w) = -6kw$, and in particular,
$$f''(\sqrt{192}) = -6\sqrt{192}\, k < 0,$$
so that $w = \pm\sqrt{192} \approx \pm13.86$ gives a relative maximum of f. Since $f''(w) < 0$ for $w > 0$, we see that the graph of f is concave downward on $(0, \infty)$ and so, $w = \sqrt{192}$ gives an absolute maximum of f. We find $h^2 = 576 - 192 = 384$ or $h \approx 19.60$. So the width and height of the log should be approximately 13.86 inches and 19.60 inches, respectively.

21. We want to minimize $C(x) = 1.50(10,000 - x) + 2.50\sqrt{3000^2 + x^2}$ subject to $0 \le x \le 10,000$. Now
$$C'(x) = -1.50 + 2.5(\tfrac{1}{2})(9,000,000 + x^2)^{-1/2}(2x) = -1.50 + \frac{2.50x}{\sqrt{9,000,000 + x^2}}$$

$$C'(x) = 0 \Rightarrow 2.5x = 1.50\sqrt{9,000,000 + x^2}$$
$$6.25x^2 = 2.25(9,000,000 + x^2) \text{ or } 4x^2 = 20250000, \ x = 2250.$$

x	0	2250	10000
$f(x)$	22500	21000	26101

From the table, we see that $x = 2250$ gives the absolute minimum.

23. The time taken for the flight is
$$T = f(x) = \frac{12 - x}{6} + \frac{\sqrt{x^2 + 9}}{4}.$$
$$f'(x) = -\frac{1}{6} + \frac{1}{4}\left(\frac{1}{2}\right)(x^2 + 9)^{-1/2}(2x) = -\frac{1}{6} + \frac{x}{4\sqrt{x^2 + 9}}$$
$$= \frac{3x - 2\sqrt{x^2 + 9}}{12\sqrt{x^2 + 9}}.$$

Setting $f'(x) = 0$ gives $3x = 2\sqrt{x^2 + 9}$, $9x^2 = 4(x^2 + 9)$ or $5x^2 = 36$. Therefore, $x = \pm 6/\sqrt{5} = \pm 6\sqrt{5}/5$. Only the critical point $x = 6\sqrt{5}/5$ is of interest. The nature of the problem suggests $x \approx 2.68$ gives an absolute minimum for T.

25. The area enclosed by the rectangular region of the racetrack is $A = (\ell)(2r) = 2r\ell$. The length of the racetrack is $2\pi r + 2\ell$, and is equal to 1760. That is,
$$2(\pi r + \ell) = 1760; \ \pi r + \ell = 880, \text{ or } \ell = 880 - \pi r.$$

Therefore, we want to maximize $A = f(r) = 2r(880 - \pi r) = 1760r - 2\pi r^2$. The restricition on r is $0 \le r \le \frac{880}{\pi}$. To maximize A, we compute

$$f'(r) = 1760 - 4\pi r. \text{ Setting } f'(r) = 0 \text{ gives } r = \frac{1760}{4\pi} = \frac{440}{\pi} \approx 140. \text{ Since}$$

$f(0) = f\left(\frac{880}{\pi}\right) = 0$, we see that the maximum rectangular area is enclosed if we take $r = \frac{440}{\pi}$ and $\ell = 880 - \pi\left(\frac{440}{\pi}\right) = 440$. So $r = 140$ and $\ell = 440$. The total area

enclosed is $2r\ell + \pi r^2 = 2\left(\frac{440}{\pi}\right)(440) + \pi\left(\frac{440}{\pi}\right)^2 = \frac{2(440)^2}{\pi} + \frac{440^2}{\pi} = \frac{580,800}{\pi} \approx 184,874$ sq ft.

27. Let x denote the number of bottles in each order. We want to minimize

$$C(x) = 200\left(\frac{2,000,000}{x}\right) + \frac{x}{2}(0.40) = \frac{400,000,000}{x} + 0.2x.$$

We compute $C'(x) = -\dfrac{400,000,000}{x^2} + 0.2$. Setting $C'(x) = 0$ gives

$$x^2 = \frac{400,000,000}{0.2} = 2,000,000,000, \text{ or } x = 44,721, \text{ a critical point of } C.$$

$C'(x) = \dfrac{800,000,000}{x^3} > 0$ for all $x > 0$, and we see that the graph of C is concave

upward and so $x = 44,721$ gives an absolute minimum of C. Therefore, there should be $2,000,000/x \approx 45$ orders per year (since we can not have fractions of an order.) Then each order should be for $2,000,000/45 \approx 44,445$ bottles.

29. a. Since the sales are assumed to be at a steady rate and D units are expected to be sold per year, the number of orders/yr is D/x. Since is costs \$$K$ per order, the ordering cost is KD/x. The purchasing cost is PD (cost per item times number purchased). Finally, the holding cost is $(x/2)h$ (the average number on hand times holding cost per item). Therefore

$$C(x) = \frac{KD}{x} + pD + \frac{hx}{2}$$

b. $\quad C'(x) = -\dfrac{KD}{x^2} + \dfrac{h}{2} = 0$

implies $\quad \dfrac{KD}{x^2} = \dfrac{h}{2}$

$$x^2 = \frac{2KD}{h}$$

$$x = \pm\sqrt{\frac{2KD}{h}}$$

We reject the negative root. So $x = \sqrt{\dfrac{2KD}{h}}$ is the only critical number. Next,

$$C''(x) = \frac{2KD}{x^3} > 0 \text{ for } x > 0$$

So $C''\left(\sqrt{\dfrac{2KD}{h}}\right) > 0$ and the second derivative test shows that $x = \sqrt{\dfrac{2KD}{h}}$ does give a relative minimum and because C is concave upward, the absolute minimum.

CHAPTER 4 CONCEPT REVIEW, page 327

1. a. $f(x_1) < f(x_2)$ b. $f(x_1) > f(x_2)$

3. a. $f(x) \le f(c)$ b. $f(x) \ge f(c)$

5. a. $f'(x)$ b. > 0 c. Concavity   d. Relative maximum; relative extremum

7. 0; 0

9. a. $f(x) \le f(c)$; absolute maximum value   b. $f(x) \ge f(c)$; open interval

CHAPTER 4 REVIEW, page 328

1. a. $f(x) = \frac{1}{3}x^3 - x^2 + x - 6$. $f'(x) = x^2 - 2x + 1 = (x - 1)^2$. $f'(x) = 0$ gives $x = 1$, the critical point of f. Now, $f'(x) > 0$ for all $x \neq 1$. Thus, f is increasing on $(-\infty, 1) \cup (1, \infty)$.
   b. Since $f'(x)$ does not change sign as we move across the critical point $x = 1$, the First Derivative Test implies that $x = 1$ does not give rise to a relative extremum of f.
   c. $f''(x) = 2(x - 1)$. Setting $f''(x) = 0$ gives $x = 1$ as a candidate for an inflection point of f. Since $f''(x) < 0$ for $x < 1$, and $f''(x) > 0$ for $x > 1$, we see that f is concave downward on $(-\infty, 1)$ and concave upward on $(1, \infty)$.
   d. The results of (c) imply that $(1, -\frac{17}{3})$ is an inflection point.

3. a. $f(x) = x^4 - 2x^2$. $f'(x) = 4x^3 - 4x = 4x(x^2 - 1) = 4x(x + 1)(x - 1)$. The sign diagram of f' shows that f is decreasing on $(-\infty, -1) \cup (0, 1)$ and increasing on $(-1, 0) \cup (1, \infty)$.

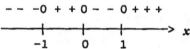

   b. The results of (a) and the First Derivative Test show that $(-1, -1)$ and $(1, -1)$ are relative minima and $(0, 0)$ is a relative maximum.

c. $f''(x) = 12x^2 - 4 = 4(3x^2 - 1) = 0$ if $x = \pm\sqrt{3}/3$. The sign diagram

```
+ + + + 0 - - - - - - - 0 + + + +
```

$$\xrightarrow{\hspace{3cm}} x$$

$$-\frac{\sqrt{3}}{3} \qquad 0 \qquad \frac{\sqrt{3}}{3}$$

shows that f is concave upward on $(-\infty, -\sqrt{3}/3) \cup (\sqrt{3}/3, \infty)$ and concave downward on $(-\sqrt{3}/3, \sqrt{3}/3)$.

d. The results of (c) show that $(-\sqrt{3}/3, -5/9)$ and $(\sqrt{3}/3, -5/9)$ are inflection points.

5. a. $f(x) = \dfrac{x^2}{x-1}$. $f'(x) = \dfrac{(x-1)(2x) - x^2(1)}{(x-1)^2} = \dfrac{x^2 - 2x}{(x-1)^2} = \dfrac{x(x-2)}{(x-1)^2}$.

The sign diagram of f'

f' is not defined here

```
+ + + + + 0 - - -    - - - 0 + + + + + +
```

$$\xrightarrow{\hspace{3cm}} x$$

$$0 \qquad 1 \qquad 2$$

shows that f is increasing on $(-\infty, 0) \cup (2, \infty)$ and decreasing on $(0,1) \cup (1,2)$.

b. The results of (a) show that $(0,0)$ is a relative maximum and $(2,4)$ is a relative minimum.

c. $f''(x) = \dfrac{(x-1)^2(2x-2) - x(x-2)2(x-1)}{(x-1)^4} = \dfrac{2(x-1)[(x-1)^2 - x(x-2)]}{(x-1)^4}$

$$= \dfrac{2}{(x-1)^3}.$$

Since $f''(x) < 0$ if $x < 1$ and $f''(x) > 0$ if $x > 1$, we see that f is concave downward on $(-\infty, 1)$ and concave upward on $(1, \infty)$.

d. Since $x = 1$ is not in the domain of f, there are no inflection points.

7. $f(x) = (1-x)^{1/3}$. $f'(x) = -\dfrac{1}{3}(1-x)^{-2/3} = -\dfrac{1}{3(1-x)^{2/3}}$.

The sign diagram for f' is

f' is not defined here

```
- - - - - - - - - -    - - - - - -
```

$$\xrightarrow{\hspace{3cm}} x$$

$$0 \qquad 1$$

a. f is decreasing on $(-\infty,1) \cup (1,\infty)$.

b. There are no relative extrema.

c. Next, we compute $f''(x) = -\dfrac{2}{9}(1-x)^{-5/3} = -\dfrac{2}{9(1-x)^{5/3}}$.

The sign diagram for f'' is

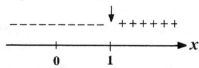

f'' is not defined here

We find f is concave downward on $(-\infty,1)$ and concave upward on $(1,\infty)$.

d. $x = 1$ is a candidate for an inflection point of f. Referring to the sign diagram for f'', we see that $(1,0)$ is an inflection point.

9. a. $f(x) = \dfrac{2x}{x+1}$. $f'(x) = \dfrac{(x+1)(2) - 2x(1)}{(x+1)^2} = \dfrac{2}{(x+1)^2} > 0$ if $x \neq -1$.

Therefore f is increasing on $(-\infty,-1) \cup (-1,\infty)$.

b. Since there are no critical points, f has no relative extrema.

c. $f''(x) = -4(x+1)^{-3} = -\dfrac{4}{(x+1)^3}$. Since $f''(x) > 0$ if $x < -1$ and $f''(x) < 0$ if $x > -1$,

we see that f is concave upward on $(-\infty,-1)$ and concave downward on $(-1,\infty)$.

d. There are no inflection points since $f''(x) \neq 0$ for all x in the domain of f.

11. $f(x) = x^2 - 5x + 5$

1. The domain of f is $(-\infty, \infty)$.

2. Setting $x = 0$ gives 5 as the y-intercept.

3. $\lim\limits_{x \to -\infty} (x^2 - 5x + 5) = \lim\limits_{x \to \infty} (x^2 - 5x + 5) = \infty$.

4. There are no asymptotes because f is a quadratic function.

5. $f'(x) = 2x - 5 = 0$ if $x = 5/2$. The sign diagram

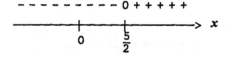

shows that f is increasing on $(\frac{5}{2},\infty)$ and decreasing on $(-\infty,\frac{5}{2})$.

6. The First Derivative Test implies that $(\frac{5}{2},-\frac{5}{4})$ is a relative minimum.

7. $f''(x) = 2 > 0$ and so f is concave upward on $(-\infty, \infty)$.

8. There are no inflection points.
The graph of f follows.

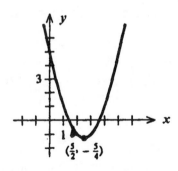

13. $g(x) = 2x^3 - 6x^2 + 6x + 1$.
1. The domain of g is $(-\infty, \infty)$.
2. Setting $x = 0$ gives 1 as the y-intercept.
3. $\lim_{x \to -\infty} g(x) = -\infty$, $\lim_{x \to \infty} g(x) = \infty$.
4. There are no vertical or horizontal asymptotes.
5. $g'(x) = 6x^2 - 12x + 6 = 6(x^2 - 2x + 1) = 6(x - 1)^2$. Since $g'(x) > 0$ for all $x \neq 1$, we see that g is increasing on $(-\infty, 1) \cup (1, \infty)$.
6. $g'(x)$ does not change sign as we move across the critical point $x = 1$, so there is no extremum.
7. $g''(x) = 12x - 12 = 12(x - 1)$. Since $g''(x) < 0$ if $x < 1$ and $g''(x) > 0$ if $x > 1$, we see that g is concave upward on $(1, \infty)$ and concave downward on $(-\infty, 1)$.
8. The point $x = 1$ gives rise to the inflection point $(1,3)$.
9. The graph of g follows.

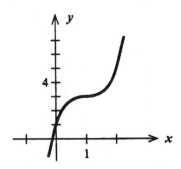

4 Applications of the Derivative

15. $h(x) = x\sqrt{x-2}$.

    1. The domain of h is $[2,\infty)$.

    2. There are no y-intercepts. Next, setting $y = 0$ gives 2 as the x-intercept.

    3. $\lim\limits_{x\to\infty} x\sqrt{x-2} = \infty$.

    4. There are no asymptotes.

    5. $h'(x) = (x-2)^{1/2} + x(\tfrac{1}{2})(x-2)^{-1/2} = \tfrac{1}{2}(x-2)^{-1/2}[2(x-2)+x]$

$$= \frac{3x-4}{2\sqrt{x-2}} > 0 \quad \text{on } [2,\infty)$$

    and so h is increasing on $[2,\infty)$.

    6. Since h has no critical points in $(2,\infty)$, there are no relative extrema.

    7. $h''(x) = \dfrac{1}{2}\left[\dfrac{(x-2)^{1/2}(3) - (3x-4)\tfrac{1}{2}(x-2)^{-1/2}}{x-2}\right]$

$$= \frac{(x-2)^{-1/2}[6(x-2)-(3x-4)]}{4(x-2)} = \frac{3x-8}{4(x-2)^{3/2}} .$$

    The sign diagram for h''

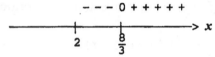

    shows that h is concave downward on $(2,\tfrac{8}{3})$ and concave upward on $(\tfrac{8}{3},\infty)$.

    8. The results of (7) tell us that $(\tfrac{8}{3}, \tfrac{8\sqrt{6}}{9})$ is an inflection point.

    The graph of h follows.

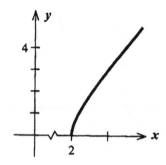

17. $f(x) = \dfrac{x-2}{x+2}$.

1. The domain of f is $(-\infty,-2) \cup (-2,\infty)$.

2. Setting $x = 0$ gives -1 as the y-intercept. Setting $y = 0$ gives 2 as the x-intercept.

3. $\displaystyle\lim_{x\to-\infty} \dfrac{x-2}{x+2} = \lim_{x\to\infty} \dfrac{x-2}{x+2} = 1.$

4. The results of (3) tell us that $y = 1$ is a horizontal asymptote. Next, observe that the denominator of $f(x)$ is equal to zero at $x = -2$, but its numerator is not equal to zero there. Therefore, $x = -2$ is a vertical asymptote.

5. $\qquad f'(x) = \dfrac{(x+2)(1)-(x-2)(1)}{(x+2)^2} = \dfrac{4}{(x+2)^2}.$

The sign diagram of f'

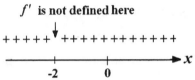

tells us that f is increasing on $(-\infty,-2) \cup (-2,\infty)$.

6. The results of (5) tells us that there are no relative extrema.

7. $f''(x) = -\dfrac{8}{(x+2)^3}.$ The sign diagram of f'' follows

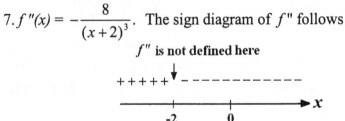

and it shows that f is concave upward on $(-\infty,-2)$ and concave downward on $(-2,\infty)$.

8. There are no inflection points.

4 *Applications of the Derivative*

The graph of f follows.

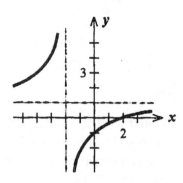

19. $\lim\limits_{x\to-\infty} \dfrac{1}{2x+3} = \lim\limits_{x\to\infty}\dfrac{1}{2x+3} = 0$ and so $y = 0$ is a horizontal asymptote. Since the denominator is equal to zero at $x = $ -3/2, but the numerator is not equal to zero there, we see that $x = $ -3/2 is a vertical asymptote.

21. $\lim\limits_{x\to-\infty} \dfrac{5x}{x^2 -2x-8} = \lim\limits_{x\to\infty}\dfrac{5x}{x^2-2x-8} = 0$ and so $y = 0$ is a horizontal asymptote. Next, note that the denominator is zero if x^2 - 2x - 8 = (x - 4)(x + 2) = 0, or $x = $ -2 or $x = 4$. Since the numerator is not equal to zero at these points, we see that $x = $ -2 and $x = 4$ are vertical asymptotes.

23. $f(x) = 2x^2 + 3x$ - 2; $f'(x) = 4x + 3$. Setting $f'(x) = 0$ gives $x = $ -3/4 as a critical point of f. Next, $f''(x) = 4 > 0$ for all x, so f is concave upward on $(-\infty, \infty)$. Therefore, $f(-\frac{3}{4}) = -\frac{25}{8}$ is an absolute minimum of f. There is no absolute maximum.

25. $g(t) = \sqrt{25 - t^2} = (25 - t^2)^{1/2}$. Differentiating $g(t)$, we have
$$g'(t) = \tfrac{1}{2}(25 - t^2)^{-1/2}(-2t) = -\dfrac{t}{\sqrt{25 - t^2}}.$$
Setting $g'(t) = 0$ gives $t = 0$ as a critical point of g. The domain of g is given by solving the inequality $25 - t^2 \geq 0$ or $(5 - t)(5 + t) \geq 0$ which implies that $t \in [-5,5]$. From the table

t	-5	0	5
$g(t)$	0	5	0

we conclude that $g(0) = 5$ is the absolute maximum of g and $g(-5) = 0$ and $g(5) = 0$ is the absolute minimum value of g.

27. $h(t) = t^3 - 6t^2$. $h'(t) = 3t^2 - 12t = 3t(t-4) = 0$ if $t = 0$ or $t = 4$, critical points of h. But only $t = 4$ lies in $(2,5)$.

t	2	4	5
$h(t)$	-16	-32	-25

From the table, we see that there is an absolute minimum at $(4,-32)$ and an absolute maximum at $(2,-16)$.

29. $f(x) = x - \dfrac{1}{x}$ on $[1,3]$. $f'(x) = 1 + \dfrac{1}{x^2}$. Since $f'(x)$ is never zero, f has no critical point.

x	1	3
$f(x)$	0	$\frac{8}{3}$

We see that $f(1) = 0$ is the absolute minimum value and $f(3) = 8/3$ is the absolute maximum value.

31. $f(s) = s\sqrt{1-s^2}$ on $[-1,1]$. The function f is continuous on $[-1,1]$ and differentiable on $(-1,1)$. Next,

$$f'(s) = (1-s^2)^{1/2} + s(\tfrac{1}{2})(1-s^2)^{-1/2}(-2s) = \frac{1-2s^2}{\sqrt{1-s^2}}.$$

Setting $f'(s) = 0$, we have $s = \pm\sqrt{2}/2$, giving the critical points of f. From the table

4 *Applications of the Derivative*

x	-1	$-\sqrt{2}/2$	$\sqrt{2}/2$	1
$f(x)$	0	-1/2	1/2	0

we see that $f(-\sqrt{2}/2) = -1/2$ is the absolute minimum value and $f(\sqrt{2}/2) = 1/2$ is the absolute maximum value of f.

33. We want to maximize $P(x) = -x^2 + 8x + 20$. Now, $P'(x) = -2x + 8 = 0$ if $x = 4$, a critical point of P. Since $P''(x) = -2 < 0$, the graph of P is concave downward. Therefore, the critical point $x = 4$ yields an absolute maximum. So, to maximize profit, the company should spend $4000 on advertising per month.

35. a. $N'(t) = 16.25t + 24.625$; Since $N'(t) > 0$ for $0 < t < 3$, we see that sales of camera phones is always increasing between 2002 and 2005.
    b. $N''(t) = 16.25$; Since $N''(t) > 0$ for $0 < t < 3$, we see that the rate of sales is increasing between 2002 and 2005.

37. $C(x) = 0.0001x^3 - 0.08x^2 + 40x + 5000$; $C'(x) = 0.0003x^2 - 0.16x + 40$; $C''(x) = 0.0006x - 0.16$. Thus, $x = 266.67$ is a candidate for an inflection point of C. The sign diagram for C'' is

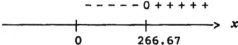

We see that the inflection point of C is (266.67, 11,874.08).

39. The revenue is $R(x) = px = x(-0.0005x^2 + 60) = -0.0005x^3 + 60x$. Therefore, the total profit is $P(x) = R(x) - C(x) = -0.0005x^3 + 0.001x^2 + 42x - 4000$. $P'(x) = -0.0015x^2 + 0.002x + 42$. Setting $P'(x) = 0$, we have $3x^2 - 4x - 84,000 = 0$. Solving for x, we find

$$x = \frac{4 \pm \sqrt{16 - 4(3)(84,000)}}{2(3)} = \frac{4 \pm 1004}{6} = 168, \text{ or } -167.$$

We reject the negative root. Next, $P''(x) = -0.003x + 0.002$ and $P''(168) = -0.003(168) + 0.002 = -0.502 < 0$. By the Second Derivative Test, $x = 168$ gives a relative maximum. Therefore, the required level of production is

168 DVDs.

41. a. $C(x) = 0.001x^2 + 100x + 4000.$

$$\overline{C}(x) = \frac{C(x)}{x} = \frac{0.001x^2 + 100x + 4000}{x} = 0.001x + 100 + \frac{4000}{x}.$$

b. $\overline{C}'(x) = 0.001 - \dfrac{4000}{x^2} = \dfrac{0.001x^2 - 4000}{x^2} = \dfrac{0.001(x^2 - 4,000,000)}{x^2}.$

Setting $\overline{C}'(x) = 0$ gives $x = \pm 2000$. We reject the negative root.

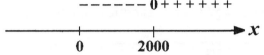

The sign diagram of $\overline{C}'$ shows that $x = 2000$ gives rise to a relative minimum of $\overline{C}$.

Since $\overline{C}''(x) = \dfrac{8000}{x^3} > 0$ if $x > 0$, we see that $\overline{C}$ is concave upward on $(0, \infty)$. So

$x = 2000$ yields an absolute minimum. So the required production level is 2000 units.

43. a. $P'(t) = -0.0006t^2 + 0.036t - 0.36$; Setting $P'(t) = 0$ gives

$$-0.0006t^2 + 0.036t - 0.36 = 0$$
$$t^2 - 60t + 600 = 0$$

So
$$t = \frac{60 \pm \sqrt{60^2 - 4(1)(600)}}{2} \approx 12.7 \text{ or } 47.3$$

We reject the root 47.3 because it lies outside $[0, 30]$. The sign diagram for P' follows.

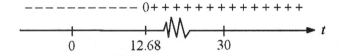

So P is decreasing on $(0, 12.7)$ and increasing on $(12.7, 30)$.

b. The absolute minimum of P occurs at $t = 12.7$ and $P(12.7) \approx 7.9$.

45. The volume is $V = f(x) = x(10 - 2x)^2$ cubic units for $0 \le x \le 5$.

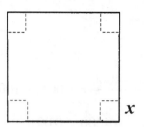

To maximize V, we compute
$$f'(x) = 12x^2 - 80x + 100 = 4(3x^2 - 20x + 25) = 4(3x - 5)(x - 5).$$
Setting $f'(x) = 0$ gives $x = 5/3$, or 5 as critical points of f. From the table

x	0	5/3	5
$f(x)$	0	2000/27≈ 74.07	0

We see that the box has a maximum volume of 74.07 cu in.

47. Refer to the following picture.

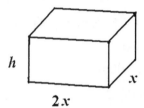

$C(x) = 30(2)(2x)(x) + 20(2)(2xh + xh) = 120x^2 + 120xh$. But $x(2x)h = 4$, or $h = \dfrac{2}{x^2}$.

Therefore, $C(x) = 120x^2 + 120x\left(\dfrac{2}{x^2}\right) = 120x^2 + \dfrac{240}{x}$

$$C'(x) = 240x - \dfrac{240}{x^2}.$$

Setting $C'(x) = 0$ gives $240x - \dfrac{240}{x^2} = 0$, or $x^3 = 1$. Therefore, $x = 1$.

$C''(x) = 240 + \dfrac{480}{x^3}$. In particular, $C''(1) > 0$. Therefore, the cost is minimized by

taking $x = 1$. The required dimensions are 1 ft × 2 ft × 2 ft.

49. $f(x) = ax^2 + bx + c$; $f'(x) = 2ax + b = 2a\left(x + \dfrac{b}{2a}\right)$. Then f' is continuous

everywhere and has a zero at $x = -\frac{b}{2a}$. The sign diagram of f' is

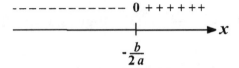

when $a > 0$, or

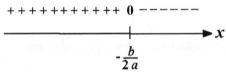

when $a < 0$. Therefore, if $a > 0$, f is decreasing on $(-\infty, -\frac{b}{2a})$ and increasing on $(-\frac{b}{2a}, \infty)$, and if $a < 0$, f is increasing on $(-\infty, -\frac{b}{2a})$ and decreasing on $(-\frac{b}{2a}, \infty)$.

CHAPTER 4, BEFORE MOVING ON, page 330

1. $f'(x) = \dfrac{(1-x)(2x) - x^2(-1)}{(1-x)^2} = \dfrac{2x - 2x^2 + x^2}{(1-x)^2} = \dfrac{x(2-x)}{(1-x)^2}$; f' is not defined at $= 1$ and has zeros at $x = 0$ and $x = 2$. The sign diagram of f follows:

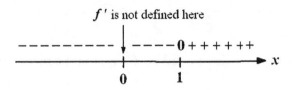

f' is not defined here

We see that f is decreasing on $(-\infty, 0) \cup (2, \infty)$ and increasing on $(0,1) \cup (1,2)$.

2. $f'(x) = 4x - 4x^{-2/3} = 4x^{-2/3}(x^{5/3} - 1) = \dfrac{4(x^{5/3} - 1)}{x^{2/3}}$; f' is discontinuous at $x = 0$ and has a zero where $x^{5/3} = 1$ or $x = 1$. Therefore, f has critical numbers at 0 and 1. The sign diagram for f' follows:

f' is not defined here

$$-\,-\,-\,-\,-\,-\,-\,-\,-\,\Big\downarrow\,-\,-\,-\,-\,0\,+\,+\,+\,+\,+\,+$$

We see that $x = 1$ gives a relative minimum. Since $f(1) = 2 - 12 = -10$, the relative

4 Applications of the Derivative

minimum is (1,-10). There are no relative maxima.

3. $f'(x) = x^2 - \frac{1}{2}x - \frac{1}{2}$; $f''(x) = 2x - \frac{1}{2}$; $f''(x) = 0$ gives $x = \frac{1}{4}$. The sign diagram of f'' follows:

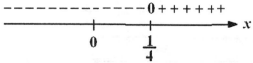

We see that f is concave downward on $(-\infty, \frac{1}{4})$ and concave upward on $(\frac{1}{4}, \infty)$. Since

$$f(\tfrac{1}{4}) = \tfrac{1}{3}(\tfrac{1}{4})^3 - \tfrac{1}{4}(\tfrac{1}{4})^2 - \tfrac{1}{2}(\tfrac{1}{4}) + 1 = \frac{83}{96}$$

the inflection point is $(\frac{1}{4}, \frac{83}{96})$.

4. $f(x) = 2x^3 - 9x^2 + 12x - 1$
   1. Domain of f is $(-\infty, \infty)$.
   2. Setting $y = f(x) = 0$ gives -1 as the y-intercept of f.
   3. $\lim\limits_{x \to -\infty} f(x) = -\infty$ and $\lim\limits_{x \to \infty} f(x) = \infty$.
   4. There are no asymptotes.
   5. $f'(x) = 6x^2 - 18x + 12 = 6(x^2 - 3x + 2) = 6(x-2)(x-1)$. The sign diagram of f'

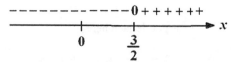

   shows that f is increasing on $(-\infty, 1) \cup (2, \infty)$ and decreasing on $(1, 2)$.
   6. We see that $(1, 4)$ is a relative maximum and $(2, 3)$ is a relative minimum.
   7. $f''(x) = 12x - 18 = 6(2x - 3)$. The sign diagram of f'' is

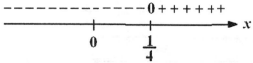

   and shows that f is concave downward on $(-\infty, \frac{3}{2})$ and concave upward on $(\frac{3}{2}, \infty)$.
   8. $f(\frac{3}{2}) = 2(\frac{3}{2})^3 - 9(\frac{3}{2})^2 + 12(\frac{3}{2}) - 1 = \frac{7}{2}$; So $(\frac{3}{2}, \frac{7}{2})$ is an inflection point of f.

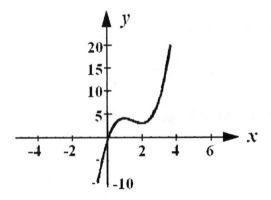

5. f is continuous on a closed interval [-2, 3]. $f'(x) = 6x^2 + 6x = 6x(x+1)$. The critical numbers of f are -1 and 0.

x	-2	-1	0	3
y	-5	0	-1	80

The absolute maximum value of f is 80; the absolute minimum value is -5.

6. The amount of material used (the surface area) is

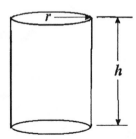

$A = \pi r^2 + 2\pi rh$. But $V = \pi r^2 h = 1$ and so $h = \dfrac{1}{\pi r^2}$. Therefore,

$A = \pi r^2 + 2\pi rh \left(\dfrac{1}{\pi r^2} \right) = \pi r^2 + \dfrac{2}{r}$; $A' = 2\pi r - \dfrac{2}{r^2} = 0$ implies

$2\pi r = \dfrac{2}{r^2}$, $r^3 = \dfrac{2}{r^2}$, $r^3 = \dfrac{1}{\pi}$, or $r = \dfrac{1}{\sqrt[3]{\pi}}$; Since $A'' = 2\pi + \dfrac{4}{r^3} > 0$ for $r > 0$, we see that $r = \dfrac{1}{\sqrt[3]{\pi}}$ does give an absolute maximum. Also

$$h = \frac{1}{\pi r^2} = \frac{1}{\pi} \cdot \pi^{2/3} = \frac{1}{\pi^{1/3}} = \frac{1}{\sqrt[3]{\pi}}$$

Therefore both the radius and height should be $\dfrac{1}{\sqrt[3]{\pi}}$ ft.

CHAPTER 5

5.1 Problem Solving Tips

1. Be careful to remember the order of operations when you work with exponents. Note that $-5^2 \neq 25$, but rather $-5^2 = -(5)^2 = -25$. On the other hand $(-5)^2 = 25$.

2. $b^{-x} = \dfrac{1}{b^x} = \left(\dfrac{1}{b}\right)^x$. If $b > 1$, then $0 < \frac{1}{b} < 1$. So the graph of b^{-x} for $b > 1$ is similar to the graph of $y = (1/2)^x$. (See Figure 3 in the text.)

5.1 CONCEPT QUESTIONS, page 335

1. $f(x) = b^x$; $a < b$, $b \neq 1$. Its domain is $(-\infty, \infty)$.

EXERCISES 5.1, page 335

1. a. $4^{-3} \times 4^5 = 4^{-3+5} = 4^2 = 16$

    b. $3^{-3} \times 3^6 = 3^{6-3} = 3^3 = 27$.

3. a. $9(9)^{-1/2} = \dfrac{9}{9^{1/2}} = \dfrac{9}{3} = 3$.

    b. $5(5)^{-1/2} = 5^{1/2} = \sqrt{5}$.

5. a. $\dfrac{(-3)^4(-3)^5}{(-3)^8} = (-3)^{4+5-8} = (-3)^1 = -3$.

    b. $\dfrac{(2^{-4})(2^6)}{2^{-1}} = 2^{-4+6+1} = 2^3 = 8$.

7. a. $\dfrac{5^{3.3} \cdot 5^{-1.6}}{5^{-0.3}} = \dfrac{5^{3.3-1.6}}{5^{-0.3}} = 5^{1.7+(0.3)} = 5^2 = 25$.

    b. $\dfrac{4^{2.7} \cdot 4^{-1.3}}{4^{-0.4}} = 4^{2.7-1.3+0.4} = 4^{1.8} \approx 12.1257$.

9. a. $(64x^9)^{1/3} = 64^{1/3}(x^{9/3}) = 4x^3$.

   b. $(25x^3y^4)^{1/2} = 25^{1/2}(x^{3/2})(y^{4/2}) = 5x^{3/2}y^2 = 5xy^2\sqrt{x}$.

11. a. $\dfrac{6a^{-5}}{3a^{-3}} = 2a^{-5+3} = 2a^{-2} = \dfrac{2}{a^2}$.

    b. $\dfrac{4b^{-4}}{12b^{-6}} = \dfrac{1}{3}b^{-4+6} = \dfrac{1}{3}b^2$.

13. a. $(2x^3y^2)^3 = 2^3 \times x^{3(3)} \times y^{2(3)} = 8x^9y^6$.

    b. $(4x^2y^2z^3)^2 = 4^2 \times x^{2(2)} \times y^{2(2)} \times z^{3(2)} = 16x^4y^4z^6$.

15. a. $\dfrac{5^0}{(2^{-3}x^{-3}y^2)^2} = \dfrac{1}{2^{-3(2)}x^{-3(2)}y^{2(2)}} = \dfrac{2^6x^6}{y^4} = \dfrac{64x^6}{y^4}$.

    b. $\dfrac{(x+y)(x-y)}{(x-y)^0} = (x+y)(x-y)$.

17. $6^{2x} = 6^4$ if and only if $2x = 4$ or $x = 2$.

19. $3^{3x-4} = 3^5$ if and only if $3x - 4 = 5$, $3x = 9$, or $x = 3$.

21. $(2.1)^{x+2} = (2.1)^5$ if and only if $x + 2 = 5$, or $x = 3$.

23. $8^x = (\frac{1}{32})^{x-2}$, $(2^3)^x = (32)^{2-x} = (2^5)^{2-x}$, so $2^{3x} = 2^{5(2-x)}$, $3x = 10 - 5x$, $8x = 10$, or $x = 5/4$.

25. Let $y = 3^x$, then the given equation is equivalent to

$$y^2 - 12y + 27 = 0$$
$$(y-9)(y-3) = 0$$

giving $y = 3$ or 9. So $3^x = 3$ or $3^x = 9$, and therefore, $x = 1$ or $x = 2$.

27. $y = 2^x$, $y = 3^x$, and $y = 4^x$

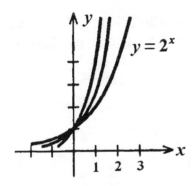

29. $y = 2^{-x}$, $y = 3^{-x}$, and $y = 4^{-x}$

31. $y = 4^{0.5x}$, $y = 4x$, and $y = 4^{2x}$

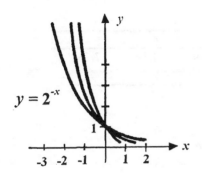

$y = 2^{-x}$

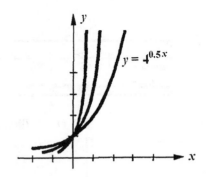

$y = 4^{0.5x}$

33. $y = e^{0.5x}$, $y = e^{x}$, $y = e^{1.5x}$

35. $y = 0.5e^{-x}$, $y = e^{-x}$, and $y = 2e^{-x}$

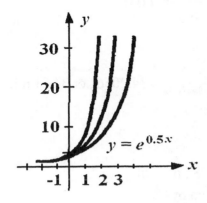

$y = e^{0.5x}$

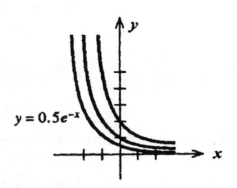

$y = 0.5e^{-x}$

37. a. $R(t) = 26.3e^{-0.016t}$; In 1982, the rate was $R(0) = 26.3\%$. In 1986, the rate was $R(4) = 24.7\%$. In 1994, the rate was $R(12) = 21.7\%$. In 2000, the rate was $R(18) = 19.7\%$.

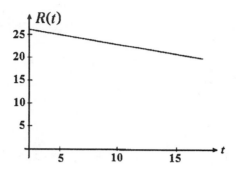

39. a.

Year	0	1	2	3	4	5
Number (billions)	0.45	0.80	1.41	2.49	4.39	7.76

b.

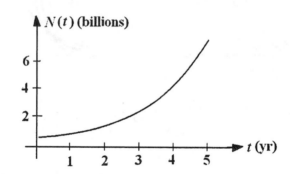

41. $N(t) = \dfrac{35.5}{1 + 6.89e^{-0.8674t}}$; $N(6) = \dfrac{35.5}{1 + 6.89e^{-0.8674(6)}} \approx 34.2056$, or 34.21 million.

43. a. The concentration initially is given by
$$N(0) = 0.08 + 0.12(1 - e^{-0.02(0)}) = 0.08 \text{, or } 0.08 \text{ g/cm}^3.$$
b. The concentration after 20 seconds is given by
$$N(20) = 0.08 + 0.12(1 - e^{-0.02(20)}) = 0.11956, \text{ or } 0.1196 \text{ g/cm}^3.$$
c. The concentration in the long run is given by

$$\lim_{t \to \infty} x(t) = \lim_{t \to \infty}[0.08 + 0.12(1 - e^{-0.02t})] = 0.2, \text{ or } 0.2 \text{ g/cm}^3.$$

d.

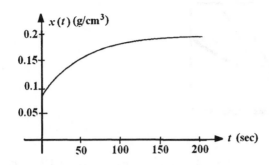

45. False. $(x^2 + 1)^3 = x^6 + 3x^4 + 3x^2 + 1$.

47. True. $f(x) = e^x$ is an increasing function and so if $x < y$, then $f(x) < f(y)$, or $e^x < e^y$.

USING TECHNOLOGY EXERCISES 5.1, page 338

1.

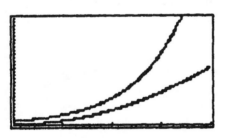

3.

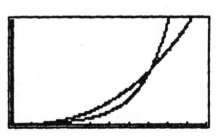

5.

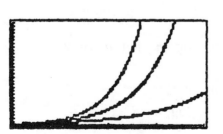

7.

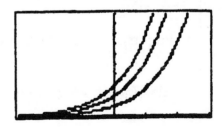

9.

11. a.

b. 0.08 g/cm^3 c. 0.12 g/cm^3
d. 0.2 g/cm^3

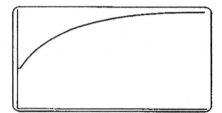

13. a.

b. 20 sec.   c. 35.1 sec

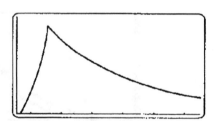

5.2 Problem Solving Tips

1. Property 1 of logarithms says that $\log_b mn = \log_b m + \log_b n$. However,

$$\log_b (m+n) \neq \log_b m + \log_b n \text{ and } \log_b \frac{m}{n} \neq \frac{\log_b m}{\log_b n}$$

2. When you work with logarithms be sure that you can distinguish between the following two operations:

$$\frac{\log 6}{\log 2} = \log 6 \div \log 2 \approx 2.585 \quad \text{and} \quad \log \frac{6}{2} = \log 6 - \log 2 \approx 0.477$$

(Property 2 of logarithms says that $\log_b \dfrac{m}{n} = \log_b m - \log_b n$.)

3. The domain of the logarithmic function is $(0, \infty)$. So the logarithm of 0 and the logarithm of negative numbers are not defined.

5.2 CONCEPT QUESTIONS, page 345

1. a. $y = \log_b x$ if and only if $x = b^y$; $b > 0$, $b \neq 1$; Its domain is $(0, \infty)$.
3. a. $e^{\ln x} = x$ b. $\ln e^x = x$

EXERCISES 5.2 , page 345

1. $\log_2 64 = 6$ 3. $\log_3 \dfrac{1}{9} = -2$ 5. $\log_{1/3} \dfrac{1}{3} = 1$

7. $\log_{32} 8 = \dfrac{3}{5}$ 9. $\log_{10} 0.001 = -3$

11. $\log 12 = \log 4 \times 3 = \log 4 + \log 3 = 0.6021 + 0.4771 = 1.0792.$

13. $\log 16 = \log 4^2 = 2 \log 4 = 2(0.6021) = 1.2042.$

15. $\log 48 = \log 3 \times 4^2 = \log 3 + 2 \log 4 = 0.4771 + 2(0.6021) = 1.6813.$

17. $2 \ln a + 3 \ln b = \ln a^2 b^3.$

19. $\ln 3 + \dfrac{1}{2} \ln x + \ln y - \dfrac{1}{3} \ln z = \ln \dfrac{3\sqrt{xy}}{\sqrt[3]{z}}$

21. $\log x(x+1)^4 = \log x + \log (x+1)^4 = \log x + 4 \log (x+1)$.

23. $\log \dfrac{\sqrt{x+1}}{x^2+1} = \log (x+1)^{1/2} - \log(x^2+1) = \frac{1}{2} \log (x+1) - \log (x^2+1)$

25. $\ln xe^{-x^2} = \ln x - x^2$.

27. $\ln \left(\dfrac{x^{1/2}}{x^2 \sqrt{1+x^2}} \right) = \ln x^{1/2} - \ln x^2 - \ln (1+x^2)^{1/2}$

$$= \frac{1}{2} \ln x - 2 \ln x - \frac{1}{2} \ln (1+x^2) = -\frac{3}{2} \ln x - \frac{1}{2} \ln (1+x^2).$$

29. $y = \log_3 x$

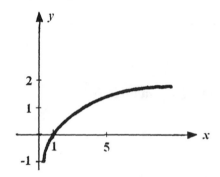

31. $y = \ln 2x$

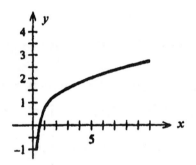

33. $y = 2^x$ and $y = \log_2 x$

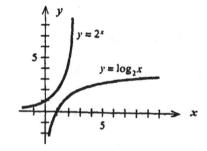

35. $e^{0.4t} = 8$, $0.4t \ln e = \ln 8$, and $0.4t = \ln 8$ $(\ln e = 1.)$ So, $t = \dfrac{\ln 8}{0.4} = 5.1986$.

37. $5e^{-2t} = 6$, $e^{-2t} = \frac{6}{5} = 1.2$. Taking the logarithm, we have

$$-2t \ln e = \ln 1.2, \text{ or } t = -\frac{\ln 1.2}{2} \approx -0.0912.$$

39. $2e^{-0.2t} - 4 = 6$, $2e^{-0.2t} = 10$. Taking the logarithm on both sides of this last equation, we have $\ln e^{-0.2t} = \ln 5$; $-0.2t \ln e = \ln 5$; $-0.2t = \ln 5$;

and $\quad t = -\dfrac{\ln 5}{0.2} \approx -8.0472.$

41. $\dfrac{50}{1+4e^{0.2t}} = 20$, $1+4e^{0.2t} = \dfrac{50}{20} = 2.5$, $4e^{0.2t} = 1.5$,

$e^{0.2t} = \dfrac{1.5}{4} = 0.375$, $\ln e^{0.2t} = \ln 0.375$, $0.2t = \ln 0.375$. So $t = \dfrac{\ln 0.375}{0.2} \approx -4.9041.$

43. Taking the logarithm on both sides, we obtain

$\ln A = \ln Be^{-t/2}$, $\ln A = \ln B + \ln e^{-t/2}$, $\ln A - \ln B = -t/2 \ln e$,

$\ln \dfrac{A}{B} = -\dfrac{t}{2}$ or $t = -2 \ln \dfrac{A}{B} = 2 \ln \dfrac{B}{A}.$

45. $p(x) = 19.4 \ln x + 18$. For a child weighing 92 lb, we find
$p(92) = 19.4 \ln 92 + 18 = 105.72$ millimeters of mercury.

47. a. $30 = 10 \log \dfrac{I}{I_0}$; $3 = \log \dfrac{I}{I_0}$; $\dfrac{I}{I_0} = 10^3 = 1000.$ So $I = 1000\, I_0.$

b. When $D = 80$, $I = 10^8 I_0$ and when $D = 30$, $I = 10^3 I_0$. Therefore, an 80–decibel sound is $10^8/10^3$ or $10^5 = 100,000$ times louder than a 30–decibel sound.
c. It is $10^{15}/10^8 = 10^7$, or $10,000,000$, times louder.

49. We solve the following equation for t. Thus,

$$\frac{160}{1+240e^{-0.2t}} = 80; \quad 1+240e^{-0.2t} = \frac{160}{80},$$

$$240e^{-0.2t} = 2-1 = 1; \quad e^{-0.2t} = \frac{1}{240}; \quad -0.2t = \ln \frac{1}{240}$$

$$t = -\frac{1}{0.2} \ln \frac{1}{240} \approx 27.40, \text{ or approximately 27.4 years old.}$$

51. We solve the following equation for t:
$$200(1-0.956e^{-0.18t})=140$$
$$1-0.956e^{-0.18t} = \frac{140}{200} = 0.7$$
$$-0.956e^{-0.18t} = 0.7-1 = -0.3$$
$$e^{-0.18t} = \tfrac{0.3}{0.956}$$
$$-0.18t = \ln\left(\tfrac{0.3}{0.956}\right)$$
$$t = -\frac{\ln\left(\tfrac{0.3}{0.956}\right)}{0.18} \approx 6.43875.$$

So, its approximate age is 6.44 years.

53. a. We solve the equation $0.08+0.12e^{-0.02t} = 0.18$.
$$0.12e^{-0.02t} = 0.1;\ e^{-0.02t} = \frac{0.1}{0.12} = \frac{1}{1.2}$$
$$\ln e^{-0.02t} = \ln\frac{1}{1.2} = \ln 1 - \ln 1.2 = -\ln 1.2$$
$$-0.02t = -\ln 1.2$$
$$t = \frac{\ln 1.2}{0.02} \approx 9.116, \quad \text{or } 9.12 \text{ sec.}$$

   b. We solve the equation $0.08+0.12e^{-0.02t} = 0.16$.
$$0.12e^{-0.02t} = 0.08;\ e^{-0.02t} = \frac{0.08}{0.12} = \frac{2}{3};\ -0.02t = \ln\frac{2}{3}$$
$$t = -\frac{\ln\left(\frac{2}{3}\right)}{0.02} \approx 20.2733, \quad \text{or } 20.27 \text{ sec.}$$

55. False. Take $x = e$. Then $(\ln e)^3 = 1^3 = 1 \neq 3 \ln e = 3$.

57. True. $g(x) = \ln x$ is continuous and greater than zero on $(1, \infty)$. Therefore,
$f(x) = \dfrac{1}{\ln x}$ is continuous on $(1, \infty)$.

59. a. Taking the logarithm on both sides gives $\ln 2^x = \ln e^{kx}$, $x \ln 2 = kx(\ln e) = kx$.
   So, $x(\ln 2 - k) = 0$ for all x and this implies that $k = \ln 2$.
   b. Tracing the same steps as done in (a), we find that $k = \ln b$.

61. Let $\log_b m = p$, then $m = b^p$. Therefore, $m^n = (b^p)^n = b^{np}$. Therefore,

$$\log_b m^n = \log_b b^{np} = np \log_b b = np \qquad (\text{Since } \log_b b = 1.)$$
$$= n \log_b m,$$

as was to be shown.

5.3 Problem Solving Tips

1. When you look at an applied problem involving interest rates, it is important to

choose the correct interest formula to solve the problem. If the problems asks for the

future value of an investment, then use the compound interest formula. If the problem

asks for the *amount of money that needs to be invested now to accumulate a certain*

sum in the future, then use the present value formula for compound interest. If the

interest in the applied problem is compounded continuously then use the corresponding

formulas for continuous compound interest.

5.3 CONCEPT QUESTIONS, page 357

1. a. In simple interest, the interest is based on the original principal. In compound
interest, interest earned is periodically added to the principal and thereafter earns
interest at the same rate.

   b. Simple interest formula: $A = P(1 + rt)$;

   Compound interest formula: $A = \left(1 + \dfrac{r}{m}\right)^{mt}$

3. $P = A\left(1 + \dfrac{r}{m}\right)^{-mt}$

1. $A = 2500\left(1 + \dfrac{0.07}{2}\right)^{20} = 4974.47$, or $4974.47.

3. $A = 150{,}000\left(1 + \dfrac{0.1}{12}\right)^{48} = 223{,}403.11$, or $223,403.11

5. a. Using the formula $r_{\mathit{eff}} = \left(1 + \dfrac{r}{m}\right)^{m} - 1$ with $r = 0.10$ and $m = 2$, we have

$$r_{\mathit{eff}} = \left(1 + \frac{0.10}{2}\right)^{2} - 1 = 0.1025, \quad \text{or } 10.25 \text{ percent/yr}$$

b. Using the formula $r_{\mathit{eff}} = \left(1 + \dfrac{r}{m}\right)^{m} - 1$ with $r = 0.09$ and $m = 4$, we have

$$r_{\mathit{eff}} = \left(1 + \frac{0.09}{4}\right)^{4} - 1 = 0.09308, \text{ or } 9.308 \text{ percent/yr.}$$

7. a. The present value is given by $P = 40{,}000\left(1 + \dfrac{0.08}{2}\right)^{-8} = 29{,}227.61$,

or $29,227.61.

b. The present value is given by $P = 40{,}000\left(1 + \dfrac{0.08}{4}\right)^{-16} = 29{,}137.83$, or

$29,137.83.

9. $A = 5000e^{0.08(4)} \approx 6885.64$, or $6,885.64.

11. We use formula (6) with $A = 7500$, $P = 5000$, $m = 12$, and $t = 3$. Thus
$$7500 = 5000\left(1 + \tfrac{r}{12}\right)^{36};$$
$$\left(1 + \tfrac{r}{12}\right)^{36} = \tfrac{7500}{5000} = \tfrac{3}{2}, \quad \ln\left(1 + \tfrac{r}{12}\right)^{36} = \ln 1.5;$$
$$36\left(1 + \tfrac{r}{12}\right) = \ln 1.5$$
$$\left(1 + \tfrac{r}{12}\right) = \tfrac{\ln 1.5}{36} = 0.0112629$$
$$1 + \tfrac{r}{12} = e^{0.0112629} = 1.011327; \tfrac{r}{12} = 0.011327;$$
$$r = 0.13592$$

So the interest rate is 13.59% per year.

13. We use formula (6) with $A = 8000$, $P = 5000$, $m = 2$, and $t = 4$. Thus
$$8000 = 5000\left(1+\tfrac{r}{2}\right)^8$$
$$\left(1+\tfrac{r}{2}\right)^8 = \tfrac{8000}{5000} = 1.6, \ \ln\left(1+\tfrac{r}{2}\right)^8 = \ln 1.6;$$
$$8\ln\left(1+\tfrac{r}{2}\right) = \ln 1.6$$
$$\ln\left(1+\tfrac{r}{2}\right) = \tfrac{\ln 1.6}{8} = 0.05875$$
$$1+\tfrac{r}{2} = e^{0.05875} = 1.06051; \ \tfrac{r}{2} = 0.06051$$
$$r = 0.1210$$
So the required interest rate is 12.1% per year.

15. We use formula (6) with $A = 4000$, $P = 2000$, $m = 1$, and $t = 5$. Thus
$$4000 = 2000\left(1+r\right)^5; \ \ \left(1+r\right)^5 = 2; 5\ln(1+r) = \ln 2; \ \ln(1+r) = \tfrac{\ln 2}{5} = 0.138629$$
$$1+r = e^{0.138629} = 1.148698; \ r = 0.1487$$
So the required interst rate is 14.87% per year.

17. We use formula (6) with $A = 6500$, $P = 5000$, $m = 12$, and $r = 0.12$. Thus
$$6500 = 5000\left(1+\frac{0.12}{12}\right)^{12t}; \ (1.01)^{12t} = \frac{6500}{5000} = 1.3; \ 12t\ln(1.01) = \ln 1.3$$
$$t = \frac{\ln 1.3}{12\ln 1.01} \approx 2.197$$
So, it will take approximately 2.2 years.

19. We use formula (6) with $A = 4000$, $P = 2000$, $m = 12$, and $r = 0.09$. Thus,
$$4000 = 2000\left(1+\tfrac{0.09}{12}\right)^{12t}$$
$$\left(1+\tfrac{0.09}{12}\right)^{12t} = 2$$
$$12t\ln\left(1+\tfrac{0.09}{12}\right) = \ln 2 \ \text{ and } \ t = \frac{\ln 2}{12\ln\left(1+\tfrac{0.09}{12}\right)} \approx 7.73.$$
So it will take approximately 7.7 years.

21. We use formula (10) with $A = 6000$, $P = 5000$, and $t = 3$. Thus,

$$6000 = 5000e^{3r}$$

$$e^{3r} = \frac{6000}{5000} = 1.2; \qquad 3r = \ln 1.2$$

$$r = \frac{\ln 1.2}{3} \approx 0.6077$$

So the interest rate is 6.08% per year.

23. We use formula (10) formula (6) with $A = 7000$, $P = 6000$, and $r = 0.075$. Thus

$$7000 = 6000e^{0.075t}; \quad e^{0.075t} = \tfrac{7000}{6000} = \tfrac{7}{6}$$

$$0.075t \ln e = \ln \tfrac{7}{6} \quad \text{and} \quad t = \frac{\ln \tfrac{7}{6}}{0.075} \approx 2.055.$$

So, it will take 2.06 years.

25. The Estradas can expect to pay $180,000(1+0.09)^4$, or approximately \$254,084.69.

27. The investment will be worth

$$A = 1.5\left(1+\frac{0.065}{2}\right)^{20} = 2.84376 \text{ , or approximately \$2.844 million.}$$

29. The present value of the \$8000 loan due in 3 years is given by

$$P = 8000\left(1+\frac{0.10}{2}\right)^{-6} = 5969.72, \text{ or } \$5969.72.$$

The present value of the \$15,000 loan due in 6 years is given by

$$P = 15{,}000\left(1+\frac{0.10}{2}\right)^{-12} = 8352.56, \text{ or } \$8352.56.$$

Therefore, the amount the proprietors of the inn will be required to pay at the end

of 5 years is given by $A = 14{,}322.28\left(1+\dfrac{0.10}{2}\right)^{10} = 23{,}329.48, \text{ or } \$23{,}329.48.$

31. He can expect the minimum revenue for 2007 to be

$$240{,}000(1.2)(1.3)(1.25)^3 \approx 731{,}250 \text{ or } \$731{,}250.$$

33. We want the value of a 2004 dollar in the year 2000. Denoting this value by x, we have

$$(1.034)(1.028)(1.016)(1.023)x = 1$$

or $x \approx 0.9051$. Thus, the purchasing power is approximately 91 cents.

35. The effective annual rate of return on his investment is found by solving the equation $(1+r)^2 = \dfrac{32100}{25250}$

$$1+r = \left(\frac{32100}{25250}\right)^{1/2}$$

$1+r \approx 1.1275$ and $r = 0.1275$, or 12.75 percent.

37. $P = Ae^{-rt} = 59673e^{-(0.08)5} \approx 40{,}000.008$, or approximately \$40,000.

39. a. If they invest the money at 10.5 percent compounded quarterly, they should set aside $P = 70{,}000\left(1+\frac{0.105}{4}\right)^{-28} \approx 33{,}885.14$, or \$33,885.14.

   b. If they invest the money at 10.5 percent compounded continuously, they should set aside $P = 70{,}000e^{-(0.105)(7)} = 33{,}565.38$, or \$33,565.38.

41. a. If inflation over the next 15 years is 6 percent, then Eleni's first year's pension will be worth $P = 40{,}000e^{-0.9} = 16{,}262.79$, or \$16,262.79.

   b. If inflation over the next 15 years is 8 percent, then Eleni's first year's pension will be worth $P = 40{,}000e^{-1.2} = 12{,}047.77$, or \$12,047.77.

   c. If inflation over the next 15 years is 12 percent, then Eleni's first year's pension will be worth $P = 40{,}000e^{-1.8} = 6611.96$, or \$6,611.96.

43. $r_{eff} = \lim\limits_{m\to\infty}\left(1+\dfrac{r}{m}\right)^m - 1 = e^r - 1.$

45. The effective rate of interest at Bank A is given by
$$R = \left(1+\tfrac{0.07}{4}\right)^4 - 1 = 0.07186,$$
or 7.186 percent. The effective rate at Bank B is given by
$$R = e^r - 1 = e^{0.07125} - 1 = 0.07385$$
or 7.385 percent. We conclude that Bank B has the higher effective rate of interest.

47. The nominal rate of interest that, when compounded continuously, yields an effective rate of interest of 10 percent per year is found by solving the equation
$$R = e^r - 1,\ 0.10 = e^r - 1,\ 1.10 = e^r,\quad \ln 1.10 = r\ln e,\ r = \ln 1.10 \approx 0.09531,$$
or 9.531 percent.

5.4 Problem Solving Tips

1. The derivative of e^x is equal to e^x. Similarly,

$$\frac{d}{dx}[e^{3x}] = 3e^{3x} \quad \text{and} \quad \frac{d}{dx}\left[e^{2x^2-1}\right] = 4xe^{2x^2-1}.$$

Note that the exponents in the original function and the derivative are the same.

2. Don't confuse functions of the type e^x with functions of the type x^r. The latter is a power function and its exponent is a constant; whereas the exponent in an exponential function such as e^x is a variable. A different rule is used to differentiate each type of function. Thus

$$\frac{d}{dx}\left[x^2 e^x\right] = x^2 \frac{d}{dx}[e^x] + \frac{d}{dx}\left[x^2\right]e^x = x^2 e^x + 2xe^x = xe^x(x+2).$$

5.4 CONCEPT QUESTIONS, page 365

1. a. $f'(x) = e^x$ b. $g'(x) = e^{f(x)} \cdot f'(x)$

EXERCISES 5.4 , page 365

1. $f(x) = e^{3x};\ f'(x) = 3e^{3x}$ 3. $g(t) = e^{-t};\ g'(t) = -e^{-t}$

5. $f(x) = e^x + x;\ f'(x) = e^x + 1$

7. $f(x) = x^3 e^x,\ f'(x) = x^3 e^x + e^x(3x^2) = x^2 e^x(x+3)$.

9. $f(x) = \dfrac{2e^x}{x},\ f'(x) = \dfrac{x(2e^x) - 2e^x(1)}{x^2} = \dfrac{2e^x(x-1)}{x^2}$.

11. $f(x) = 3(e^x + e^{-x});\ f'(x) = 3(e^x - e^{-x}).$

13. $f(w) = \dfrac{e^w + 1}{e^w} = 1 + \dfrac{1}{e^w} = 1 + e^{-w}.\ f'(w) = -e^{-w} = -\dfrac{1}{e^w}.$

15. $f(x) = 2e^{3x-1},\ f'(x) = 2e^{3x-1}(3) = 6e^{3x-1}.$

17. $h(x) = e^{-x^2};\ h'(x) = e^{-x^2}(-2x) = -2xe^{-x^2}.$

19. $f(x) = 3e^{-1/x};\ f'(x) = 3e^{-1/x} \cdot \dfrac{d}{dx}\left(-\dfrac{1}{x}\right) = 3e^{-1/x}\left(\dfrac{1}{x^2}\right) = \dfrac{3e^{-1/x}}{x^2}.$

21. $f(x) = (e^x + 1)^{25},\ f'(x) = 25(e^x + 1)^{24}e^x = 25e^x(e^x + 1)^{24}.$

23. $f(x) = e^{\sqrt{x}};\ f'(x) = e^{\sqrt{x}}\dfrac{d}{dx}x^{1/2} = e^{\sqrt{x}}\dfrac{1}{2}x^{-1/2} = \dfrac{e^{\sqrt{x}}}{2\sqrt{x}}.$

25. $f(x) = (x - 1)e^{3x+2};\ f'(x) = (x - 1)(3)e^{3x+2} + e^{3x+2} = e^{3x+2}(3x - 3 + 1) = e^{3x+2}(3x - 2).$

27. $f(x) = \dfrac{e^x - 1}{e^x + 1};\ f'(x) = \dfrac{(e^x + 1)(e^x) - (e^x - 1)(e^x)}{(e^x + 1)^2} = \dfrac{e^x(e^x + 1 - e^x + 1)}{(e^x + 1)^2} = \dfrac{2e^x}{(e^x + 1)^2}.$

29. $f(x) = e^{-4x} + 2e^{3x};\ f'(x) = -4e^{-4x} + 6e^{3x}$ and
$f''(x) = 16e^{-4x} + 18e^{3x} = 2(8e^{-4x} + 9e^{3x}).$

31. $f(x) = 2xe^{3x};\ f'(x) = 2e^{3x} + 2xe^{3x}(3) = 2(3x + 1)e^{3x}.$
$f''(x)\quad = 6e^{3x} + 2(3x + 1)e^{3x}(3) = 6(3x + 2)e^{3x}.$

33. $y = f(x) = e^{2x-3}.\ f'(x) = 2e^{2x-3}.$ To find the slope of the tangent line to the graph
of f at $x = 3/2,$ we compute $f'(\tfrac{3}{2}) = 2e^{3-3} = 2.$ Next, using the point–slope form of
the equation of a line, we find that
$$y - 1 = 2(x - \tfrac{3}{2})$$
$$= 2x - 3, \qquad \text{or} \qquad y = 2x - 2.$$

35. $f(x) = e^{-x^2/2},\ f'(x) = e^{-x^2/2}(-x) = -xe^{-x^2/2}.$ Setting $f'(x) = 0,$ gives $x = 0$ as the only
critical point of $f.$ From the sign diagram,

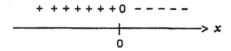

we conclude that f is increasing on $(-\infty,0)$ and decreasing on $(0,\infty)$.

37. $f(x) = \frac{1}{2}e^x - \frac{1}{2}e^{-x}$, $f'(x) = \frac{1}{2}(e^x + e^{-x})$, $f''(x) = \frac{1}{2}(e^x - e^{-x})$. Setting $f''(x) = 0$, gives $e^x = e^{-x}$ or $e^{2x} = 1$, and $x = 0$. From the sign diagram for f'',

$$+ + + + + + + + + + + 0 \quad - - - - - - - -$$
$$\xrightarrow{\hspace{2cm}}\;x$$
$$0$$

we conclude that f is concave upward on $(0,\infty)$ and concave downward on $(-\infty,0)$.

39. $f(x) = xe^{-2x}$. $f'(x) = e^{-2x} + xe^{-2x}(-2) = (1 - 2x)e^{-2x}$.
    $f''(x) = -2e^{-2x} + (1 - 2x)e^{-2x}(-2) = 4(x - 1)e^{-2x}$.
    Observe that $f''(x) = 0$ if $x = 1$. The sign diagram of f''

$$- - - - - - - - - - - - 0 + + + + + + + + + + +$$
$$\xrightarrow{\hspace{2cm}}\;x$$
$$1$$

shows that $(1, e^{-2})$ is an inflection point.

41. $f(x) = e^{-x^2}$, $f'(x) = -2xe^{-x^2}$;
    $f''(x) = -2e^{-x^2} - 2xe^{-x^2} \cdot (-2x) = -2e^{-x^2(1-2x^2)=0}$ implies $x = \pm\frac{\sqrt{2}}{2}$. The sign diagram of f'' follows:

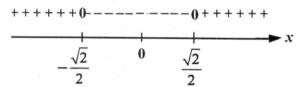

We see that the graph of f has inflection points at $(-\frac{\sqrt{2}}{2}, e^{-1/2})$ and $(\frac{\sqrt{2}}{2}, e^{-1/2})$. The slope of the tangent line at $(-\frac{\sqrt{2}}{2}, e^{-1/2})$ is $f'(-\frac{\sqrt{2}}{2}) = \sqrt{2}e^{-1/2}$. The tangent line has equation

$$y - e^{-1/2} = \sqrt{2}e^{-1/2}(x + \frac{\sqrt{2}}{2}) \text{ or } y = \sqrt{\frac{2}{e}}x + \frac{2}{\sqrt{e}} \text{ or } e^{-1/2}(\sqrt{2}x + 2)$$

The slope of the tangent line at $(\frac{\sqrt{2}}{2}, e^{-1/2})$ is $f'(\frac{\sqrt{2}}{2}) = -\sqrt{2}e^{-1/2}$. The tangent line has equation $\quad y - e^{-1/2} = -\sqrt{2}e^{-1/2}(x - \frac{\sqrt{2}}{2})$ or $y = e^{-1/2}(-\sqrt{2}x + 2)$

43. $f(x) = e^{-x^2}$. $f'(x) = -2xe^{-x^2} = 0$ if $x = 0$, the only critical point of f.

x	-1	0	1
$f(x)$	e^{-1}	1	e^{-1}

From the table, we see that f has an absolute minimum value of e^{-1} attained at $x = -1$ and $x = 1$. It has an absolute maximum at $(0,1)$.

45. $g(x) = (2x - 1)e^{-x}$; $g'(x) = 2e^{-x} + (2x - 1)e^{-x}(-1) = (3 - 2x)e^{-x} = 0$, if $x = 3/2$. The graph of g shows that $(\frac{3}{2}, 2e^{-3/2})$ is an absolute maximum, and $(0,-1)$ is an absolute minimum.

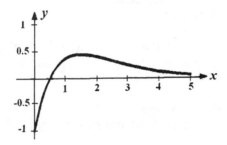

47. $f(t) = e^t - t$;
   We first gather the following information on f.
   1. The domain of f is $(-\infty, \infty)$.
   2. Setting $t = 0$ gives 1 as the y-intercept.
   3. $\lim\limits_{t \to -\infty} (e^t - t) = \infty$ and $\lim\limits_{t \to \infty} (e^t - t) = \infty$.
   4. There are no asymptotes.
   5. $f'(t) = e^t - 1$ if $t = 0$, a critical point of f. From the sign diagram for f'

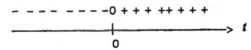

   we see that f is decreasing on $(-\infty, 0)$ and increasing on $(0, \infty)$.
   6. From the results of (5), we see that $(0,1)$ is a relative minimum of f.
   7. $f''(t) = e^t > 0$ for all t in $(-\infty, \infty)$. So the graph of f is concave upward on

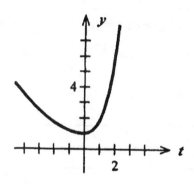

$(-\infty,\infty)$.
8. There are no inflection points.
The graph of f follows.

49. $f(x) = 2 - e^{-x}$.
    We first gather the following information on f.
    1. The domain of f is $(-\infty,\infty)$.
    2. Setting $x = 0$ gives 1 as the y-intercept.
    3. $\lim\limits_{x \to -\infty} (2 - e^{-x}) = -\infty$ and $\lim\limits_{x \to \infty} (2 - e^{-x}) = 2$,
    4. From the results of (3), we see that $y = 2$ is a horizontal asymptote of f.
    5. $f'(x) = e^{-x}$. Observe that $f'(x) > 0$ for all x in $(-\infty,\infty)$ and so f is increasing on $(-\infty,\infty)$.
    6. Since there are no critical points, f has no relative extrema.
    7. $f''(x) = -e^{-x} < 0$ for all x in $(-\infty,\infty)$ and so the graph of f is concave downward on $(-\infty,\infty)$.
    8. There are no inflection points
    The graph of f follows.

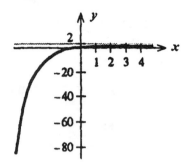

51. $P'(t) = 20.6(-0.009)e^{-0.009t} = -0.1854e^{-0.009t}$
    $P'(10) = -0.1694$, $P'(20) = -0.1549$, and $P'(30) = -0.1415$,
    and this tells us that the percentage of the total population relocating was

decreasing at the rate of 0.17% in 1970, 0.15% in 1980, and 0.14% in 1990.

53. a. The population at the beginning of 2000 was $P(0) = 0.07$, or 70,000. The
population at the beginning of 2030 will be $P(3) = 0.3537$, or approximately
353,700.
b. $P'(t) = 0.0378e^{0.54t}$; The population was changing at the rate of $P'(0) = 0.0378$
or 37,800/decade at the beginning of 2000. At the beginning of 2030, it was
changing at the rate of $P'(3) \approx 0.191$, or approximately 191,000/decade.

55. a. The total loans outstanding in 1998 were $L(0) = 4.6$, or \$4.6 trillion. The total
loans outstanding in 2004 were $L(6) = 3.6$, or \$3.6 trillion.
b. $L'(t) = -0.184e^{-0.04t}$; The total loans outstanding were changing at the rate of
$L'(0) = -0.184$, that is, they were declining at the rate of \$0.18 trillion/yr in 1998.
In 2004, they were changing at $L'(16) \approx -0.145$ or declining at the rate of \$0.15
trillion/yr.
c. $L''(t) = 0.00736e^{-0.04t}$; Since $L''(t) > 0$ on the interval $(0, 6)$, we see L is
decreasing but at a slower rate and this proves the assertion.

57. a. $S(t) = 20,000(1 + e^{-0.5t})$

$S'(t) = 20,000(-0.5e^{-0.5t}) = -10,000e^{-0.5t}$;
$S'(1) = -10,000e^{-0.5} = -6065$, or $-\$6065$/day.
$S'(2) = -10,000e^{-1} = -3679$, or $-\$3679$/day.
$S'(3) = -10,000(e^{-1.5}) = -2231$, or $-\$2231$/day.
$S'(4) = -10,000e^{-2} = -1353$, or $-\$1353$/day.
b. $S(t) = 20,000(1+e^{-0.5t}) = 27,400$

$$1+e^{-0.5t} = \frac{27,400}{20,000}$$

$$e^{-0.5t} = \frac{274}{200} - 1$$

$$-0.5t = \ln\left(\frac{274}{200} - 1\right)$$

$$t = \frac{\ln\left(\frac{274}{200} - 1\right)}{-0.5} \approx 2, \text{ or 2 days}$$

59. $N(t) = 5.3e^{0.095t^2 - 0.85t}$.

a. $N'(t) = 5.3e^{0.095t^2 - 0.85t}(0.19t - 0.85)$. Since $N'(t)$ is negative for $(0 \le t \le 4)$, we see that $N(t)$ is decreasing over that interval.

b. To find the rate at which the number of polio cases was decreasing at the beginning of 1959, we compute
$$N'(0) = 5.3e^{0.095(0^2) - 0.85(0)}(0.85) = 5.3(-0.85) = -4.505$$
(t is measured in thousands), or 4,505 cases per year. To find the rate at which the number of polio cases was decreasing at the beginning of 1962, we compute
$$N'(3) = 5.3e^{0.095(9) - 0.85(3)}(0.57 - 0.85)$$
$$= (-0.28)(0.9731) \approx -0.273, \text{ or 273 cases per year.}$$

61. From the results of Exercise 60, we see that $R'(x) = 100(1 - 0.0001x)e^{-0.0001x}$.
Setting $R'(x) = 0$ gives $x = 10,000$, a critical point of R. From the graph of R

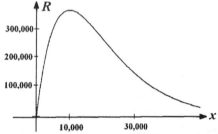

we see that the revenue is maximized when $x = 10,000$. So 10,000 pairs must be sold, yielding a maximum revenue of $R(10,000) = 367,879.44$, or \$367,879.

63. $p = 240\left(1 - \dfrac{3}{3 + e^{-0.0005x}}\right) = 240[1 - 3(3 + e^{-0.0005x})^{-1}]$.

$p' = 720(3 + e^{-0.0005x})^{-2}(-0.0005e^{-0.0005x})$
$p'(1000) = 720(3 + e^{-0.0005(1000)})^{-2}(-0.0005e^{-0.0005(1000)})$
$$= -\frac{0.36(0.606531)}{(3 + 0.606531)^2} \approx -0.0168, \text{ or } -1.68 \text{ cents per case.}$$

$$p(1000) = 240(1 - \frac{3}{3.606531}) \approx 40.36, \text{ or } \$40.36/\text{case.}$$

65. a. $W = 2.4e^{1.84h}$; $W = 2.4e^{1.84(16)} \approx 45.58$, or approximately 45.6 kg.

b. $\Delta W \approx dW = (2.4)(1.84)e^{1.84h} dh$. With $h = 1.6$ and $dh = \Delta h = 1.65 - 1.6 = 0.05$, we find $\Delta W \approx (2.4)(1.84)e^{1.84(1.6)} \cdot (0.05) \approx 4.19$, or approximately 4.2 kg.

67. $P(t) = 80{,}000\, e^{\sqrt{t}/2 - 0.09t} = 80{,}000\, e^{\frac{1}{2}t^{1/2} - 0.09t}$.

$P'(t) = 80{,}000(\frac{1}{4}t^{-1/2} - 0.09)e^{\frac{1}{2}t^{1/2} - 0.09t}$.

Setting $P'(t) = 0$, we have

$$\tfrac{1}{4}t^{-1/2} = 0.09, \quad t^{-1/2} = 0.36, \quad \frac{1}{\sqrt{t}} = 0.36, \quad t = \left(\frac{1}{0.36}\right)^2 \approx 7.72.$$

Evaluating $P(t)$ at each of its endpoints and at the point $t = 7.72$, we find

t	$P(t)$
0	80,000
7.72	160,207.69
8	160,170.71

We conclude that P is optimized at $t = 7.72$. The optimal price is $160,207.69.

69. $f(t) = 1.5 + 1.8te^{-1.2t}$

$f'(t) = 1.8\dfrac{d}{dt}(te^{-1.2t}) = 1.8[e^{-1.2t} + te^{-1.2t}(-1.2)] = 1.8e^{-1.2t}(1 - 1.2t)$.

$f'(0) = 1.8$, $f'(1) = -0.11$, $f'(2) = -0.23$, and $f'(3) = -0.13$,

and this tells us that the rate of change of the amount of oil used is 1.8 barrels per $1000 of output per decade in 1965; it is decreasing at the rate of 0.11 barrels per $1000 of output per decade in 1966, and so on.

71. a. The price at $t = 0$ is $8 + 4$, or 12, dollars per unit.

b. $\dfrac{dp}{dt} = -8e^{-2t} + e^{-2t} - 2te^{-2t}$.

$\left.\dfrac{dp}{dt}\right|_{t=0} = -8e^{-2t} + e^{-2t} - 2te^{-2t}\Big|_{t=0} = -8 + 1 = -7.$

That is, the price is decreasing at the rate of $7/week.

c. The equilibrium price is $\lim_{t\to\infty}(8+4e^{-2t}+te^{-2t}) = 8+0+0$, or $8 per unit.

73. We are given that

$$c(1-e^{-at/V}) < m$$

$$1-e^{-at/V} < \frac{m}{c}$$

$$-e^{-at/V} < \frac{m}{c}-1 \quad \text{and} \quad e^{-at/V} > 1-\frac{m}{c}.$$

Taking the log of both sides of the inequality, we have

$$-\frac{at}{V}\ln e > \ln\frac{c-m}{c}$$

$$-\frac{at}{V} > \ln\frac{c-m}{c}$$

$$-t > \frac{V}{a}\ln\frac{c-m}{c} \quad \text{or} \quad t < \frac{V}{a}\left(-\ln\frac{c-m}{c}\right) = \frac{V}{a}\ln\left(\frac{c}{c-m}\right).$$

Therefore the liquid must not be allowed to enter the organ for a time longer than

$$t = \frac{V}{a}\ln\left(\frac{c}{c-m}\right) \text{ minutes.}$$

75. $C'(t) = \begin{cases} 0.3+18e^{-t/60}(-\frac{1}{60}) & 0 \le t \le 20 \\ -\frac{18}{60}e^{-t/60}+\frac{12}{60}e^{-(t-20)/60} & t > 20 \end{cases} = \begin{cases} 0.3(1-e^{-t/60}) & 0 \le t \le 20 \\ -0.3e^{-t/60}+0.2e^{-(t-20)/60} & t > 20 \end{cases}$

a. $C'(10) = 0.3(1-e^{-10/60}) \approx 0.05$ or 0.05 g/cm^3/sec.

b. $C'(30) = -0.3e^{-30/60}+0.2e^{-10/60} \approx -0.01$, or decreasing at the rate of 0.01 g/ cm^3/sec.

c. On the interval $(0, 20)$, $C'(t) = 0$ implies $1-e^{-t/60} = 0$, or $t = 0$.
Therefore, C attains its absolute maximum value at an endpoint. In this case, at $t = 20$. On the interval $[20,\infty)$, $C'(t) = 0$ implies

$$-0.3e^{-t/60} = -0.2e^{-(t-20)/60}$$

$$\frac{e^{-\left(\frac{t-20}{60}\right)}}{e^{-t/60}} = \frac{3}{2}; \text{ or } e^{1/3} = \frac{3}{2},$$

which is not possible. Therefore $C'(t) \ne 0$ on $[20,\infty)$. Since $C(t)\to 0$ as $t\to\infty$, the absolute maximum of c occurs at $t = 20$. Thus, the concentration of the drug reaches a maximum at $t = 20$.

d. The maximum concentration is $C(20) = 0.90$ g/cm^3.

77. False. $f(x) = 3^x = e^{x \ln 3}$ and so $f'(x) = e^{x \ln 3} \cdot \dfrac{d}{dx}(x \ln 3) = (\ln 3)e^{x \ln 3} = (\ln 3)3^x$.

79. False. $f'(x) = (\ln \pi)\pi^x$. See Exercise 74.

USING TECHNOLOGY EXERCISES 5.4, page 370

1. 5.4366   3. 12.3929   5. 0.1861

7. a. The initial population of crocodiles is $P(0) = \frac{300}{6} = 50$.

   b. $\displaystyle\lim_{t \to 0} P(t) = \lim_{t \to 0} \dfrac{300e^{-0.024t}}{5e^{-0.024t} + 1} = \dfrac{0}{0+1} = 0$.

   c.

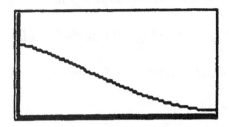

9. a.   b. 4.2720 billion/half century

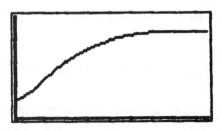

11. a. Using the function evaluation capabilities of a graphing utility, we find
          $f(11) = 153.024$ and $g(11) = 235.180977624$
    and this tells us that the number of violent-crime arrests will be 153,024 at the
    beginning of the year 2000, but if trends like inner-city drug use and wider
    availability of guns continue, then the number of arrests will be 235,181.
    b. Using the differentiation capability of a graphing utility, we find
          $f'(11) = -0.634$ and $g'(11) = 18.4005596893$

and this tells us that the number of violent-crime arrests will be decreasing at the rate of 634 per year at the beginning of the year 2000. But if the trends like inner-city drug use and wider availability of guns continues, then the number of arrests will be increasing at the rate of 18,401 per year at the beginning of the year 2000.

13. a. $P(10) = \dfrac{74}{1 + 2.6e^{-0.166(10) + 0.04536(10)^2 - 0.0066(10)^3}} \approx 69.63$ percent.

b. $P'(10) = 5.09361$, or 5.09361%/decade

5.5 Problem Solving Tips

1. If you trying to find the derivative of a complicated function involving products, quotients, or powers, check to see if you can use logarithmic differentiation to simplify the process. Look at problems 37-46, and try to become familiar with the type of problems this method is especially suitable for.

2. Example 7 provided us with a method for finding the derivative of the function $y = x^x$ ($x > 0$). Note that we use the *Power rule* ($\dfrac{d}{dx}[x^n] = nx^{n-1}$) to differentiate functions of the form $y = x^n$, where the base is a variable and the exponent is a constant, the *rule for differentiating exponential functions* to differentiate functions of the form $y = e^x$, where the base is the constant e and the exponent is a variable, and *logarithmic differentiation* to differentiate functions of the form $y = x^x$ where both the base and the exponent of the function are variables. Be sure that you can distinguish between these functions and the rule to be applied in each of these cases.

5.5 CONCEPT QUESTIONS, page 376

1. a. $f'(x) = \dfrac{1}{x}$; $g'(x) = \dfrac{f'(x)}{f(x)}$

EXERCISES 5.5, page 379

1. $f(x) = 5 \ln x; f'(x) = 5\left(\dfrac{1}{x}\right) = \dfrac{5}{x}$.

3. $f(x) = \ln (x + 1); f'(x) = \dfrac{1}{x+1}$.

5. $f(x) = \ln x^8; f'(x) = \dfrac{8x^7}{x^8} = \dfrac{8}{x}$.

7. $f(x) = \ln x^{1/2}$; $f'(x) = \dfrac{\frac{1}{2}x^{-1/2}}{x^{1/2}} = \dfrac{1}{2x}$.

9. $f(x) = \ln \left(\dfrac{1}{x^2}\right) = \ln x^{-2} = -2 \ln x$; $f'(x) = -\dfrac{2}{x}$.

11. $f(x) = \ln (4x^2 - 6x + 3)$; $f'(x) = \dfrac{8x-6}{4x^2-6x+3} = \dfrac{2(4x-3)}{4x^2-6x+3}$.

13. $f(x) = \ln \left(\dfrac{2x}{x+1}\right) = \ln 2x - \ln (x + 1)$.

$f'(x) = \dfrac{2}{2x} - \dfrac{1}{x+1} = \dfrac{2(x+1)-2x}{2x(x+1)} = \dfrac{2x+2-2x}{2x(x+1)} = \dfrac{2}{2x(x+1)} = \dfrac{1}{x(x+1)}$.

15. $f(x) = x^2 \ln x$; $f'(x) = x^2\left(\dfrac{1}{x}\right) + (\ln x)(2x) = x + 2x \ln x = x(1 + 2 \ln x)$

17. $f(x) = \dfrac{2 \ln x}{x}$. $f'(x) = \dfrac{x\left(\frac{2}{x}\right) - 2 \ln x}{x^2} = \dfrac{2(1-\ln x)}{x^2}$.

19. $f(u) = \ln (u - 2)^3$; $f'(u) = \dfrac{3(u-2)^2}{(u-2)^3} = \dfrac{3}{u-2}$.

21. $f(x) = (\ln x)^{1/2}$ and $f'(x) = \dfrac{1}{2}(\ln x)^{-1/2}\left(\dfrac{1}{x}\right) = \dfrac{1}{2x\sqrt{\ln x}}$.

23. $f(x) = (\ln x)^3$; $f'(x) = 3(\ln x)^2\left(\dfrac{1}{x}\right) = \dfrac{3(\ln x)^2}{x}$.

25. $f(x) = \ln (x^3 + 1)$; $f'(x) = \dfrac{3x^2}{x^3+1}$.

27. $f(x) = e^x \ln x.$ $f'(x) = e^x \ln x + e^x \left(\dfrac{1}{x}\right) = \dfrac{e^x(x \ln x + 1)}{x}.$

29. $f(t) = e^{2t} \ln (t + 1)$

$f'(t) = e^{2t}\left(\dfrac{1}{t+1}\right) + \ln(t+1) \cdot (2e^{2t}) = \dfrac{[2(t+1)\ln(t+1)+1]e^{2t}}{t+1}.$

31. $f(x) \dfrac{\ln x}{x}.$ $f'(x) = \dfrac{x(\frac{1}{x}) - \ln x}{x^2} = \dfrac{1 - \ln x}{x^2}.$

33. $f(x) = \ln 2 + \ln x;$ So $f'(x) = \dfrac{1}{x}$ and $f''(x) = -\dfrac{1}{x^2}.$

35. $f(x) = \ln (x^2 + 2);$ $f'(x) = \dfrac{2x}{(x^2 + 2)}$ and

$f''(x) = \dfrac{(x^2 + 2)(2) - 2x(2x)}{(x^2 + 2)^2} = \dfrac{2(2 - x^2)}{(x^2 + 2)^2}.$

37. $y = (x + 1)^2(x + 2)^3$

$\ln y = \ln (x + 1)^2(x + 2)^3 = \ln (x + 1)^2 + \ln (x + 2)^3$

$\quad = 2 \ln (x + 1) + 3 \ln (x + 2).$

$\dfrac{y'}{y} = \dfrac{2}{x+1} + \dfrac{3}{x+2} = \dfrac{2(x+2)+3(x+1)}{(x+1)(x+2)} = \dfrac{5x+7}{(x+1)(x+2)}$

$y' = \dfrac{(5x+7)(x+1)^2(x+2)^3}{(x+1)(x+2)} = (5x+7)(x+1)(x+2)^2.$

39. $y = (x - 1)^2(x + 1)^3(x + 3)^4$

$\ln y = 2 \ln (x - 1) + 3 \ln (x + 1) + 4 \ln (x + 3)$

$\dfrac{y'}{y} = \dfrac{2}{x-1} + \dfrac{3}{x+1} + \dfrac{4}{x+3}$

$\quad = \dfrac{2(x+1)(x+3) + 3(x-1)(x+3) + 4(x-1)(x+1)}{(x-1)(x+1)(x+3)}$

$\quad = \dfrac{2x^2 + 8x + 6 + 3x^2 + 6x - 9 + 4x^2 - 4}{(x-1)(x+1)(x+3)} = \dfrac{9x^2 + 14x - 7}{(x-1)(x+1)(x+3)}.$

Therefore,

$$y' = \frac{9x^2 + 14x - 7}{(x-1)(x+1)(x+3)} \cdot y$$

$$= \frac{(9x^2 + 14x - 7)(x-1)^2(x+1)^3(x+3)^4}{(x-1)(x+1)(x+3)}$$

$$= (9x^2 + 14x - 7)(x-1)(x+1)^2(x+3)^3.$$

41. $y = \dfrac{(2x^2 - 1)^5}{\sqrt{x+1}}.$

$$\ln y = \ln \frac{(2x^2 - 1)^5}{(x+1)^{1/2}} = 5 \ln(2x^2 - 1) - \frac{1}{2}\ln(x+1)$$

So $\dfrac{y'}{y} = \dfrac{20x}{2x^2 - 1} - \dfrac{1}{2(x+1)} = \dfrac{40x(x+1) - (2x^2 - 1)}{2(2x^2 - 1)(x+1)}$

$$= \frac{38x^2 + 40x + 1}{2(2x^2 - 1)(x+1)}.$$

$$y' = \frac{38x^2 + 40x + 1}{2(2x^2 - 1)(x+1)} \cdot \frac{(2x^2 - 1)^5}{\sqrt{x+1}} = \frac{(38x^2 + 40x + 1)(2x^2 - 1)^4}{2(x+1)^{3/2}}.$$

43. $y = 3^x$; $\quad \ln y = x \ln 3$; $\quad \dfrac{1}{y} \cdot \dfrac{dy}{dx} = \ln 3$; $\quad \dfrac{dy}{dx} = y \ln 3 = 3^x \ln 3.$

45. $y = (x^2 + 1)^x$; $\ln y = \ln (x^2 + 1)^x = x \ln (x^2 + 1)$. So

$$\frac{y'}{y} = \ln(x^2 + 1) + x\left(\frac{2x}{x^2 + 1}\right) = \frac{(x^2 + 1)\ln(x^2 + 1) + 2x^2}{x^2 + 1}.$$

$$y' = \frac{[(x^2 + 1)\ln(x^2 + 1) + 2x^2](x^2 + 1)^x}{x^2 + 1}.$$

47. $y = x \ln x$. The slope of the tangent line at any point is

$$y' = \ln x + x\left(\tfrac{1}{x}\right) = \ln x + 1.$$

In particular, the slope of the tangent line at (1,0) where $x = 1$ is $m = \ln 1 + 1 = 1$. So, an equation of the tangent line is $y - 0 = 1(x - 1)$ or $y = x - 1$.

49. $f(x) = \ln x^2 = 2 \ln x$ and so $f'(x) = 2/x$. Since $f'(x) < 0$ if $x < 0$, and $f'(x) > 0$ if

$x > 0$, we see that f is decreasing on $(-\infty, 0)$ and increasing on $(0, \infty)$.

51. $f(x) = x^2 + \ln x^2$; $f'(x) = 2x + \dfrac{2x}{x^2} = 2x + \dfrac{2}{x}$; $f''(x) = 2 - \dfrac{2}{x^2}$.

To find the intervals of concavity for f, we first set $f''(x) = 0$ giving

$$2 - \frac{2}{x^2} = 0, \quad 2 = \frac{2}{x^2}, \quad 2x^2 = 2$$

or $\qquad\qquad x^2 = 1$ and $x = \pm 1$.

Next, we construct the sign diagram for f''

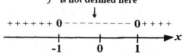

and conclude that f is concave upward on $(-\infty, -1) \cup (1, \infty)$ and concave downward on $(-1, 0) \cup (0, 1)$.

53. $f(x) = \ln(x^2 + 1)$. $f'(x) = \dfrac{2x}{x^2 + 1}$; $f''(x) = \dfrac{(x^2 + 1)(2) - (2x)(2x)}{(x^2 + 1)^2} = -\dfrac{2(x^2 - 1)}{(x^2 + 1)^2}$.

Setting $f''(x) = 0$ gives $x = \pm 1$ as candidates for inflection points of f.

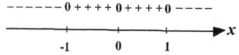

From the sign diagram for f'', we see that $(-1, \ln 2)$ and $(1, \ln 2)$ are inflection points of f.

55. $f(x) = x^2 + 2\ln x$, $f'(x) = 2x + \dfrac{2}{x}$, $f''(x) = 2 - \dfrac{2}{x^2} = 0$ implies

$2 - \dfrac{2}{x^2} = 0$, $x^2 = 1$, or $x = \pm 1$. We reject the negative root because the domain of f

is $(0, \infty)$. The sign diagram of f'' follows:

We see that $(1, 1)$ is an inflection point of the graph of f. $f'(1) = 4$. So, an equation of the required tangent line is
$$y - 1 = 4(x - 1) \quad \text{or} \quad y = 4x - 3$$

57. $f(x) = x - \ln x$; $f'(x) = 1 - \dfrac{1}{x} = \dfrac{x-1}{x} = 0$ if $x = 1$, a critical point of f.

x	1/2	1	3
$f(x)$	$1/2 + \ln 2$	1	3 - $\ln 3$

From the table, we see that f has an absolute minimum at $(1,1)$ and an absolute maximum at $(3, 3 - \ln 3)$.

59. $f(x) = 7.2956 \ln(0.0645012 x^{0.95} + 1)$

$$f'(x) = 7.2956 \cdot \frac{\frac{d}{dx}(0.0645012 x^{0.95} + 1)}{0.0645012 x^{0.95} + 1} = \frac{7.2956(0.0645012)(0.95 x^{-0.05})}{0.0645012 x^{0.95} + 1}$$

$$= \frac{0.4470462}{x^{0.05}(0.0645012 x^{0.95} + 1)}$$

So $f'(100) = 0.05799$, or approximately 0.0580 percent/kg and
$f'(500) = 0.01329$, or approximately 0.0133 percent/kg.

61. a. The projected number at the beginning of 2005 will be
$$N(1) = 34.68 + 23.88 \ln(6.35) \approx 78.82 \text{ million}$$

b. $N'(t) = 23.88 \dfrac{1.05}{1.05t + 5.3} = \dfrac{25.074}{1.05t + 5.3}$

The projected number will be changing at the rate of
$$N'(1) = \frac{25.074}{1.05 + 5.3} \approx 3.95 \text{ (million/yr)}$$

63. a. If $0 < r < 100$, then $c = 1 - \frac{r}{100}$ sastisfies $0 < c < 1$. It suffices to show that

$A_1(n) = -(1 - \frac{r}{100})^n$ is increasing, (why?), or equivalently $A_2(n) = -A_1(n) = (1 - \frac{r}{100})^n$

is decreasing. Let $y = (1 - \frac{r}{100})^n$. Then $\ln y = \ln(1 - \frac{r}{100})^n = \ln c^n = n \ln c$.
Differentiating both sides with respect to n, we find

$\dfrac{y'}{y} = \ln c$ and so $y' = (\ln c)(1 - \frac{r}{100})^n < 0$

since $\ln c < 0$ and $(1 - \frac{r}{m})^n > 0$ for $0 < r < 100$. Therefore, A is an increasing
function of n on $(0, \infty)$.

b.

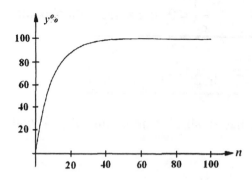

c. $\lim\limits_{n\to\infty} A(n) = \lim\limits_{n\to\infty} 100\left[1 - \left(1 - \tfrac{r}{100}\right)^n\right] = 100$

65. a. $R = \log\dfrac{10^6 I_0}{I_0} = \log 10^6 = 6.$

b. $I = I_0 10^R$ by definition. Taking the natural logarithm on both sides, we find
$$\ln I = \ln I_0 10^R = \ln I_0 + \ln 10^R = \ln I_0 + R\ln 10.$$
Differentiating implicitly with respect to R, we obtain
$$\frac{I'}{I} = \ln 10 \quad\text{or}\quad \frac{dI}{dR} = (\ln 10)I .$$
Therefore, $\Delta I \approx dI = \dfrac{dI}{dR}\Delta R = (\ln 10)I\Delta R.$ With $|\Delta R| \le (0.02)(6) = 0.12$ and
$I = 1,000,000 I_0$, (see part a), we have
$$|\Delta I| \le (\ln 10)(1,000,000 I_0)(0.12) = 276310.21 I_o$$
So the error is at most 276,310 times the standard reference intensity.

67. $f(x) = \ln(x - 1).$

1. The domain of f is obtained by requiring that $x - 1 > 0$. We find the domain to be $(1,\infty)$.

2. Since $x \ne 0$, there are no y-intercepts. Next, setting $y = 0$ gives $x - 1 = 1$ or $x = 2$ as the x-intercept.

3. $\lim\limits_{x\to 1^+} \ln(x - 1) = -\infty.$

4. There are no horizontal asymptotes. Observe that $\lim\limits_{x\to 1^+} \ln(x - 1) = -\infty$ so $x = 1$ is a vertical asymptote.

5. $f'(x) = \dfrac{1}{x-1}$.

The sign diagram for f' is

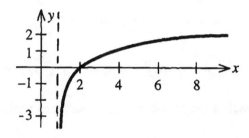

f' is not defined here $\downarrow$ + + + + + + + + + +

$$\begin{array}{ccc} & & \\ \hline 0 & 1 & \end{array} \quad x$$

We conclude that f is increasing on $(1,\infty)$.

6. The results of (5) show that f is increasing on $(1,\infty)$.

7. $f''(x) = -\dfrac{1}{(x-1)^2}$. Since $f''(x) < 0$ for $x > 1$, we see that f is concave

downward on $(1,\infty)$.

8. From the results of (7), we see that f has no inflection points.

The graph of f follows.

69. False. ln 5 is a constant function and $f'(x) = 0$.

71. If $x \le 0$, then $|x| = -x$. Therefore, $\ln|x| = \ln(-x)$. Writing $f(x) = \ln|x|$ we have

$|x| = -x = e^{f(x)}$. Differentiating both sides with respect to x and using the Chain

Rule, we have $\quad -1 = e^{f(x)} \cdot f'(x) \quad$ or $\quad f'(x) = -\dfrac{1}{e^{f(x)}} = -\dfrac{1}{-x} = \dfrac{1}{x}$.

5.6 Problem Solving Tips

Four mathematical models were introduced in this section:

Exponential growth: $Q(t) = Q_0 e^{kt}$ that describes a quantity $Q(t)$ that is initially present

in the amount of $Q(0) = Q_0$ and whose rate of growth at any time t is directly

proportional to the amount of the quantity present at time t.

Exponential decay: $Q(t) = Q_0 e^{-kt}$ that describes a quantity $Q(t)$ that is initially present

in the amount of $Q(0) = Q_0$ and decreases at a rate that is directly proportional to its

size.

Learning curves: $Q(t) = C - Ae^{-kt}$ that describes a quantity $Q(t)$, where $Q(0) = C - A$,

and $Q(t)$ increases and approaches the number C as t increases without bound.

Logistic growth functions: $Q(t) = \dfrac{A}{1 + Be^{-kt}}$ that describes a quantity $Q(t)$, where

$Q(0) = \dfrac{A}{1 + B}$. Note that $Q(t)$ increases rather rapidly for small values of t but the rate

of growth of $Q(t)$ decreases quite rapidly as t increases and $Q(t)$ approaches the

number A as t increases without bound.

Try to familiarize yourself with the examples and graphs for each of these models

before you work through the applied problems in this section.

5.6 CONCEPT QUESTIONS, page 386

1. $Q(t) = Q_0 e^{kt}$ where $k > 0$ is exponential growth and $k < 0$ is exponential decay. The larger the magnitude of k the faster the former grows and the faster the latter decays.

3. $Q(t) = \dfrac{A}{1 + Be^{-kt}}$, where A, B, and k are positive constants. Q increases rapidly for small values of t but the rate of increase slows down as Q (always increasing) approaches the number A.

EXERCISES 5.6 , page 386

1. a. The growth constant is $k = 0.05$.
   b. Initially, the quantity present is 400 units.
   c.

t	0	10	20	100	1000
Q	400	660	1087	59365	2.07×10^{24}

3. a. $Q(t) = Q_0 e^{kt}$. Here $Q_0 = 100$ and so $Q(t) = 100e^{kt}$. Since the number of cells doubles in 20 minutes, we have
$$Q(20) = 100e^{20k} = 200, \ e^{20k} = 2, \ 20k = \ln 2, \ \text{or} \ k = \tfrac{1}{20} \ln 2 \approx 0.03466.$$
$$Q(t) = 100e^{0.03466t}$$
   b. We solve the equation $100e^{0.03466t} = 1,000,000$. We obtain
$$e^{0.03466t} = 10000 \ \text{or} \ 0.03466t = \ln 10000,$$
$$t = \frac{\ln 10,000}{0.03466} \approx 266, \ \text{or } 266 \text{ minutes.}$$
   c. $Q(t) = 1000e^{0.03466t}$.

5. a. We solve the equation
$$5.3e^{0.0198t} = 3(5.3) \ \text{or} \ e^{0.0198t} = 3,$$
   or $\qquad 0.0198t = \ln 3$ and $t = \dfrac{\ln 3}{0.0198} \approx 55.5.$

   So the world population will triple in approximately 55.5 years.
   b. If the growth rate is 1.8 percent, then proceeding as before, we find
$$1.018(5.3) = 5.3e^{k} , \ \text{and} \ \ k = \ln 1.018 \approx 0.0178.$$

So $N(t) = 5.3e^{0.0178t}$. If $t = 55.5$, the population would be
$$N(55.5) = 5.3e^{0.0178(55.5)} \approx 14.23, \text{ or approximately 14.23 billion.}$$

7. $P(h) = p_0 e^{-kh}$, $P(0) = 15$, therefore, $p_0 = 15$.
$$P(4000) = 15e^{-4000k} = 12.5; \quad e^{-4000k} = \frac{12.5}{15},$$
$$-4000k = \ln\left(\frac{12.5}{15}\right) \quad \text{and } k = 0.00004558.$$

Therefore, $P(12,000) = 15e^{-0.00004558(12,000)} = 8.68$, or 8.7 lb/sq in.
The rate of change of the atmospheric pressure with respect to altitude is given by
$$P'(h) = \frac{d}{dh}(15e^{-0.00004558h}) = -0.0006837e^{-0.00004558h}.$$

So, the rate of change of the atmospheric pressure with respect to altitude when the

altitude is 12,000 feet is $P'(12,000) = -0.0006837e^{-0.00004558(12,000)} \approx -0.00039566$.
That is, it is dropping at the rate of approximately 0.0004 lbs per square inch/foot.

9. Suppose the amount of phosphorus 32 at time t is given by
$$Q(t) = Q_0 e^{-kt}$$
where Q_0 is the amount present initially and k is the decay constant. Since this
element has a half–life of 14.2 days, we have
$$\tfrac{1}{2}Q_0 = Q_0 e^{-14.2k}, \quad e^{-14.2k} = \tfrac{1}{2}, \; -14.2k = \ln\tfrac{1}{2}, \; k = -\frac{\ln\frac{1}{2}}{14.2} \approx 0.0488.$$
Therefore, the amount of phosphorus 32 present at any time t is given by
$$Q(t) = 100e^{-0.0488t}$$
The amount left after 7.1 days is given by
$$Q(7.1) = 100e^{-0.0488(7.1)} = 100e^{-0.3465}$$
$$= 70.617, \text{ or } 70.617 \text{ grams.}$$
The rate at which the phosphorus 32 is decaying when $t = 7.1$ is given by
$$Q'(t) = \frac{d}{dt}[100e^{-0.0488t}] = 100(-0.0488)e^{-0.0488t} = -4.88e^{-0.0488t}.$$

Therefore, $Q'(7.1) = -4.88e^{-0.0488(7.1)} \approx -3.451$; that is, it is changing at the rate of
3.451 gms/day.

11. We solve the equation $0.2Q_0 = Q_0 e^{-0.00012t}$
obtaining
$$t = \frac{\ln 0.2}{-0.00012} \approx 13,412, \text{ or approximately } 13,412 \text{ years.}$$

13. The graph of $Q(t)$ follows.

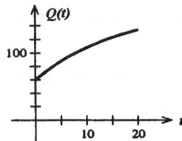

a. $Q(0) = 120(1 - e^0) + 60 = 60$, or 60 w.p.m.
b. $Q(10) = 120(1 - e^{-0.5}) + 60 = 107.22$, or approximately 107 w.p.m.
c. $Q(20) = 120(1 - e^{-1}) + 60 = 135.65$, or approximately 136 w.p.m.

15. The graph of $D(t)$ follows.

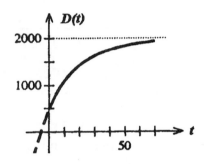

a. After one month, the demand is $D(1) = 2000 - 1500e^{-0.05} \approx 573$.
After twelve months, the demand is $D(12) = 2000 - 1500e^{-0.6} \approx 1177$.
After twenty-four months the demand is $D(24) = 2000 - 1500e^{-1.2} \approx 1548$.
After sixty months, the demand is $D(60) = 2000 - 1500e^{-3} \approx 1925$.
b. $$\lim_{t \to \infty} D(t) = \lim_{t \to \infty} 2000 - 1500e^{-0.05t} = 2000$$

and we conclude that the demand is expected to stabilize at 2000 computers per month.
c. $D'(t) = -1500e^{-0.05t}(-0.05) = 75e^{-0.05t}$. Therefore, the rate of growth after ten months is given by $D'(10) = 75e^{-0.5} \approx 45.49$, or approximately 46 computers per month.

17. a. The length is given by $f(5) = 200(1 - 0.956e^{-0.18(5)}) \approx 122.26$, or approximately 122.3 cm.
b. $f'(t) = 200(-0.956)e^{-0.18t}(-0.18) = 34.416e^{-0.18t}$. So, a 5-yr old is growing at the

rate of $f'(5) = 34.416e^{-0.18(5)} \approx 13.9925$, or approximately 14 cm/yr.

c. The maximum length is given by $\lim_{t \to \infty} 200(1 - 0.956e^{-0.18t}) = 200$, or 200 cm.

19. a. The percent of lay teachers is $f(3) = \dfrac{98}{1 + 2.77e^{-3}} \approx 86.1228$, or 86.12%.

b. $f'(t) = \dfrac{d}{dt}[98(1 + 2.77e^{-t})^{-1}] = 98(-1)(1 + 2.77e^{-t})^{-2}(2.77e^{-t})(-1)$

$= \dfrac{271.46e^{-t}}{(1 + 2.77e^{-t})^2}$

$f'(3) = \dfrac{271.46e^{-3}}{(1 + 2.77e^{-3})^2} \approx 10.4377.$

So it is increasing at the rate of 10.44%/yr.

c. $f''(t) = 271.46 \left[\dfrac{(1 + 2.77e^{-t})^2(-e^{-t}) - e^{-t} \cdot 2(1 + 2.77e^{-t})(-2.77e^{-t})}{(1 + 2.77e^{-t})^4} \right]$

$= \dfrac{271.46[-(1 + 2.77e^{-t} + 5.54e^{-t}]}{e^t(1 + 2.77e^{-t})^3} = \dfrac{271.46(2.77e^{-t} - 1)}{e^t(1 + 2.77e^{-t})^3}.$

Setting $f''(t) = 0$ gives $2.77e^{-t} = 1$

$$e^{-t} = \dfrac{1}{2.77}; \quad -t = \ln\left(\tfrac{1}{2.77}\right), \quad \text{and} \quad t = 1.0188.$$

The sign diagram of f'' shows that $t = 1.02$ gives an inflection point of P. So, the

```
        + + + +0 - - - - - - - -
      ┼─────────┼──────────────▶ t
      0        1.02
```

percent of lay teachers was increasing most rapidly in 1970.

21. The projected population of citizens 45-64 in 2010 is

$$P(20) = \dfrac{197.9}{1 + 3.274e^{-0.0361(20)}} \approx 76.3962$$

or 76.4 million.

23. The first of the given conditions implies that $f(0) = 300$, that is,

$$300 = \dfrac{3000}{1 + Be^0} = \dfrac{3000}{1 + B}.$$

So $1 + B = 10$, or $B = 9$. Therefore, $f(t) = \dfrac{3000}{1 + 9e^{-kt}}$. Next, the condition

$f(2) = 600$ gives the equation

$$600 = \frac{3000}{1+9e^{-2k}}, \quad 1+9e^{-2k} = 5, \quad e^{-2k} = \frac{4}{9}, \quad \text{or } k = -\frac{1}{2}\ln\left(\frac{4}{9}\right).$$

Therefore, $f(t) = \dfrac{3000}{1+9e^{(1/2)t \cdot \ln(4/9)}} = \dfrac{3000}{1+9\left(\frac{4}{9}\right)^{t/2}}$.

The number of students who had heard about the policy four hours later is given by

$$f(4) = \frac{3000}{1+9\left(\frac{4}{9}\right)^2} = 1080, \quad \text{or } 1080 \text{ students.}$$

To find the rate at which the rumor was spreading at any time time, we compute

$$f'(t) = \frac{d}{dt}\left[3000(1+9e^{-0.405465t})^{-1}\right]$$

$$= (3000)(-1)(1+9e^{-0.405465})^{-2}\frac{d}{dt}(9e^{-0.405465t})$$

$$= -3000(9)(-0.405465)e^{-0.405465t}(1+9e^{-0.405465t})^{-2}$$

$$= \frac{10947.555\,e^{-0.405465t}}{(1+9e^{-0.405465t})^2}$$

In particular, the rate at which the rumor was spreading 4 hours after the ceremony is given by $f'(4) = \dfrac{10947.555e^{-0.405465(4)}}{(1+9e^{-0.405465(4)})^2} \approx 280.25737$.

So , the rumor is spreading at the rate of 280 students per hour.

25. $x(t) = \dfrac{15\left(1-\left(\frac{2}{3}\right)^{3t}\right)}{1-\frac{1}{4}\left(\frac{2}{3}\right)^{3t}}$; $\quad \lim_{t\to\infty} x(t) = \lim_{t\to\infty}\dfrac{15\left(1-\left(\frac{2}{3}\right)^{3t}\right)}{1-\frac{1}{4}\left(\frac{2}{3}\right)^{3t}} = \dfrac{15(1-0)}{1-0} = 15$

or 15 lb.

27. a. $C(t) = \dfrac{k}{b-a}\left(e^{-at}-e^{-bt}\right)$;

$C'(t) = \dfrac{k}{b-a}(-ae^{-at}+be^{-bt}) = \dfrac{kb}{b-a}\left[e^{-bt}-\left(\dfrac{a}{b}\right)e^{-at}\right]$

$\qquad = \dfrac{kb}{b-a}e^{-bt}\left[1-\dfrac{a}{b}e^{(b-a)t}\right]$

$C'(t) = 0$ implies that $1 = \dfrac{a}{b}e^{(b-a)t}$ or $t = \dfrac{\ln\left(\frac{b}{a}\right)}{b-a}$.

The sign diagram of C'

$$0+ + + + + - - - - -$$

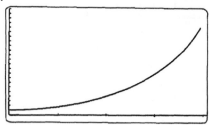

$$0 \qquad \dfrac{\ln \left(\frac{b}{a}\right)}{b-a}$$

shows that this value of t gives a minimum.

b. $\displaystyle\lim_{t\to\infty} C(t) = \dfrac{k}{b-a}$.

29. a. We solve $Q_0 e^{-kt} = \dfrac{1}{2} Q_0$ for t. Proceeding, we have

$$e^{-kt} = \dfrac{1}{2}, \ \ln e^{-kt} = \ln\dfrac{1}{2} = \ln 1 = \ln 2 = -\ln 2;$$

$$-kt = -\ln 2;$$

So $\qquad \overline{t} = \dfrac{\ln 2}{k}$

b. $\qquad \overline{t} = \dfrac{\ln 2}{0.0001238} \approx 5598.927,$ or approximately 5599 years.

USING TECHNOLOGY EXERCISES 5.6, page 390

1. a.

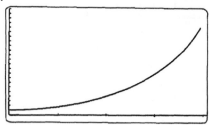

b. 12.146%/yrc. 9.474%/yr

3. a.

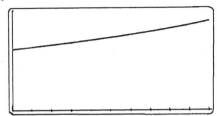

b. 666 million, 818.8 million   c. 33.8 million/yr

5. a.

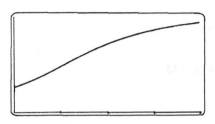

b. 86.12%/yr   c. 10.44%/yr   d. 1970

7. a.

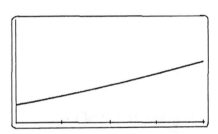

b. 325 million   c. 76.84 million/decade

9. a.

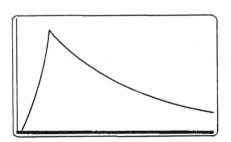

b. 0 c. 0.237 g/cm^3 d. 0.760 g/cm^3 e. 0

CHAPTER 5 CONCEPT REVIEW, page 392

1. Power; 0; 1; exponential
3. a. $(0,\infty)$; $(-\infty,\infty)$; $(1,0)$ b. < 1; >1
5. Accumulated amount; principal; nominal interest rate; number of conversion periods; term
7. Pe^{rt}
9. a. Initially; growth b. Decay c. Time; one-half

CHAPTER 5 REVIEW EXERCISES, page 393

1. a-b

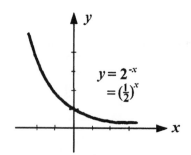

$$y = 2^{-x} = \left(\tfrac{1}{2}\right)^x$$

Since $y = \left(\dfrac{1}{2}\right)^x = \dfrac{1}{2^x} = 2^{-x}$, it has the same graph as that of $y = 2^{-x}$.

3. $16^{-3/4} = 0.125$ is equivalent to $-\dfrac{3}{4} = \log_{16} 0.125$.

5.
$$\ln (x - 1) + \ln 4 = \ln (2x + 4) - \ln 2$$
$$\ln (x - 1) - \ln (2x + 4) = -\ln 2 - \ln 4 = -(\ln 2 + \ln 4)$$
$$\ln \left(\frac{x-1}{2x+4}\right) = -\ln 8 = \ln \tfrac{1}{8} \cdot .$$
$$\left(\frac{x-1}{2x+4}\right) = \frac{1}{8}$$
$$8x - 8 = 2x + 4$$

$$6x = 12, \text{ or } x = 2.$$

CHECK: l.h.s. $\ln (2 - 1) + \ln 4 = \ln 4$

r.h.s $\ln (4 + 4) - \ln 2 = \ln 8 - \ln 2 = \ln \frac{8}{2} = \ln 4.$

7. $\ln 3.6 = \ln \frac{36}{10} = \ln 36 - \ln 10 = \ln 6^2 - \ln 2 \cdot 5 = 2 \ln 6 - \ln 2 - \ln 5$
$= 2(\ln 2 + \ln 3) - \ln 2 - \ln 5 = 2(x + y) - x - z = x + 2y - z.$

9. We first sketch the graph of $y = 2^{x-3}$. Then we take the reflection of this graph with respect to the line $y = x$.

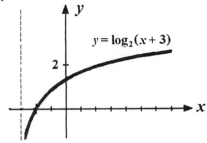

11. $f(x) = xe^{2x}; f'(x) = e^{2x} + xe^{2x}(2) = (1 + 2x)e^{2x}.$

13. $g(t) = \sqrt{t}e^{-2t}$; $g'(t) = \frac{1}{2}t^{-1/2}e^{-2t} + \sqrt{t}e^{-2t}(-2) = \frac{1 - 4t}{2\sqrt{t}e^{2t}}.$

15. $y = \dfrac{e^{2x}}{1 + e^{-2x}}$; $y' = \dfrac{(1 + e^{-2x})e^{2x}(2) - e^{2x} \cdot e^{-2x}(-2)}{(1 + e^{-2x})^2} = \dfrac{2(e^{2x} + 2)}{(1 + e^{-2x})^2}.$

17. $f(x) = xe^{-x^2}; f'(x) = e^{-x^2} + xe^{-x^2}(-2x) = (1 - 2x^2)e^{-x^2}.$

19. $f(x) = x^2e^x + e^x;$
$f'(x) = 2xe^x + x^2e^x + e^x = (x^2 + 2x + 1)e^x = (x + 1)^2e^x.$

21. $f(x) = \ln (e^{x^2} + 1); f'(x) = \dfrac{e^{x^2}(2x)}{e^{x^2} + 1} = \dfrac{2xe^{x^2}}{e^{x^2} + 1}.$

23. $f(x) = \dfrac{\ln x}{x+1}$. $f'(x) = \dfrac{(x+1)\left(\dfrac{1}{x}\right) - \ln x}{(x+1)^2} = \dfrac{1 + \dfrac{1}{x} - \ln x}{(x+1)^2} = \dfrac{x - x\ln x + 1}{x(x+1)^2}$.

25. $y = \ln(e^{4x} + 3)$; $y' = \dfrac{e^{4x}(4)}{e^{4x} + 3} = \dfrac{4e^{4x}}{e^{4x} + 3}$.

27. $f(x) = \dfrac{\ln x}{1 + e^x}$;

$$f'(x) = \dfrac{(1+e^x)\dfrac{d}{dx}\ln x - \ln x \dfrac{d}{dx}(1+e^x)}{(1+e^x)^2} = \dfrac{(1+e^x)\left(\dfrac{1}{x}\right) - (\ln x)e^x}{(1+e^x)^2}$$

$$= \dfrac{1 + e^x - xe^x \ln x}{x(1+e^x)^2} = \dfrac{1 + e^x(1 - x\ln x)}{x(1+e^x)^2}.$$

29. $y = \ln(3x + 1)$; $y' = \dfrac{3}{3x+1}$;

$$y'' = 3\dfrac{d}{dx}(3x+1)^{-1} = -3(3x+1)^{-2}(3) = -\dfrac{9}{(3x+1)^2}.$$

31. $h'(x) = g'(f(x))f'(x)$. But $g'(x) = 1 - \dfrac{1}{x^2}$ and $f'(x) = e^x$.

So $f(0) = e^0 = 1$ and $f'(0) = e^0 = 1$. Therefore,

$$h'(0) = g'(f(0))f'(0) = g'(1)f'(0) = 0 \cdot 1 = 0.$$

33. $y = (2x^3 + 1)(x^2 + 2)^3$. $\ln y = \ln(2x^3 + 1) + 3\ln(x^2 + 2)$.

$$\dfrac{y'}{y} = \dfrac{6x^2}{2x^3 + 1} + \dfrac{3(2x)}{x^2 + 2} = \dfrac{6x^2(x^2 + 2) + 6x(2x^3 + 1)}{(2x^3 + 1)(x^2 + 2)}$$

$$= \dfrac{6x^4 + 12x^2 + 12x^4 + 6x}{(2x^3 + 1)(x^2 + 2)} = \dfrac{18x^4 + 12x^2 + 6x}{(2x^3 + 1)(x^2 + 2)}.$$

Therefore, $y' = 6x(3x^3 + 2x + 1)(x^2 + 2)^2$.

35. $y = e^{-2x}$. $y' = -2e^{-2x}$ and this gives the slope of the tangent line to the graph of $y = e^{-2x}$ at any point (x, y). In particular, the slope of the tangent line at $(1, e^{-2})$ is

$y'(1) = -2e^{-2}$. The required equation is $y - e^{-2} = -2e^{-2}(x - 1)$ or $y = \dfrac{1}{e^2}(-2x + 3)$.

37. $f(x) = xe^{-2x}$.

We first gather the following information on f.
1. The domain of f is $(-\infty, \infty)$.
2. Setting $x = 0$ gives 0 as the y-intercept.
3. $\lim\limits_{x \to -\infty} xe^{-2x} = -\infty$ and $\lim\limits_{x \to \infty} xe^{-2x} = 0$.
4. The results of (3) show that $y = 0$ is a horizontal asymptote.

5. $f'(x) = e^{-2x} + xe^{-2x}(-2) = (1 - 2x)e^{-2x}$. Observe that $f'(x) = 0$ if $x = 1/2$, a critical point of f. The sign diagram of f'

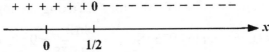

shows that f is increasing on $(-\infty, \frac{1}{2})$ and decreasing on $(\frac{1}{2}, \infty)$.

6. The results of (5) show that $(\frac{1}{2}, \frac{1}{2}e^{-1})$ is a relative maximum.

7. $f''(x) = -2e^{-2x} + (1 - 2x)e^{-2x}(-2) = 4(x - 1)e^{-2x}$ and is equal to zero if $x = 1$. The sign diagram of f''

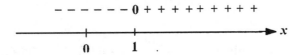

shows that the graph of f is concave downward on $(-\infty, 1)$ and concave upward on $(1, \infty)$.

The graph of f follows.

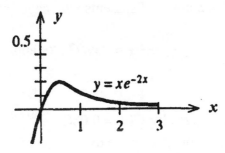

39. $f(t) = te^{-t}$. $f'(t) = e^{-t} + t(-e^{-t}) = e^{-t}(1 - t)$. Setting $f'(t) = 0$ gives $t = 1$ as the only

critical point of f. From the sign diagram of f'

$$\cdot + + + + + + + + + 0 - - - - - - - - -$$

$$\begin{array}{c} \\ \hline \quad | \quad \quad | \qquad\qquad\qquad \rightarrow t \\ 0 \qquad 1 \end{array}$$

we see that $f(1) = e^{-1} = 1/e$ is the absolute maximum value of f.

41. We want to find r where r satisfies the equation $8.2 = 4.5 \, e^{r(5)}$. We have

$$e^{5r} = \frac{8.2}{4.5} \quad \text{or} \quad r = \frac{1}{5}\ln\left(\frac{8.2}{4.5}\right) \approx 0.12$$

and so the annual rate of return is 12 percent per year.

43. We solve the equation $2 = 1(1+0.075)^t$ for t. Taking the logarithm on both sides, we have $\ln 2 = \ln(1.075)^t \approx t \ln 1.075$. So $t = \dfrac{\ln 2}{\ln 1.075} \approx 9.58$, or 9.6 years.

45. We have $Q(t) = Q_0 e^{-kt}$, where Q_0 is the amount of radium present initially. Since the half-life of radium is 1600 years, we have $\frac{1}{2}Q_0 = Q_0 e^{-1600k}$, $e^{-1600k} = \frac{1}{2}$, $-1600k = \ln\frac{1}{2} = -\ln 2$, and $k = \dfrac{\ln 2}{1600} \approx 0.0004332$.

47. We have $Q(10) = 90$ and this gives $\dfrac{3000}{1+499e^{-10k}} = 90$, $1 + 499e^{-10k} = \dfrac{3000}{90}$,

$$499e^{-10k} = \frac{2910}{90}, \quad e^{-10k} = \frac{2910}{90(499)}, \quad \text{and } k = -\frac{1}{10}\ln\frac{2910}{90(499)} \approx 0.2737.$$

So $N(t) = \dfrac{3000}{1+499e^{-0.2737t}}$. The number of students who have contracted the flu by the 20th day is $N(20) = \dfrac{3000}{1+499e^{-0.2737(20)}} \approx 969.92$, or approximately 970 students.

49. a. The concentration initially is given by $C(0) = 0.08(1-e^{-0.02(0)}) = 0$, or 0 g/cm^3.
   b. The concentration after 30 seconds is given by
      $$C(30) = 0.08(1 - e^{-0.02(30)}) = 0.03609, \text{ or } 0.0361 \ \text{g/cm}^3.$$
   c. The concentration in the long run is given by
      $$\lim_{t\to\infty} 0.08(1-e^{-0.02t}) = 0.08 \text{ or } 0.08 \ \text{g/cm}^3.$$

d.

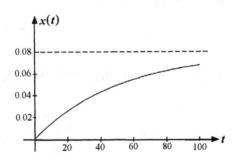

CHAPTER 5 BEFORE MOVING ON, page 394

1. $\dfrac{100}{1+2e^{0.3t}} = 40,\ 1+2e^{0.3t} = \dfrac{100}{40} = 2.5$;

$2e^{0.3t} = 1.5,\ e^{0.3t} = \dfrac{1.5}{2} = 0.75,\ 0.3t = \ln 0.75,\ t = \dfrac{\ln 0.75}{0.3} \approx -0.959$

2. $A = 3000\left(1+\dfrac{0.08}{52}\right)^{4(52)} = 4130.37$, or \$4130.37.

3. $f'(x) = \dfrac{d}{dx}e^{x^{1/2}} = e^{x^{1/2}}\dfrac{d}{dx}(x^{1/2}) = e^{x^{1/2}}(\dfrac{1}{2}x^{-1/2}) = \dfrac{e^{\sqrt{x}}}{2\sqrt{x}}$

4. $\dfrac{dy}{dx} = x\dfrac{d}{dx}\ln(x^2+1) + \ln(x^2+1)\cdot\dfrac{d}{dx}(x) = x\cdot\dfrac{2x}{x^2+1} + \ln(x^2+1) = \dfrac{2x^2}{x^2+1} + \ln(x^2+1)$

$\left.\dfrac{dy}{dx}\right|_{x=1} = \dfrac{1}{1+1} + \ln 2 = 1 + \ln 2$

5. $y' = e^{2x}\dfrac{d}{dx}\ln 3x + \ln 3x\cdot\dfrac{d}{dx}e^{2x} = \dfrac{e^{2x}}{x} + 2e^{2x}\ln 3x$

$y'' = \dfrac{d}{dx}(x^{-1}e^{2x}) + 2e^{2x}\dfrac{d}{dx}\ln 3x + (\ln 3x)\dfrac{d}{dx}(2e^{2x})$

$= -x^{-2}e^{2x} + 2x^{-1}e^{2x} + 2e^{2x}\left(\dfrac{1}{x}\right) + 4e^{2x}\cdot\ln 3x$

$= -\dfrac{1}{x^2}e^{2x} + \dfrac{4e^{2x}}{x} + 4(\ln 3x)e^{2x} = e^{2x}\left(\dfrac{4x^2\ln 3x + 4x - 1}{x^2}\right)$

6. $T(0) = 200$ gives $70 + ce^0 = 70 + C = 200,$ so $C = 130.$ So $T(t) = 70 + 130e^{-kt}.$

$T(3) = 180$ implies $70 + 130e^{-3k} = 180,$ $130e^{-3k} = 110,$ $e^{-3k} = \dfrac{110}{130},$ $-3k = \ln\dfrac{11}{13},$

$k = -\dfrac{1}{3}\ln\dfrac{11}{13} \approx 0.0557.$ Therefore, $T(t) = 70 + 130e^{-0.0557t}.$ So when $T(t) = 150,$

we have

$$70 + 130e^{-0.0557t} = 150; \quad 130e^{-0.0557t} = 80; \quad e^{-0.0557t} = \frac{80}{130} = \frac{8}{13}; -0.0557t = \ln\frac{8}{13}$$

Thus, $t = -\dfrac{\ln\dfrac{8}{13}}{0.0557} \approx 8.716,$ or approximately 8.7 minutes.

CHAPTER 6

6.1 Problem Solving Tips

1. Get into the habit of using the correct notation for integration. The indefinite integral of $f(x)$ with respect to x is written $\int x\, dx$. It is incorrect to write $\int x$ without indicating that you are integrating with respect to x. You will appreciate how important the correct notation is if you use CAS or a graphic calculator with the capability to do symbolic integration. If you don't enter this information (the variable with respect to which you are performing the integration) into your calculator or computer, the integration cannot be completed.

2. If you are finding an indefinite integral, be sure to include a constant of integration in your answer. Remember $\int f(x)\, dx$ is the family of functions given by $F(x) + C$, where $F'(x) = f(x)$.

3. It's very easy to check your answer if you are finding an indefinite integral. Just take the derivative of your answer and you should get the integrand. (You are just verifying that $F'(x) = f(x)$.) If not, you know immediately that you have made an error.

6.1 CONCEPT QUESTIONS, page 404

1. An antiderivative of a continuous function f on an interval I is a function F such that $F'(x) = f(x)$ for every x in I. For example, an antiderivative of $f(x) = x^2$ on $(-\infty, \infty)$ is the function $F(x) = \frac{1}{3}x^3$ on $(-\infty, \infty)$.

3. The indefinite integral of f is the family of functions $F(x) + C$, where F is an antiderivative of f, and C is an arbitrary constant.

EXERCISES 6.1, page 404

1. $F(x) = \frac{1}{3}x^3 + 2x^2 - x + 2$; $F'(x) = x^2 + 4x - 1 = f(x)$.

3. $F(x) = (2x^2 - 1)^{1/2}$; $F'(x) = \frac{1}{2}(2x^2 - 1)^{-1/2}(4x) = 2x(2x^2 - 1)^{-1/2} = f(x)$.

5. a. $G'(x) = \dfrac{d}{dx}(2x) = 2 = f(x)$ b. $F(x) = G(x) + C = 2x + C$

   c.

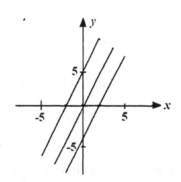

7. a. $G'(x) = \dfrac{d}{dx}(\frac{1}{3}x^3) = x^2 = f(x)$ b. $F(x) = G(x) + C = \frac{1}{3}x^3 + C$

c.

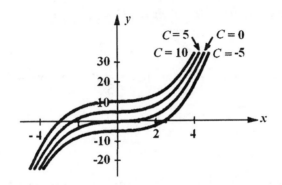

9. $\int 6\,dx = 6x + C.$

11. $\int x^3 dx = \frac{1}{4}x^4 + C$

13. $\int x^{-4} dx = -\frac{1}{3}x^{-3} + C$

15. $\int x^{2/3} dx = \frac{3}{5}x^{5/3} + C$

17. $\int x^{-5/4} dx = -4x^{-1/4} + C$

19. $\int \frac{2}{x^2}\,dx = 2\int x^{-2} dx = 2(-1x^{-1}) + C = -\frac{2}{x} + C$

21. $\int \pi\sqrt{t}\,dt = \pi\int t^{1/2} dt = \pi(\frac{2}{3}t^{3/2}) + C = \frac{2\pi}{3}t^{3/2} + C$

23. $\int (3-2x)\,dx = \int 3\,dx - 2\int x\,dx = 3x - x^2 + C$

25. $\int (x^2 + x + x^{-3})\,dx = \int x^2\,dx + \int x\,dx + \int x^{-3}\,dx = \frac{1}{3}x^3 + \frac{1}{2}x^2 - \frac{1}{2}x^{-2} + C$

27. $\int 4e^x\,dx = 4e^x + C$

29. $\int (1 + x + e^x)\,dx = x + \frac{1}{2}x^2 + e^x + C$

31. $\int (4x^3 - \frac{2}{x^2} - 1)\,dx = \int (4x^3 - 2x^{-2} - 1)\,dx = x^4 + 2x^{-1} - x + C = x^4 + \frac{2}{x} - x + C$

33. $\int (x^{5/2} + 2x^{3/2} - x)\,dx = \frac{2}{7}x^{7/2} + \frac{4}{5}x^{5/2} - \frac{1}{2}x^2 + C$

35. $\int (x^{1/2} + 3x^{-1/2})\,dx = \frac{2}{3}x^{3/2} + 6x^{1/2} + C$

37. $\int \left(\frac{u^3 + 2u^2 - u}{3u}\right)du = \frac{1}{3}\int (u^2 + 2u - 1)\,du = \frac{1}{9}u^3 + \frac{1}{3}u^2 - \frac{1}{3}u + C$

6 Integration

39. $\int (2t+1)(t-2)\,dt = \int (2t^2-3t-2)\,dt = \frac{2}{3}t^3 - \frac{3}{2}t^2 - 2t + C$

41. $\int \frac{1}{x^2}(x^4 - 2x^2 + 1)\,dx = \int (x^2 - 2 + x^{-2})\,dx = \frac{1}{3}x^3 - 2x - x^{-1} + C$

$$= \frac{1}{3}x^3 - 2x - \frac{1}{x} + C$$

43. $\int \frac{ds}{(s+1)^{-2}} = \int (s+1)^2\,ds = \int (s^2 + 2s + 1)\,ds = \frac{1}{3}s^3 + s^2 + s + C$

45. $\int (e^t + t^e)\,dt = e^t + \frac{1}{e+1}t^{e+1} + C$

47. $\int \left(\frac{x^3 + x^2 - x + 1}{x^2} \right)dx = \int \left(x + 1 - \frac{1}{x} + \frac{1}{x^2} \right)dx = \frac{1}{2}x^2 + x - \ln|x| - x^{-1} + C$

49. $\int \left(\frac{(x^{1/2} - 1)^2}{x^2} \right)dx = \int \left(\frac{x - 2x^{1/2} + 1}{x^2} \right)dx = \int (x^{-1} - 2x^{-3/2} + x^{-2})\,dx$

$$= \ln|x| + 4x^{-1/2} - x^{-1} + C = \ln|x| + \frac{4}{\sqrt{x}} - \frac{1}{x} + C$$

51. $\int f'(x)\,dx = \int (2x+1)\,dx = x^2 + x + C$. The condition $f(1) = 3$ gives $f(1) = 1 + 1 + C = 3$, or $C = 1$. Therefore, $f(x) = x^2 + x + 1$.

53. $f'(x) = 3x^2 + 4x - 1$; $f(x) = x^3 + 2x^2 - x + C$. Using the given initial condition, we have $f(2) = 8 + 2(4) - 2 + C = 9$, so $16 - 2 + C = 9$, or $C = -5$. Therefore, $f(x) = x^3 + 2x^2 - x - 5$.

55. $f(x) = \int f'(x)\,dx = \int \left(1 + \frac{1}{x^2} \right)dx = \int (1 + x^{-2})\,dx = x - \frac{1}{x} + C$.

Using the given initial condition, we have $f(1) = 1 - 1 + C = 2$, or $C = 2$.

Therefore, $f(x) = x - \dfrac{1}{x} + 2$.

57. $f(x) = \displaystyle\int \dfrac{x+1}{x}\,dx = \int \left(1 + \dfrac{1}{x}\right)dx = x + \ln|x| + C$. Using the initial condition, we

have $f(1) = 1 + \ln 1 + C = 1 + C = 1$, or $C = 0$. So $f(x) = x + \ln|x|$.

59. $f(x) = \displaystyle\int f'(x)\,dx = \int \tfrac{1}{2}x^{-1/2}\,dx = \tfrac{1}{2}(2x^{1/2}) + C = x^{1/2} + C$; $f(2) = \sqrt{2} + C = \sqrt{2}$

implies $C = 0$. So $f(x) = \sqrt{x}$.

61. $f'(x) = e^x + x$; $f(x) = e^x + \tfrac{1}{2}x^2 + C$; $f(0) = e^0 + \tfrac{1}{2}(0) + C = 1 + C$

So $3 = 1 + C$ or $2 = C$. Therefore, $f(x) = e^x + \tfrac{1}{2}x^2 + 2$.

63. The net amount on deposit in Branch A is given by the area under the graph of f from $t = 0$ to $t = 180$. On the other hand, the net amount on deposit in Branch B is given by the area under the graph of g over the same interval. Evidently the first area is larger than the second. Therefore, we see that Branch A has the larger net deposit.

65. The position of the car is

$s(t) = \displaystyle\int f(t)\,dt = \int 2\sqrt{t}\,dt = \int 2t^{1/2}\,dt = 2(\tfrac{2}{3}t^{3/2}) + C = \tfrac{4}{3}t^{3/2} + C$.

$s(0) = 0$ implies $s(0) = C = 0$. So $s(t) = \tfrac{4}{3}t^{3/2}$.

67. $C(x) = \displaystyle\int C'(x)\,dx = \int (0.000009x^2 - 0.009x + 8)\,dx$

$= 0.000003x^3 - 0.0045x^2 + 8x + k$.

$C(0) = k = 120$ and so $C(x) = 0.000003x^3 - 0.0045x^2 + 8x + 120$.

$C(500) = 0.000003(500)^3 - 0.0045(500)^2 + 8(500) + 120$, or $\$3370$.

69. $P'(x) = -0.004x + 20$, $P(x) = -0.002x^2 + 20x + C$. Since $C = -16{,}000$, we find that $P(x) = -0.002x^2 + 20x - 16{,}000$. The company realizes a maximum profit when $P'(x) = 0$, that is, when $x = 5000$ units. Next,

$P(5000) = -0.002(5000)^2 + 20(5000) - 16{,}000 = 34{,}000$.

6   *Integration*

Thus, a maximum profit of $34,000 is realized at a production level of 5000 units.

71. a. $N(t) = \int N'(t)\,dt = \int (-3t^2 + 12t + 45)\,dt = -t^3 + 6t^2 + 45t + C$. But $N(0) = C = 0$
    and so $N(t) = -t^3 + 6t^2 + 45t$.
    b. The number is $N(4) = -4^3 + 6(4)^2 + 45(4) = 212$.

73. a. The number of subscribers in year t is given by
    $$N(t) = \int r(t)\,dt = \int (-0.375t^2 + 2.1t + 2.45)\,dt$$
    $$= -0.125t^3 + 1.05t^2 + 2.45t + C$$
    To find C, note that $N(0) = 1.5$. This gives
    $$N(0) = C = 1.5.$$
    Therefore,
    $$N(t) = -0.125t^3 + 1.05t^2 + 2.45t + 1.5$$
    b. $N(5) = -0.125(5^3) + 1.05(5^2) + 2.45(5) + 1.5 = 24.375$, or 24.375 million subscribers.

75. a. The approximate average credit card debt per U.S. household in year t is
    $$A(t) = \int D(t)\,dt = \int (-4.479t^2 + 69.8t + 279.5)\,dt$$
    $$= -1.493t^3 + 34.9t^2 + 279.5t + C$$
    Using the condition $A(0) = 2917$, we find
    $$A(0) = C = 2917. \text{ Therefore,}$$
    $$A(t) = -1.493t^3 + 34.9t^2 + 279.5t + 2917$$
    b. The average credit card debt per U.S. household in 2003 was
    $$A(13) = -1.493(13^3) + 34.9(13^2) + 279.5(13) + 2917 \approx 9168.479,$$
    or approximately $9168.

77. a. The number of gastric bypass surgeries performed in year t is
    $$N(t) = \int R(t)\,dt = \int (9.399t^2 - 13.4t + 14.07)\,dt$$
    $$= 3.133t^3 - 6.7t^2 + 14.07t + C$$
    Using the condition $N(0) = 36.7$, we find
    $$N(0) = C = 36.7. \text{ Therefore, } N(t) = 3.133t^3 - 6.7t^2 + 14.07t + 36.7.$$
    b. The number of bypass surgeries performed in 2003 was

$$N(3) = 3.133(3^3) - 6.7(3^2) + 14.07(3) + 36.7 = 103,201,$$

or approximately 103,201.

79. a. We have the initial-value problem:
$$C'(t) = 12.288t^2 - 150.5594t + 695.23$$
$$C(0) = 3142$$

Integrating, we find
$$C(t) = \int C'(t)dt = \int (12.288t^2 - 150.5594t + 695.23)dt$$
$$= 4.096t^3 - 75.2797t^2 + 695.23t + k$$

Using the initial condition, we find
$$C(0) = 0 + k = 3142, \quad \text{and so } k = 3142.$$

Therefore, $C(t) = 4.096t^3 - 75.2797t^2 + 695.23t + 3142.$

b. The projected average out-of-pocket costs for beneficiaries is 2010 is
$$C(2) = 4.096(2^3) - 75.2797(2^2) + 695.23(2) + 3142 = 4264.1092$$

or \$4264.11.

81. The number of new subscribers at any time is
$$N(t) = \int (100 + 210t^{3/4})dt = 100t + 120t^{7/4} + C.$$

The given condition implies that $N(0) = 5000$. Using this condition, we find $C = 5000$. Therefore, $N(t) = 100t + 120t^{7/4} + 5000$. The number of subscribers 16 months from now is
$$N(16) = 100(16) + 120(16)^{7/4} + 5000, \quad \text{or } 21,960.$$

83. $A(t) = \int A'(t)dt = \int (3.2922t^2 - 0.366t^3)dt = 1.0974t^3 - 0.0915t^4 + C$

Now, $A(0) = C = 0$. So, $A(t) = 1.0974t^3 - 0.0915t^4.$

85. $h(t) = \int h'(t)dt = \int (-3t^2 + 192t + 120)dt = -t^3 + 96t^2 + 120t + C$
$$= -t^3 + 96t^2 + 120t + C.$$
$h(0) = C = 0$ implies $h(t) = -t^3 + 96t^2 + 120t.$
The altitude 30seconds after lift-off is
$$h(30) = -30^3 + 96(30)^2 + 120(30) = 63,000 \text{ ft.}$$

87. a. The number of health-care agencies in year t is
$$N(t) = \int -0.186te^{-0.02t}\, dt = 9.3e^{-0.02t} + C$$
Using the condition $N(0) = 9.3$, we find
$$N(0) = 9.3 + C = 9.3 \quad \text{or} \quad C = 0.$$
Therefore,
$$N(t) = 9.3e^{-0.02t}$$
b. The number of health-care agencies in 2002 was
$$N(14) = 9.3e^{-0.02(14)} \approx 7.03, \text{ or } 7.03 \text{ thousand units.}$$
c. The number of health-care agencies in 2005 would be
$$N(17) = 9.3e^{-0.02(17)} \approx 6.62, \text{ or } 6.62 \text{ thousand units.}$$

89. $v(r) = \int v'(r)\, dr = \int -kr\, dr = -\frac{1}{2}kr^2 + C.$

But $v(R) = 0$ and so $v(R) = -\frac{1}{2}kR^2 + C = 0$, or $C = \frac{1}{2}kR^2$. Therefore,
$$v(R) = -\frac{1}{2}kr^2 + \frac{1}{2}kR^2 = \frac{1}{2}k(R^2 - r^2).$$

91. Denote the constant deceleration by k (ft/sec^2). Then $f''(t) = -k$, so $f'(t) = v(t) = -kt + C_1$. Next, the given condition implies that $v(0) = 88$. This gives $C_1 = 88$, or $f'(t) = -kt + 88$.
$$s = f(t) = \int f'(t)\, dt = \int (-kt + 88)\, dt = -\frac{1}{2}kt^2 + 88t + C_2.$$

Also, $f(0) = 0$ gives $s = f(t) = -\frac{1}{2}kt^2 + 88t$. Since the car is brought to rest in 9 seconds, we have $v(9) = -9k + 88 = 0$, or $k = \frac{88}{9}$, or $9\frac{7}{9}$. So the deceleration is $9\frac{7}{9}$ ft/sec^2. The distance covered is
$$s = f(9) = -\frac{1}{2}\left(\frac{88}{9}\right)(81) + 88(9) = 396.$$
So the stopping distance is 396 ft.

93. The time taken by runner A to cross the finish line is
$$t = \frac{200}{22} = \frac{100}{11} \text{ sec}$$
Let a be the constant acceleration of runner B as he begins to spurt. Then
$$\frac{dv}{dt} = a \text{ or } v = \int a\, dt = at + c,$$
the velocity of runner B as he runs towards the finish line. At $t = 0$, $v = 20$ and so

$v = at = 20$. Now, $\dfrac{ds}{dt} = v = at + 20$, so $s = \displaystyle\int (at + 20)dt = \tfrac{1}{2}at^2 + 20t + k$

(k, constant of integration) . Next, $s(0) = 0$ gives $s = \tfrac{1}{2}at^2 + 20t = (\tfrac{1}{2}at + 20)t$. In order for runner B to cover 220 ft in $\tfrac{100}{11}$ sec, we must have

$$[\tfrac{1}{2}a(\tfrac{100}{11}) + 20](\tfrac{100}{11}) = 220$$

$\tfrac{50}{11}a + 20 = \tfrac{(220)(11)}{100} = \tfrac{121}{5}$; $\tfrac{50}{11}a = \tfrac{121}{5} - 20 = \tfrac{21}{5}$ or $a = \tfrac{21}{5} \cdot \tfrac{11}{50} = 0.924 \ (\text{ft}/\sec^2)$

So B must have a minimum acceleration of 0.924 ft/sec^2 .

95. a. We have the initial-value problem $R'(t) = \dfrac{8}{(t+4)^2}$ and $R(0) = 0$.

Integrating, we find $R(t) = \displaystyle\int \dfrac{8}{(t+4)^2}\, dt = 8\int (t+4)^{-2}\, dt = -\dfrac{8}{t+4} + C$

$R(0) = 0$ implies $-\dfrac{8}{4} + C = 0$ or $C = 2$.

Therefore, $R(t) = -\dfrac{8}{t+4} + 2 = \dfrac{-8 + 2t + 8}{t+4} = \dfrac{2t}{t+4}$.

b. After 1 hr, $R(1) = \dfrac{2}{5} = 0.4$, or 0.0.4" had fallen. After 2 hr, $R(2) = \dfrac{4}{6} = \dfrac{2}{3}$, or

$\dfrac{2}{3}$" had fallen.

97. True. See proof in Section 6.1 in the text.

99. True. Use the Sum Rule followed by the Constant Multiple Rule.

6.2 Problem Solving Tips

1. Here are some hints for using the method of substitution.

   a. The idea here is to replace the given integral by a simpler integral. So look for a

   substitution $u = g(x)$ that simplifies the integral.

b. Check to see that $du = g'(x)\,dx$ appears in the integral.

2. Look through Problems 1-50, so that you familiarize yourself with the types of functions that can be integrated using the method of substitution. Even if you don't complete every problem, check to see if you can set up the given integral so that the method of substitution can be used to complete the integration.

6.2 CONCEPT QUESTIONS, page 417

1. To find $I = \int f(g(x))g'(x)\,dx$ by the Method of Substitution, let $u = g(x)$, so that $du = g'(x)dx$. Making the substitution, we obtain $I = \int f(u)\,du$, which can be integrated with respect to u. Finally, replace u by $u = g(x)$ to obtain the integral.

EXERCISES 6.2, page 417

1. Put $u = 4x + 3$ so that $du = 4\,dx$, or $dx = \frac{1}{4}du$. Then
$$\int 4(4x+3)^4\,dx = \int u^4\,du = \tfrac{1}{5}u^5 + C = \tfrac{1}{5}(4x+3)^5 + C.$$

3. Let $u = x^3 - 2x$ so that $du = (3x^2 - 2)\,dx$. Then
$$\int (x^3 - 2x)^2(3x^2 - 2)\,dx = \int u^2\,du = \tfrac{1}{3}u^3 + C = \tfrac{1}{3}(x^3 - 2x)^3 + C.$$

5. Let $u = 2x^2 + 3$ so that $du = 4x\,dx$. Then
$$\int \frac{4x}{(2x^2+3)^3}\,dx = \int \frac{1}{u^3}\,du = \int u^{-3}\,du = -\tfrac{1}{2}u^{-2} + C = -\frac{1}{2(2x^2+3)^2} + C.$$

7. Put $u = t^3 + 2$ so that $du = 3t^2\,dt$ or $t^2\,dt = \frac{1}{3}du$. Then
$$\int 3t^2\sqrt{t^3 + 2}\,dt = \int u^{1/2}\,du = \tfrac{2}{3}u^{3/2} + C = \tfrac{2}{3}(t^3 + 2)^{3/2} + C$$

9. Let $u = x^2 - 1$ so that $du = 2x\,dx$ and $x\,dx = \frac{1}{2}du$. Then,

$$\int (x^2 - 1)^9 \, x\,dx = \int \tfrac{1}{2}u^9 \, du = \tfrac{1}{20}u^{10} + C = \tfrac{1}{20}(x^2 - 1)^{10} + C.$$

11. Let $u = 1 - x^5$ so that $du = -5x^4\,dx$ or $x^4\,dx = -\frac{1}{5}du$. Then

$$\int \frac{x^4}{1-x^5}\,dx = -\frac{1}{5}\int \frac{du}{u} = -\frac{1}{5}\ln|u| + C = -\frac{1}{5}\ln\left|1 - x^5\right| + C.$$

13. Let $u = x - 2$ so that $du = dx$. Then

$$\int \frac{2}{x-2}\,dx = 2\int \frac{du}{u} = 2\ln|u| + C = \ln u^2 + C = \ln(x-2)^2 + C.$$

15. Let $u = 0.3x^2 - 0.4x + 2$. Then $du = (0.6x - 0.4)\,dx = 2(0.3x - 0.2)\,dx$.

$$\int \frac{0.3x - 0.2}{0.3x^2 - 0.4x + 2}\,dx = \int \frac{1}{2u}\,du = \frac{1}{2}\ln|u| + C = \frac{1}{2}\ln(0.3x^2 - 0.4x + 2) + C.$$

17. Let $u = 3x^2 - 1$ so that $du = 6x\,dx$, or $x\,dx = \frac{1}{6}du$. Then

$$\int \frac{x}{3x^2 - 1}\,dx = \frac{1}{6}\int \frac{du}{u} = \frac{1}{6}\ln|u| + C = \frac{1}{6}\ln\left|3x^2 - 1\right| + C.$$

19. Let $u = -2x$ so that $du = -2\,dx$ or $dx = -\frac{1}{2}du$. Then

$$\int e^{-2x}\,dx = -\frac{1}{2}\int e^u\,du = -\frac{1}{2}e^u + C = -\frac{1}{2}e^{-2x} + C.$$

21. Let $u = 2 - x$ so that $du = -\,dx$ or $dx = -\,du$. Then

$$\int e^{2-x}\,dx = -\int e^u\,du = -e^u + C = -e^{2-x} + C.$$

23. Let $u = -x^2$, then $du = -2x\,dx$ or $x\,dx = -\frac{1}{2}du$.

$$\int xe^{-x^2}\,dx = \int -\tfrac{1}{2}e^u\,du = -\tfrac{1}{2}e^u + C = -\tfrac{1}{2}e^{-x^2} + C.$$

25. $\displaystyle \int (e^x - e^{-x})\,dx = \int e^x\,dx - \int e^{-x}\,dx = e^x - \int e^{-x}\,dx.$

To evaluate the second integral on the right, let $u = -x$ so that $du = -dx$ or $dx = -du.$ Therefore,

$$\int (e^x - e^{-x})\,dx = e^x + \int e^u\,du = e^x + e^u + C = e^x + e^{-x} + C.$$

27. Let $u = 1 + e^x$ so that $du = e^x\,dx.$ Then

$$\int \frac{e^x}{1+e^x}\,dx = \int \frac{du}{u} = \ln|u| + C = \ln(1+e^x) + C.$$

29. Let $u = \sqrt{x} = x^{1/2}.$ Then $du = \tfrac{1}{2}x^{-1/2}\,dx$ or $2\,du = x^{-1/2}\,dx.$

$$\int \frac{e^{\sqrt{x}}}{\sqrt{x}}\,dx = \int 2e^u\,du = 2e^u + C = 2e^{\sqrt{x}} + C.$$

31. Let $u = e^{3x} + x^3$ so that $du = (3e^{3x} + 3x^2)\,dx = 3(e^{3x} + x^2)\,dx$ or $(e^{3x} + x^2)\,dx = \tfrac{1}{3}\,du.$ Then

$$\int \frac{e^{3x} + x^2}{(e^{3x}+x^3)^3}\,dx = \frac{1}{3}\int \frac{du}{u^3} = \frac{1}{3}\int u^{-3}\,du = -\frac{1}{6}u^{-2} + C = -\frac{1}{6(e^{3x}+x^3)^2} + C.$$

33. Let $u = e^{2x} + 1,$ so that $du = 2e^{2x}\,dx$, or $\tfrac{1}{2}\,du = e^{2x}\,dx.$

$$\int e^{2x}(e^{2x} + 1)^3\,dx = \int \tfrac{1}{2}u^3\,du = \tfrac{1}{8}u^4 + C = \tfrac{1}{8}(e^{2x} + 1)^4 + C.$$

35. Let $u = \ln 5x$ so that $du = \dfrac{1}{x}\,dx.$ Then

$$\int \frac{\ln 5x}{x}\,dx = \int u\,du = \tfrac{1}{2}u^2 + C = \tfrac{1}{2}(\ln 5x)^2 + C.$$

37. Let $u = \ln x$ so that $du = \tfrac{1}{x}\,dx.$ Then

$$\int \frac{1}{x\ln x}\,dx = \int \frac{du}{u} = \ln|u| + C = \ln|\ln x| + C.$$

39. Let $u = \ln x$ so that $du = \frac{1}{x}\,dx$. Then

$$\int \frac{\sqrt{\ln x}}{x}\,dx = \int \sqrt{u}\,du = \frac{2}{3}u^{3/2} + C = \frac{2}{3}(\ln x)^{3/2} + C.$$

41. $\displaystyle \int \left(xe^{x^2} - \frac{x}{x^2+2} \right)dx = \int xe^{x^2}\,dx - \int \frac{x}{x^2+2}\,dx.$

To evaluate the first integral, let $u = x^2$ so that $du = 2x\,dx$, or $x\,dx = \frac{1}{2}\,du$. Then

$$\int xe^{x^2}\,dx = \frac{1}{2}\int e^u\,du + C_1 = \frac{1}{2}e^u + C_1 = \frac{1}{2}e^{x^2} + C_1.$$

To evaluate the second integral, let $u = x^2 + 2$ so that $du = 2x\,dx$, or $x\,dx = \frac{1}{2}\,du$. Then

$$\int \frac{x}{x^2+2}\,dx = \frac{1}{2}\int \frac{du}{u} = \frac{1}{2}\ln|u| + C_2 = \frac{1}{2}\ln(x^2+2) + C_2.$$

Therefore, $\displaystyle \int \left(xe^{x^2} - \frac{x}{x^2+2} \right)dx = \frac{1}{2}e^{x^2} - \frac{1}{2}\ln(x^2+2) + C.$

43. Let $u = \sqrt{x} - 1$ so that $du = \frac{1}{2}x^{-1/2}\,dx = \frac{1}{2\sqrt{x}}\,dx$ or $dx = 2\sqrt{x}\,du$.

Also, we have $\sqrt{x} = u + 1$, so that $x = (u+1)^2 = u^2 + 2u + 1$ and $dx = 2(u+1)\,du$. So

$$\int \frac{x+1}{\sqrt{x}-1}\,dx = \int \frac{u^2+2u+2}{u}\cdot 2(u+1)\,du = 2\int \frac{(u^3+3u^2+4u+2)}{u}\,du$$

$$= 2\int \left(u^2 + 3u + 4 + \frac{2}{u} \right)du = 2\left(\frac{1}{3}u^3 + \frac{3}{2}u^2 + 4u + 2\ln|u| \right) + C$$

$$= 2\left[\frac{1}{3}(\sqrt{x}-1)^3 + \frac{3}{2}(\sqrt{x}-1)^2 + 4(\sqrt{x}-1) + 2\ln|\sqrt{x}-1| \right] + C.$$

45. Let $u = x - 1$ so that $du = dx$. Also, $x = u + 1$ and so

$$\int x(x-1)^5\, dx = \int (u+1)u^5\, du = \int (u^6 + u^5)\, du$$

$$= \frac{1}{7}u^7 + \frac{1}{6}u^6 + C = \frac{1}{7}(x-1)^7 + \frac{1}{6}(x-1)^6 + C$$

$$= \frac{(6x+1)(x-1)^6}{42} + C.$$

47. Let $u = 1 + \sqrt{x}$ so that $du = \frac{1}{2}x^{-1/2}\, dx$ and $dx = 2\sqrt{x} = 2(u-1)\, du$

$$\int \frac{1-\sqrt{x}}{1+\sqrt{x}}\, dx = \int \left(\frac{1-(u-1)}{u}\right) \cdot 2(u-1)\, du = 2\int \frac{(2-u)(u-1)}{u}\, du$$

$$= 2\int \frac{-u^2 + 3u - 2}{u}\, du = 2\int \left(-u + 3 - \frac{2}{u}\right)\, du = -u^2 + 6u - 4\ln|u| + C$$

$$= -(1+\sqrt{x})^2 + 6(1+\sqrt{x}) - 4\ln(1+\sqrt{x}) + C$$

$$= -1 - 2\sqrt{x} - x + 6 + 6\sqrt{x} - 4\ln(1+\sqrt{x}) + C$$

$$= -x + 4\sqrt{x} + 5 - 4\ln(1+\sqrt{x}) + C.$$

49. $I = \int v^2(1-v)^6\, dv$. Let $u = 1 - v$, then $du = -dv$. Also, $1 - u = v$, and $(1-u)^2 = v^2$. Therefore,

$$I = \int -(1 - 2u + u^2)u^6\, du = \int -(u^6 - 2u^7 + u^8)\, du = -\left(\frac{u^7}{7} - \frac{2u^8}{8} + \frac{u^9}{9}\right) + C$$

$$= -u^7\left(\frac{1}{7} - \frac{1}{4}u + \frac{1}{9}u^2\right) + C = -\frac{1}{252}(1-v)^7[36 - 63(1-v) + 28(1 - 2v + v^2)]$$

$$= -\frac{1}{252}(1-v)^7[36 - 63 + 63v + 28 - 56v + 28v^2]$$

$$= -\frac{1}{252}(1-v)^7(28v^2 + 7v + 1) + C.$$

51. $f(x) = \int f'(x)\, dx = 5\int (2x-1)^4\, dx$. Let $u = 2x - 1$ so that $du = 2x - 1$ so that $du = 2\, dx$, or $dx = \frac{1}{2}\, du$. Then

$$f(x) = \frac{5}{2} \int u^4 \, du = \frac{1}{2} u^5 + C = \frac{1}{2}(2x-1)^5 + C.$$

Next, $f(1) = 3$ implies $\frac{1}{2} + C = 3$ or $C = \frac{5}{2}$. Therefore,

$$f(x) = \frac{1}{2}(2x-1)^5 + \frac{5}{2}.$$

53. $f(x) = \int -2xe^{-x^2+1} \, dx$. Let $u = -x^2 + 1$ so that $du = -2x \, dx$. Then

$f(x) = \int e^u \, du = e^u + C = e^{-x^2+1} + C$. The condition $f(1) = 0$ implies

$f(1) = 1 + C = 0$, or $C = -1$. Therefore, $f(x) = e^{-x^2+1} - 1$.

55. The number of subscribers at time t is
$$N(t) = \int R(t) \, dt = \int 3.36(t+1)^{0.05} \, dt$$
Let $u = t+1$, so that $du = dt$. Thus
$$N = 3.36 \int u^{0.05} \, du = 3.2u^{1.05} + C = 3.2(t+1)^{1.05} + C$$
To find C, use the condition $N(0) = 3.2$ giving
$$N(0) = 3.2 + C = 3.2,$$
so $C = 0$. Therefore,
$$N(t) = 3.2(t+1)^{1.05}$$
If the projection holds true, then the number of subscribers at the beginning of 2008 will be $N(4) = 3.2(4+1)^{1.05} \approx 17.341$, or 17.341 million.

57. $N'(t) = 2000(1+0.2t)^{-3/2}$. Let $u = 1 + 0.2t$. Then $du = 0.2 \, dt$ and $5 \, du = dt$.
Therefore, $N(t) = (5)(2000)$
$$\int u^{-3/2} \, du = -20{,}000u^{-1/2} + C = -20{,}000(1+0.2t)^{-1/2} + C.$$
Next, $N(0) = -20{,}000(1)^{-1/2} + C = 1000$. Therefore, $C = 21{,}000$ and
$$N(t) = -\frac{20{,}000}{\sqrt{1+0.2t}} + 21{,}000. \text{ In particular, } N(5) = -\frac{20{,}000}{\sqrt{2}} + 21{,}000 \approx 6{,}858.$$

6   *Integration*

59. $p(x) = \int -\dfrac{250x}{(16+x^2)^{3/2}}\,dx = -250\int \dfrac{x}{(16+x^2)^{3/2}}\,dx.$

Let $u = 16 + x^2$ so that $du = 2x\,dx$ and $x\,dx = \tfrac{1}{2}\,du.$

Then $p(x) = -\dfrac{250}{2}\int u^{-3/2}\,du = (-125)(-2)u^{-1/2} + C = \dfrac{250}{\sqrt{16+x^2}} + C.$

$p(3) = \dfrac{250}{\sqrt{16+9}} + C = 50$ implies $C = 0$ and $p(x) = \dfrac{250}{\sqrt{16+x^2}}.$

61. Let $u = 2t + 4$, so that $du = 2\,dt.$ Then

$r(t) = \int \dfrac{30}{\sqrt{2t+4}}\,dt = 30\int \dfrac{1}{2}u^{-1/2}\,du = 30u^{1/2} + C = 30\sqrt{2t+4} + C.$

$r(0) = 60 + C = 0$, and $C = -60.$ Therefore, $r(t) = 30\left(\sqrt{2t+4} - 2\right).$ Then

$r(16) = 30\left(\sqrt{36} - 2\right) = 120$ ft. Therefore, the polluted area is

$\pi r^2 = \pi(120)^2 = 14{,}400\pi$, or $14{,}400\pi$ sq ft.

63. Let $u = 1 + 2.449e^{-0.3277t}$ so that $du = -0.802537e^{-0.3277t}\,dt$ and $e^{-0.3277t}\,dt = -1.24605\,du.$

Then $h(t) = \int \dfrac{52.8706e^{-0.3277t}}{(1+2.449e^{-0.3277t})^2}\,dt = (52.8706)(-1.24605)\int \dfrac{du}{u^2}$

$= 65.8794u^{-1} + C = \dfrac{65.8794}{1+2.449e^{-0.3277t}} + C.$

$h(0) = \dfrac{65.8794}{1+2.449} + C = 19.4,$ and $C = 0.3.$

Therefore, $h(t) = \dfrac{65.8794}{1+2.449e^{-0.3277t}} + 0.3,$

and $h(8) = \dfrac{65.8794}{1+2.449e^{-0.3277(8)}} + 0.3 \approx 56.22$, or 56.22 inches.

65. $A(t) = \int A'(t)\,dt = r\int e^{-at}\,dt.$ Let $u = -at$ so that $du = -a\,dt,$ or $dt = -\tfrac{1}{a}\,du.$

$A(t) = r\left(-\tfrac{1}{a}\right)\int e^u\,du = -\tfrac{r}{a}e^u + C = -\tfrac{r}{a}e^{-at} + C$

$A(0)$ implies $-\dfrac{r}{a} + C = 0,$ or $C = \dfrac{r}{a}.$ So, $A(t) = -\dfrac{r}{a}e^{-at} + \dfrac{r}{a} = \dfrac{r}{a}(1 - e^{-at}).$

6.3 Problem Solving Tips

1. In Sections 6.1 and 6.2, we found the indefinite integral of a function , that is,

$\displaystyle\int f(x)\,dx = F(x) + C$. Note that our answer ($F(x) + C$) was a *family of functions*

where $F'(x) = f(x)$. In this section, we found the definite integral of a function,

$\displaystyle\int_a^b f(x)\,dx = \lim_{n \to \infty}[f(x_1)\Delta x + f(x_2)\Delta x + \cdots + f(x_n)\Delta x])$. Note that the answer here is a

number.

2. The geometric interpretation of a definite integral follows: If f is continuous on $[a,b]$,

then $\displaystyle\int_a^b f(x)\,dx$ is equal to the area of the region above $[a, b]$ minus the area of the

region below $[a,b]$.

6.3 CONCEPT QUESTIONS, page 427

1. See text page 424.

EXERCISES 6.3, page 427

1. $\frac{1}{3}(1.9 + 1.5 + 1.8 + 2.4 + 2.7 + 2.5) = \frac{12.8}{3} \approx 4.27.$

3. a. $A = \frac{1}{2}(2)(6) = 6$ sq units.

   b. $\Delta x = \frac{2}{4} = \frac{1}{2}$; $x_1 = 0$, $x_2 = \frac{1}{2}$, $x_3 = 1$, $x_4 = \frac{3}{2}$.

   $$A \approx \frac{1}{2}[3(0) + 3(\tfrac{1}{2}) + 3(1) + 3(\tfrac{3}{2})] = \frac{9}{2}$$
   $$= 4.5 \text{ sq units.}$$

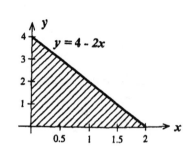

   c. $\Delta x = \frac{2}{8} = \frac{1}{4}$. $x_1 = 0, \dots, x_8 = \frac{7}{4}$.

   $$A \approx \frac{1}{4}\left[3(0) + 3(\tfrac{1}{4}) + 3(\tfrac{1}{2}) + 3(\tfrac{3}{4}) + 3(1) + 3(\tfrac{5}{4}) + 3(\tfrac{3}{2}) + 3(\tfrac{7}{4})\right]$$
   $$= \frac{21}{4} = 5.25 \text{ sq units.}$$

   d. Yes.

5. a. $A = 4$

   b. $\Delta x = \frac{2}{5} = 0.4$; $x_1 = 0$, $x_2 = 0.4$, $x_3 = 0.8$, $x_4 = 1.2$

   $x_5 = 1.6$,
   $$A \approx 0.4\{[4 - 2(0)] + [4 - 2(0.4)] + [4 - 2(0.8)]$$
   $$+ [4 - 2(1.2)] + [4 - 2(1.6)]\}$$
   $$= 4.8$$

   c. $\Delta x = \frac{2}{10} = 0.2$, $x_1 = 0$, $x_2 = 0.2$, $x_3 = 0.4$, ..., $x_{10} = 1.8$.
   $$A \approx 0.2\{[4 - 2(0)] + [4 - 2(0.2)] + [4 - 2(0.4)]$$
   $$+ \cdots + [4 - 2(1.8)]\} = 4.4$$

   d. Yes.

7. a. $\Delta x = \dfrac{4-2}{2} = 1$; $x_1 = 2.5$, $x_2 = 3.5$;   The Riemann sum is $[(2.5)^2 + (3.5)^2] = 18.5$.

   b. $\Delta x = \dfrac{4-2}{5} = 0.4$; $x_1 = 2.2$, $x_2 = 2.6$, $x_3 = 3.0$, $x_4 = 3.4$, $x_5 = 3.8$.

   The Riemann sum is $0.4[2.2^2 + 2.6^2 + 3.0^2 + 3.4^2 + 3.8^2] = 18.64$.

   c. $\Delta x = \dfrac{4-2}{10} = 0.2$; $x_1 = 2.1$, $x_2 = 2.3$, $x_2 = 2.5$, ..., $x_{10} = 3.9$

   The Riemann sum is $0.2[2.1^2 + 2.3^2 + 2.5^2 + \cdots + 3.9^2] = 18.66$.
   The area seems to be $18\frac{2}{3}$ sq units.

9. a. $\Delta x = \dfrac{4-2}{2} = 1$; $x_1 = 3$, $x_2 = 4$. The Riemann sum is $(1)[3^2 + 4^2] = 25$.

   b. $\Delta x = \dfrac{4-2}{5} = 0.4$; $x_1 = 2.4$, $x_2 = 2.8$, $x_3 = 3.2$, $x_4 = 3.6$, $x_5 = 4$.

   The Riemann sum is $0.4[2.4^2 + 2.8^2 + \cdots + 4^2] = 21.12$.

   c. $\Delta x = \dfrac{4-2}{10} = 0.2$; $x_1 = 2.2$, $x_2 = 2.4$, $x_3 = 2.6$, ..., $x_{10} = 4$.

   The Riemann sum is $0.2[2.2^2 + 2.4^2 + 2.6^2 + \cdots + 4^2] = 19.88$.

   d. 19.9 sq units.

11. a. $\Delta x = \dfrac{1}{2}$, $x_1 = 0$, $x_2 = \dfrac{1}{2}$. The Riemann sum is

   $f(x_1)\Delta x + f(x_2)\Delta x = \left[(0)^3 + (\tfrac{1}{2})^3\right]\tfrac{1}{2} = \tfrac{1}{16} = 0.0625$.

   b. $\Delta x = \dfrac{1}{5}$, $x_1 = 0$, $x_2 = \dfrac{1}{5}$, $x_3 = \dfrac{2}{5}$, $x_4 = \dfrac{3}{5}$, $x_5 = \dfrac{4}{5}$. The Riemann sum

   is $f(x_1)\Delta x + f(x_2)\Delta x + \cdots + f(x_5)\Delta x = \left[(\tfrac{1}{5})^3 + (\tfrac{2}{5})^3 + \cdots + (\tfrac{4}{5})^3\right]\tfrac{1}{5} = \tfrac{100}{625} = 0.16$.

   c. $\Delta x = \dfrac{1}{10}$; $x_1 = 0$, $x_2 = \dfrac{1}{10}$, $x_3 = \dfrac{2}{10}$, ..., $x_{10} = \dfrac{9}{10}$.

   The Riemann sum is

   $f(x_1)\Delta x + f(x_2)\Delta x + \cdots + f(x_{10})\Delta x = \left[(\tfrac{1}{10})^3 + (\tfrac{2}{10})^3 + \cdots + (\tfrac{9}{10})^3\right]\tfrac{1}{10}$

   $\qquad\qquad\qquad = \tfrac{2025}{10,000} = 0.2025 \approx 0.2$ sq units.

   The Riemann sum seems to approach 0.2.

13. $\Delta x = \dfrac{2-0}{5} = \dfrac{2}{5}$; $x_1 = \dfrac{1}{5}$, $x_2 = \dfrac{3}{5}$, $x_3 = \dfrac{5}{5}$, $x_4 = \dfrac{7}{5}$, $x_5 = \dfrac{9}{5}$.

   $A \approx \left\{\left[(\tfrac{1}{5})^2 + 1\right] + \left[(\tfrac{3}{5})^2 + 1\right] + \left[(\tfrac{5}{5})^2 + 1\right] + \left[(\tfrac{7}{5})^2 + 1\right] + \left[(\tfrac{9}{5})^2 + 1\right](\tfrac{2}{5})\right\}$

   $= \tfrac{580}{125} = 4.64$ sq units.

15. $\Delta x = \dfrac{3-1}{4} = \dfrac{1}{2}$; $x_1 = \dfrac{3}{2}$, $x_2 = \dfrac{4}{2}$, $x_3 = \dfrac{5}{2}$, $x_4 = 3$.

$$A \approx \left[\frac{1}{\frac{3}{2}} + \frac{1}{\frac{4}{2}} + \frac{1}{\frac{5}{2}} + \frac{1}{3} \right] \frac{1}{2} \approx 0.95 \text{ sq units.}$$

17. $A = 20[f(10) + f(30) + f(50) + f(70) + f(90)]$
     $= 20(80 + 100 + 110 + 100 + 80) = 9400 \text{ sq ft.}$

6.4 CONCEPT QUESTIONS, page 437

1. See the Fundamental of Calculus Theorem on page 429 in the text.

EXERCISES 6.4, page 437

1. $A = \int_{1}^{4} 2\,dx = 2x\big|_{1}^{4} = 2(4-1) = 6$, or 6

    square units. The region is a rectangle
    whose area is

    $3 \cdot 2$, or 6, square units.

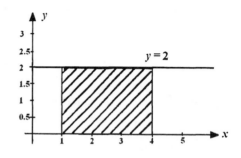

3. $A = \int_{1}^{3} 2x\,dx = x^2\big|_{1}^{3} = 9 - 1 = 8$, or 8 sq units.

    The region is a parallelogram of area
    $(1/2)(3 - 1)(2 + 6) = 8$ sq units.

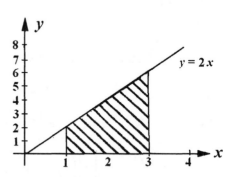

5. $A = \int_{-1}^{2} (2x + 3)\,dx = x^2 + 3x\big|_{-1}^{2} = (4+6) - (1-3) = 12$, or 12 sq units.

7. $A = \int_{-1}^{2} (-x^2 + 4)\,dx = -\frac{1}{3}x^3 + 4x\big|_{-1}^{2} = \left(-\frac{8}{3} + 8 \right) - \left(\frac{1}{3} - 4 \right) = 9$, or 9 sq units.

9. $A = \int_1^2 \frac{1}{x}\,dx = \ln|x|\Big|_1^2 = \ln 2 - \ln 1 = \ln 2$, or $\ln 2$ sq units.

11. $A = \int_1^9 \sqrt{x}\,dx = \frac{2}{3}x^{3/2}\Big|_1^9 = \frac{2}{3}(27-1) = \frac{52}{3}$, or $17\frac{1}{3}$ sq units.

13. $A = \int_{-8}^{-1}(1-x^{1/3})\,dx = x - \frac{3}{4}x^{4/3}\Big|_{-8}^{-1} = (-1-\frac{3}{4})-(-8-12) = 18\frac{1}{4}$, or $18\frac{1}{4}$ sq units.

15. $A = \int_0^2 e^x\,dx = e^x\Big|_0^2 = (e^2 - 1)$, or approximately 6.39 sq units.

17. $\int_2^4 3\,dx = 3x\Big|_2^4 = 3(4-2) = 6.$

19. $\int_1^3 (2x+3)\,dx = x^2 + 3x\Big|_1^3 = (9+9)-(1+3) = 14.$

21. $\int_{-1}^3 2x^2\,dx = \frac{2}{3}x^3\Big|_{-1}^3 = \frac{2}{3}(27) - \frac{2}{3}(-1) = \frac{56}{3}.$

23. $\int_{-2}^2 (x^2 - 1)\,dx = \frac{1}{3}x^3 - x\Big|_{-2}^2 = (\frac{8}{3}-2)-(-\frac{8}{3}+2) = \frac{4}{3}.$

25. $\int_1^8 4x^{1/3}\,dx = (4)(\frac{3}{4})x^{4/3}\Big|_1^8 = 3(16-1) = 45.$

27. $\int_0^1 (x^3 - 2x^2 + 1)\,dx = \frac{1}{4}x^4 - \frac{2}{3}x^3 + x\Big|_0^1 = \frac{1}{4} - \frac{2}{3} + 1 = \frac{7}{12}$

29. $\int_2^4 \frac{1}{x}\,dx = \ln|x|\Big|_2^4 = \ln 4 - \ln 2 = \ln(\frac{4}{2}) = \ln 2.$

31. $\int_0^4 x(x^2 - 1)\,dx = \int_0^4 (x^3 - x)\,dx = \frac{1}{4}x^4 - \frac{1}{2}x^2\Big|_0^4 = 64 - 8 = 56.$

33. $\int_1^3 (t^2 - t)^2\,dt = \int_1^3 t^4 - 2t^3 + t^2)\,dt = \frac{1}{5}t^5 - \frac{1}{2}t^4 + \frac{1}{3}t^3\Big|_1^3$

$$= \left(\frac{243}{5} - \frac{81}{2} + \frac{27}{3}\right) - \left(\frac{1}{5} - \frac{1}{2} + \frac{1}{3}\right) = \frac{512}{30} = \frac{256}{15}.$$

35. $\int_{-3}^{-1} x^{-2} \, dx = -\frac{1}{x}\Big|_{-3}^{-1} = 1 - \frac{1}{3} = \frac{2}{3}.$

37. $\int_{1}^{4}\left(\sqrt{x} - \frac{1}{\sqrt{x}}\right) dx = \int_{1}^{4}(x^{1/2} - x^{-1/2}) \, dx = \frac{2}{3}x^{3/2} - 2x^{1/2}\Big|_{1}^{4}$

$$= \left(\frac{16}{3} - 4\right) - \left(\frac{2}{3} - 2\right) = \frac{8}{3}.$$

39. $\int_{1}^{4} \frac{3x^3 - 2x^{2.} + 4}{x^2} \, dx = \int_{1}^{4}(3x - 2 + 4x^{-2}) \, dx = \frac{3}{2}x^2 - 2x - \frac{4}{x}\Big|_{1}^{4}$

$$= (24 - 8 - 1) - (\tfrac{3}{2} - 2 - 40 = \tfrac{39}{2}.$$

41. a. $C(300) - C(0) = \int_{0}^{300}(0.0003x^2 - 0.12x + 20) \, dx = 0.0001x^3 - 0.06x^2 + 20x\Big|_{0}^{300}$

$$= 0.0001(300)^3 - 0.06(300)^2 + 20(300) = 3300.$$

Therefore $C(300) = 3300 + C(0) = 3300 + 800 = 4100$, or \$4100.

b. $\int_{200}^{300} C'(x) \, dx = (0.0001x^3 - 0.06x^2 + 20x)\Big|_{200}^{300}$

$$= [0.0001(300)^3 - 0.06(300)^2 + 20(300)]$$
$$-[0.0001(200)^3 - 0.06(200)^2 + 20(200)]$$
$$= 900 \text{ or } \$900.$$

43. a. The profit is $\int_{0}^{200}(-0.0003x^2 + 0.02x + 20) \, dx + P(0)$

$$= -0.0001x^3 + 0.01x^2 + 20x\Big|_{0}^{200} + P(0)$$
$$= 3600 + P(0) = 3600 - 800, \text{ or } \$2800.$$

b. $\int_{200}^{220} P'(x) \, dx = P(220) - P(200) = -0.0001x^3 + 0.01x^2 + 20x\Big|_{200}^{220}$

$$= 219.20, \text{ or } \$219.20.$$

45. The distance is

$$\int_0^{20} v(t)\,dt = \int_0^{20} (-t^2 + 20t + 440)\,dt = -\tfrac{1}{3}t^3 + 10t^2 + 440t\Big|_0^{20} \approx 10{,}133\tfrac{1}{3}\ \text{ft}.$$

47. a. The percent of these households in decade t is
$$P(t) = \int R(t)\,dt = \int (0.8499t^2 - 3.872t + 5)\,dt$$
$$= 0.2833t^3 - 1.936t^2 + 15t + C$$
The condition $P(0) = 5.6$ gives
$$P(0) = C = 5.6$$
Therefore,
$$P(t) = 0.2833t^3 - 1.936t^2 + 15t + 5.6$$
b. The percent of these households in 2010 will be
$$P(4) = 0.2833(4^3) - 1.936(4^2) + 5(4) + 5.6 = 12.7552$$
or approximately 12.8%.
c. The percent of these households in 2000 was
$$P(3) = 0.2833(3^3) - 1.936(3^2) + 5(3) + 5.6 = 10.8251.$$
Therefore, the net increase in the percent of these households from 1970 ($t = 0$) to 2000 ($t = 3$) is $P(3) - P(0) = 10.825 - 5.6 = 5.2$, or approximately 5.2%.

49. The average population over the period in question is
$$A = \tfrac{1}{3} \int \frac{85}{1 + 1.859e^{-0.66t}}\,dt$$
Multiplying the integrand by $e^{0.66t}/e^{0.66t}$ gives
$$A = \frac{85}{3} \int_0^3 \frac{e^{0.66t}}{e^{0.66t} + 1.859}\,dt$$
Let $u = 1.859 + e^{0.66t}$, $du = 0.66e^{0.66t}\,dt$, or $e^{0.66t}\,dt = \dfrac{du}{0.66}$.

If $t = 0$, then $u = 2.859$. If $t = 3$, then $u = 9.1017$
Substituting
$$A = \frac{85}{3} \int_{2.859}^{9.1017} \frac{du}{(0.66)u} = \frac{85}{3(0.66)} \ln u \Big|_{2.859}^{9.1017}$$
$$= \frac{85}{3(0.66)}(\ln 9.1017 - \ln 2.859) = \frac{85}{3(0.66)} \ln \frac{9.1017}{2.859} \approx 49.712,$$
or approximately 49.7 million people.

51. $f(x) = x^4 - 2x^2 + 2$. $f'(x) = 4x^3 - 4x = 4x(x^2 - 1) = 4x(x+1)(x-1)$

Setting $f'(x) = 0$ gives $x = -1, 0$, and 1 as critical numbers.

$f''(x) = 12x^2 - 4 = 4(3x^2 - 1)$

Using the second derivative test, we find

$f''(-1) = 8 > 0$ and so $(-1,1)$ is a relative minimum. $f''(0) = -4 < 0$ and so $(0, 2)$ is a relative maximum. $f''(1) = 8 > 0$ and so $(1, 1)$ is a relative minimum. The graph of f is symmetric with respect to the y-axis because

$$f(-x) = (-x)^4 - 2(-x)^2 + 2 = x^4 - 2x^2 + 2.$$

So, the required area is the area under the graph of f between $x = 0$ and $x = 1$. Thus,

$$A = \int_0^1 (x^4 - 2x^2 + 2)\, dx = \tfrac{1}{5}x^5 - \tfrac{2}{3}x^3 + 2x\Big|_0^1 = \tfrac{1}{5} - \tfrac{2}{3} + 2 = \tfrac{23}{15} \text{ sq units.}$$

53. False. The integrand $f(x) = \dfrac{1}{x^3}$ is discontinuous at $x = 0$.

55. False. $f(x)$ is not nonnegative on $[0, 2]$.

USING TECHNOLOGY EXERCISES 6.4, page 440

1. 6.1787   3. 0.7873   5. −0.5888

7. 2.7044   9. 3.9973

11. 46 %; 24%   13. 333,209   15. 903,213

6.5 Problem Solving Tips

1. If you use the method of substitution to evaluate a definite integral, make sure that you change the corresponding limits of integration to reflect the fact that the integration is being performed with respect to the new variable u.

6.5 CONCEPT QUESTIONS, page 447

1. Approach I: We first find the indefinite integral. Let $u = x^3 + 1$ so that $du = 3x^2\, dx$ or $x^2\, dx = \frac{1}{3} du$. Then

$$\int x^2 (x^3 + 1)^2\, dx = \frac{1}{3} \int u^2\, du = \frac{1}{9} u^3 + C = \frac{1}{9}(x^3 + 1)^3 + C.$$

Therefore,

$$\int_0^1 x^2 (x^3 + 1)^2\, dx = \frac{1}{9}(x^3 + 1)^3 \Big|_0^1 = \frac{1}{9}(8 - 1) = \frac{7}{9}.$$

Approach II: Transform the definite integral in x into an integral in u: Let $u = x^3 + 1$, so that $du = 3x^2\, dx$ or $x^2\, dx = \frac{1}{3} du$. Next, find the limits of integration with respect to u: If $x = 0$, then $u = 0^3 + 1 = 1$ and if $x = 1$, then $u = 1^3 + 1 = 2$. Therefore,

$$\int_0^1 x^2 (x^3 + 1)^2\, dx = \frac{1}{3} \int_1^2 u^2\, du = \frac{1}{9} u^3 \Big|_1^2 = \frac{1}{9}(8 - 1) = \frac{7}{9}.$$

EXERCISES 6.5, page 449

1. Let $u = x^2 - 1$ so that $du = 2x\, dx$ or $x\, dx = \frac{1}{2}\, du$. Also, if $x = 0$, then $u = -1$ and if $x = 2$, then $u = 3$. So

$$\int_0^2 x(x^2 - 1)^3\, dx = \frac{1}{2} \int_{-1}^3 u^3\, du = \frac{1}{8} u^4 \Big|_{-1}^3 = \frac{1}{8}(81) - \frac{1}{8}(1) = 10.$$

3. Let $u = 5x^2 + 4$ so that $du = 10x\, dx$ or $x\, dx = \frac{1}{10}\, du$. Also, if $x = 0$, then $u = 4$, and if $x = 1$, then $u = 9$. So

$$\int_0^1 x\sqrt{5x^2 + 4}\, dx = \frac{1}{10} \int_4^9 u^{1/2}\, du = \frac{1}{15} u^{3/2} \Big|_4^9 = \frac{1}{15}(27) - \frac{1}{15}(8) = \frac{19}{15}.$$

5. Let $u = x^3 + 1$ so that $du = 3x^2\, dx$ or $x^2\, dx = \frac{1}{3}\, du$. Also, if $x = 0$, then $u = 1$, and if $x = 2$, then $u = 9$. So,

$$\int_0^2 x^2 (x^3 + 1)^{3/2}\, dx = \frac{1}{3} \int_1^9 u^{3/2}\, du = \frac{2}{15} u^{5/2} \Big|_1^9 = \frac{2}{15}(243) - \frac{2}{15}(1) = \frac{484}{15}.$$

7. Let $u = 2x + 1$ so that $du = 2\,dx$ or $dx = \frac{1}{2}\,du$. Also, if $x = 0$,
    then $u = 1$ and if $x = 1$ then $u = 3$. So

    $$\int_0^1 \frac{1}{\sqrt{2x+1}}\,dx = \frac{1}{2}\int_1^3 \frac{1}{\sqrt{u}}\,du = \frac{1}{2}\int_1^3 u^{-1/2}\,du = u^{1/2}\Big|_1^3 = \sqrt{3} - 1.$$

9. $\int_1^2 (2x-1)^4\,dx$. Put $u = 2x - 1$ so that $du = 2\,dx$ or $dx = \frac{1}{2}\,du$. Then if $x = 1$, $u = 1$
    and if $x = 2$, then $u = 3$. Then

    $$\int_1^2 (2x-1)^4\,dx = \frac{1}{2}\int_1^3 u^4\,du = \frac{1}{10}u^5\Big|_1^3 = \frac{1}{10}(243 - 1) = \frac{121}{5} = 24\frac{1}{5}.$$

11. Let $u = x^3 + 1$ so that $du = 3x^2\,dx$ or $x^2\,dx = \frac{1}{3}\,du$. Also, if $x = -1$,
    then $u = 0$ and if $x = 1$, then $u = 2$. So

    $$\int_{-1}^1 x^2(x^3+1)^4\,dx = \frac{1}{3}\int_0^2 u^4\,du = \frac{1}{15}u^5\Big|_0^2 = \frac{32}{15}.$$

13. Let $u = x - 1$ so that $du = dx$. Then if $x = 1$, $u = 0$, and if $x = 5$, then $u = 4$.

    $$\int_1^5 x\sqrt{x-1}\,dx = \int_0^4 (u+1)u^{1/2}\,du = \int_0^4 (u^{3/2} + u^{1/2})\,du$$

    $$= \frac{2}{5}u^{5/2} + \frac{2}{3}u^{3/2}\Big|_0^4 = \frac{2}{5}(32) + \frac{2}{3}(8) = 18\frac{2}{15}.$$

15. Let $u = x^2$ so that $du = 2x\,dx$ or $x\,dx = \frac{1}{2}\,du$. If $x = 0$, $u = 0$ and if
    $x = 2$, $u = 4$. So

    $$\int_0^2 xe^{x^2}\,dx = \frac{1}{2}\int_0^4 e^u\,du = \frac{1}{2}e^u\Big|_0^4 = \frac{1}{2}(e^4 - 1).$$

17. $\int_0^1 (e^{2x} + x^2 + 1)\,dx = \frac{1}{2}e^{2x} + \frac{1}{3}x^3 + x\Big|_0^1 = (\frac{1}{2}e^2 + \frac{1}{3} + 1) - \frac{1}{2}$

    $$= \frac{1}{2}e^2 + \frac{5}{6}.$$

19. Put $u = x^2 + 1$ so that $du = 2x\,dx$ or $x\,dx = \frac{1}{2}\,du$. Then

    $$\int_{-1}^1 xe^{x^2+1}\,dx = \frac{1}{2}\int_2^2 e^u\,du = \frac{1}{2}e^u\Big|_2^2 = 0$$

(Since the upper and lower limits are equal.)

21. Let $u = x - 2$ so that $du = dx$. If $x = 3$, $u = 1$ and if $x = 6$, $u = 4$. So

$$\int_3^6 \frac{2}{x-2}\,dx = 2\int_1^4 \frac{du}{u} = 2\ln|u|\Big|_1^4 = 2\ln 4.$$

23. Let $u = x^3 + 3x^2 - 1$ so that $du = (3x^2 + 6x)dx = 3(x^2 + 2x)dx$. If $x = 1$, $u = 3$, and if $x = 2$, $u = 19$. So

$$\int_1^2 \frac{x^2 + 2x}{x^3 + 3x^2 - 1}\,dx = \frac{1}{3}\int_3^{19} \frac{du}{u} = \frac{1}{3}\ln u\Big|_3^{19} = \frac{1}{3}(\ln 19 - \ln 3).$$

25. $\displaystyle\int_1^2 \left(4e^{2u} - \frac{1}{u}\right)du = 2e^{2u} - \ln u\Big|_1^2 = (2e^4 - \ln 2) - (2e^2 - 0) = 2e^4 - 2e^2 - \ln 2.$

27. $\displaystyle\int_1^2 (2e^{-4x} - x^{-2})dx = -\frac{1}{2}e^{-4x} + \frac{1}{x}\Big|_1^2 = (-\frac{1}{2}e^{-8} + \frac{1}{2}) - (-\frac{1}{2}e^{-4} + 1)$

$$= -\frac{1}{2}e^{-8} + \frac{1}{2}e^{-4} - \frac{1}{2} = \frac{1}{2}(e^{-4} - e^{-8} - 1).$$

29. $\displaystyle \text{AV} = \frac{1}{2}\int_0^2 (2x+3)\,dx = \frac{1}{2}(x^2 + 3x)\Big|_0^2 = \frac{1}{2}(10) = 5.$

31. $\displaystyle \text{AV} = \frac{1}{2}\int_1^3 (2x^2 - 3)\,dx = \frac{1}{2}(\frac{2}{3}x^3 - 3x)\Big|_1^3 = \frac{1}{2}(9 + \frac{7}{3}) = \frac{17}{3}.$

33. $\displaystyle \text{AV} = \frac{1}{3}\int_{-1}^2 (x^2 + 2x - 3)\,dx = \frac{1}{3}(\frac{1}{3}x^3 + x^2 - 3x)\Big|_{-1}^2$

$$= \frac{1}{3}[(\frac{8}{3} + 4 - 6) - (-\frac{1}{3} + 1 + 3)] = \frac{1}{3}(\frac{8}{3} - 2 + \frac{1}{3} - 4) = -1.$$

35. $\displaystyle \text{AV} = \frac{1}{4}\int_0^4 (2x+1)^{1/2}\,dx = (\frac{1}{4})(\frac{1}{2})(\frac{2}{3})(2x+1)^{3/2}\Big|_0^4 = \frac{1}{12}(27-1) = \frac{13}{6}$

37. $AV = \dfrac{1}{2}\displaystyle\int_0^2 xe^{x^2}\,dx = \dfrac{1}{4}e^{x^2}\Big|_0^2 = \dfrac{1}{4}(e^4 - 1).$

39. The amount produced was

$$\int_0^{20} 3.5e^{0.05t}\,dt = \dfrac{3.5}{0.05}e^{u}\Big|_0^{20} \qquad\qquad (\text{Use the substitution } u = 0.05t.)$$
$$= 70(e-1) \approx 120.3, \quad \text{or } 120.3 \text{ billion metric tons.}$$

41. The amount is $\displaystyle\int_1^2 t(\tfrac{1}{2}t^2 + 1)^{1/2}\,dt$. Let $u = \tfrac{1}{2}t^2 + 1$, so that $du = t\,dt$. Therefore,

$$\int_1^2 t(\tfrac{1}{2}t^2 + 1)^{1/2}\,dt = \int_{3/2}^3 u^{1/2}\,du = \tfrac{2}{3}u^{3/2}\Big|_{3/2}^3 = \tfrac{2}{3}[(3)^{3/2} - (\tfrac{3}{2})^{3/2}]$$
$$\approx 2.24 \text{ million dollars.}$$

43. The tractor will depreciate

$$\int_0^5 13388.61e^{-0.22314t}\,dt = \dfrac{13388.61}{-0.22314}e^{-0.22314t}\Big|_0^5$$
$$= -60{,}000.94e^{-0.22314t}\Big|_0^5 = -60{,}000.94(-0.672314)$$
$$= 40{,}339.47, \quad \text{or } \$40{,}339.$$

45. $\bar{A} = \tfrac{1}{5}\displaystyle\int (\tfrac{1}{12}t^2 + 2t + 44)\,dt = \tfrac{1}{5}\left[\tfrac{1}{36}t^3 + t^2 + 44t\Big|_0^5\right]$

$$= \tfrac{1}{5}\left[\tfrac{125}{36} + 25 + 220\right] = \dfrac{125+900+7920}{5(36)} \approx 49.69, \text{ or } 49.7 \text{ ft/sec.}$$

47. The average whale population will be

$$\tfrac{1}{10}\int_0^{10}(3t^3 + 2t^2 - 10t + 600)\,dt = \tfrac{1}{10}(\tfrac{3}{4}t^4 + \tfrac{2}{3}t^3 - 5t^2 + 600t\Big|_0^{10}$$
$$\approx \tfrac{1}{10}(7500 + 666.67 - 500 + 6000) \approx 1367 \text{ whales.}$$

49. The average yearly sales of the company over its first 5 years of operation is given

by $\qquad \tfrac{1}{5-0}\displaystyle\int_0^5 t(0.2t^2 + 4)^{1/2}\,dt = \tfrac{1}{5}[(\tfrac{5}{2})(\tfrac{2}{3})(0.2t^2 + 4)^{3/2}\Big|_0^5 \qquad$ [Let $u = -0.2t^2 + 4$.]

$$= \tfrac{1}{5}[\tfrac{5}{3}(5+4)^{3/2} - \tfrac{5}{3}(4)^{3/2}] = \tfrac{1}{3}(27-8) = \tfrac{19}{3}, \quad \text{or } 6\tfrac{1}{3} \text{ million dollars.}$$

51. The average velocity is $\tfrac{1}{4}\int_0^4 3t\sqrt{16-t^2}\, dt = \tfrac{1}{4}(64) = 16$, or 16 ft/sec.

    (Using the results of Exercise 44.)

53. $\int_0^5 p\, dt = \int_0^5 (18 - 3e^{-2t} - 6e^{-t/3})\, dt = \tfrac{1}{5}\left[18t + \tfrac{3}{2}e^{-2t} + 18e^{-t/3}\right]_0^5$

    $= \tfrac{1}{5}\left[18(5) + \tfrac{3}{2}e^{-10} + 18e^{-5/3} - \tfrac{3}{2} - 18\right] = 14.78$, or \$14.78.

55. The average content of oxygen in the pond over the first 10 days is

$$\frac{1}{10-0}\int_0^{10} 100\left(\frac{t^2 + 10t + 100}{t^2 + 20t + 100}\right) dt = \frac{100}{10}\int_0^{10}\left[1 - \frac{10}{t+10} + \frac{100}{(t+10)^2}\right] dt$$

$$= 10\int_0^{10}\left[1 - \frac{10}{t+10} + 100(t+10)^{-2}\right]$$

$$= 10\left[t - 10\ln(t+10) - \frac{100}{t+10}\right]_0^{10} \quad \text{[Use the substitution } u = t + 10 \text{ for the}$$

third integral.]

$$= 10\{[10 - 10\ln 20 - \frac{100}{2p}] - [-10\ln 10 - 10]\}$$

$$= 10[10 - 10\ln 20 - 5 + 10\ln 10 + 10]$$

$$\approx 80.6853, \quad \text{or approximately } 80.7\%.$$

57. $\int_a^a f(x)\, dx = F(x)\Big|_a^a = F(a) - F(a) = 0$, where $F'(x) = f(x)$.

59. $\int_1^3 x^2\, dx = \tfrac{1}{3}x^3\Big|_1^3 = 9 - \tfrac{1}{3} = \tfrac{26}{3} = -\int_3^1 x^2\, dx = -\tfrac{1}{3}x^3\Big|_3^1 = -\tfrac{1}{3} + 9 = \tfrac{26}{3}$.

61. $\int_1^9 2\sqrt{x}\, dx = \tfrac{4}{3}x^{3/2}\Big|_1^9 = \tfrac{4}{3}(27-1) = \tfrac{104}{3} = 2\int_1^9 \sqrt{x}\, dx = (2)(\tfrac{2}{3}x^{3/2})\Big|_1^9 = \tfrac{104}{3}$.

63. $\int_0^3 (1+x^3)\, dx = (x + \tfrac{1}{4}x^4)\Big|_0^3 = 3 + \tfrac{81}{4} = \tfrac{93}{4}$.

315

$$\int_0^1 (1+x^3)\,dx + \int_1^2 (1+x^3)\,dx + \int_2^3 (1+x^3)\,dx$$
$$= (x+\tfrac{1}{4}x^4)\Big|_0^1 + (x+\tfrac{1}{4}x^4)\Big|_1^2 + (x+\tfrac{1}{4}x^4)\Big|_2^3$$
$$= (1+\tfrac{1}{4})+(2+4)-(1+\tfrac{1}{4})+(3+\tfrac{81}{4})-(2+4) = \tfrac{93}{4}. \quad \text{[Property 5.]}$$

65. $\int_3^3 (1+\sqrt{x})e^{-x}\,dx = 0$ by Property 1 of definite integrals.

67. a. $\int_{-1}^2 [2f(x)+g(x)]\,dx = 2\int_{-1}^2 f(x)\,dx + \int_{-1}^2 g(x)\,dx = 2(-2) + 3 = -1.$

   b. $\int_{-1}^2 [g(x)-f(x)]\,dx = \int_{-1}^2 g(x)\,dx - \int_{-1}^2 f(x)\,dx = 3 - (-2) = 5.$

   c. $\int_{-1}^2 [2f(x)-3g(x)]\,dx = 2\int_{-1}^2 f(x)\,dx - 3\int_{-1}^2 g(x)\,dx = 2(-2) - 3(3) = -13.$

69. True. This follows from Property 1 of the definite integral.

71. False. Only a constant can be "moved out" of the integral sign.

73. True. This follows from Properties 3 and 4 of the definite integral.

USING TECHNOLOGY EXERCISES 6.5, page 451

1. 7.71667   3. 17.5649   5. 10,140   7. 60.5mg/day

6.6 Problem Solving Tips

1. Note that the formula for the area between two curves where f and g are continuous

functions and $f(x) \geq g(x)$ on $[a,b]$ is given by $\int_a^b [f(x)-g(x)]\,dx$. The condition

$f(x) \geq g(x)$ on $[a,b]$ tells us that we cannot interchange $f(x)$ and $g(x)$ in this formula,

as that would yield a negative answer, and area cannot be negative

---

6.6 CONCEPT QUESTIONS, page 458

1. $\int_a^b [f(x) - g(x)]\, dx$

EXERCISES 6.6, page 458

1. $-\int_0^6 (x^3 - 6x^2)\, dx = -\frac{1}{4}x^4 + 2x^3\Big|_0^6 = -\frac{1}{4}(6^4) + 2(6^3) = 108$ sq units.

3. $A = -\int_{-1}^0 x\sqrt{1-x^2}\, dx + \int_0^1 x\sqrt{1-x^2}\, dx = 2\int_0^1 x(1-x^2)^{1/2}\, dx$ (by symmetry). Let
   $u = 1 - x^2$ so that $du = -2x\, dx$ or $x\, dx = -\frac{1}{2}\, du$. Also, if $x = 0$, then $u = 1$ and
   if $x = 1$, $u = 0$. So $A = (2)(-\frac{1}{2})\int_0^1 u^{1/2}\, du = -\frac{2}{3}u^{3/2}\Big|_1^0 = \frac{2}{3}$, or $\frac{2}{3}$ sq unit.

5. $A = -\int_0^4 (x - 2\sqrt{x})\, dx = \int_0^4 (-x + 2x^{1/2})\, dx = -\frac{1}{2}x^2 + \frac{4}{3}x^{3/2}\Big|_0^4$
   $= 8 + \frac{32}{3} = \frac{8}{3}$ sq units.

7. The required area is given by
   $$\int_{-1}^0 (x^2 - x^{1/3})\, dx + \int_0^1 (x^{1/3} - x^2)\, dx = \frac{1}{3}x^3 - \frac{3}{4}x^{4/3}\Big|_{-1}^0 + \frac{3}{4}x^{4/3} - \frac{1}{3}x^3\Big|_0^1$$
   $$= -\left(-\frac{1}{3} - \frac{3}{4}\right) + \left(\frac{3}{4} - \frac{1}{3}\right) = 1\frac{1}{2}\quad \text{sq units.}$$

9. The required area is given by
   $-\int_{-1}^2 -x^2\, dx = \frac{1}{3}x^3\Big|_{-1}^2 = \frac{8}{3} + \frac{1}{3} = 3$ sq units.

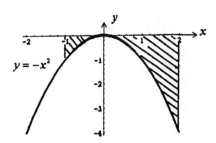

$y = -x^2$

11. $y = x^2 - 5x + 4 = (x-4)(x-1) = 0$
    if $x = 1$ or 4. These give the x-intercepts.

$$A = -\int_1^3 (x^2 - 5x + 4)\, dx = -\tfrac{1}{3}x^3 + \tfrac{5}{2}x^2 - 4x\Big|_1^3$$

$$= (-9 + \tfrac{45}{2} - 12) - (-\tfrac{1}{3} + \tfrac{5}{2} - 4) = \tfrac{10}{3} = 3\tfrac{1}{3}.$$

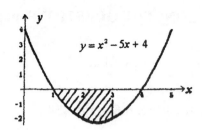

13. The required area is given by

$$-\int_0^9 -(1 + \sqrt{x})\, dx = x + \tfrac{2}{3}x^{3/2}\Big|_0^9 = 9 + 18 = 27.$$

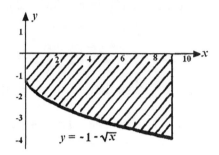

15. $-\int_{-2}^4 -e^{(1/2)x}\, dx = 2e^{(1/2)x}\Big|_{-2}^4$

$$= 2(e^2 - e^{-1})\text{ sq units.}$$

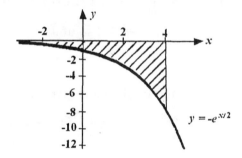

17. $A = \int_1^3 [(x^2 + 3) - 1]\, dx$

$$= \int_1^3 (x^2 + 2)\, dx = \tfrac{1}{3}x^3 + 2x\Big|_1^3$$

$$= (9 + 6) - (\tfrac{1}{3} + 2) = \tfrac{38}{3}.$$

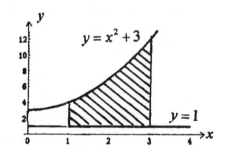

19. $A = \int_0^2 (-x^2 + 2x + 3 + x - 3)\, dx$

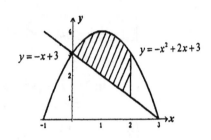

$\qquad = \int_0^2 (-x^2 + 3x)\, dx$

$\qquad = -\frac{1}{3}x^3 + \frac{3}{2}x^2 \Big|_0^2 = -\frac{1}{3}(8) + \frac{3}{2}(4)$

21. $A = \int_{-1}^2 \left[(x^2 + 1) - \frac{1}{3}x^3 \right] dx$

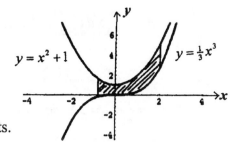

$\qquad = \int_{-1}^2 \left(-\frac{1}{3}x^3 + x^2 + 1 \right) dx$

$\qquad = -\frac{1}{12}x^4 + \frac{1}{3}x^3 + x \Big|_{-1}^2$

$\qquad = \left(-\frac{4}{3} + \frac{8}{3} + 2 \right) - \left(-\frac{1}{12} - \frac{1}{3} - 1 \right) = 4\frac{3}{4}$ sq units.

23. $A = \int_1^4 \left[(2x - 1) - \frac{1}{x} \right] dx = \int_1^4 \left(2x - 1 - \frac{1}{x} \right) dx$

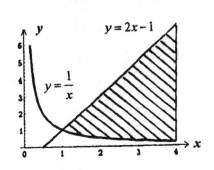

$\qquad = (x^2 - x - \ln x) \Big|_1^4$

$\qquad = (16 - 4 - \ln 4) - (1 - 1 - \ln 1)$

$\qquad = 12 - \ln 4 \approx 10.6$ sq units.

25. $A = \int_{1}^{2}\left(e^x - \dfrac{1}{x}\right)dx = e^x - \ln x\Big|_{1}^{2}$

$= (e^2 - \ln 2) - e = (e^2 - e - \ln 2)$ sq units.

$= \left(\tfrac{1}{2}e^6 - \tfrac{9}{2}\right) - \left(\tfrac{1}{2}e^2 - \tfrac{1}{2}\right)$

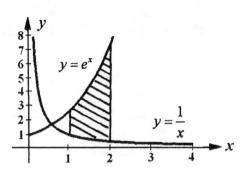

27.

$A = -\int_{-1}^{0} x\,dx + \int_{0}^{2} x\,dx$

$= -\tfrac{1}{2}x^2\Big|_{-1}^{0} + \tfrac{1}{2}x^2\Big|_{0}^{2}$

$= \tfrac{1}{2} + 2 = 2\tfrac{1}{2}$ sq units.

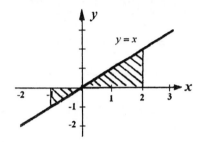

29. The x–intercepts are found by solving
$x^2 - 4x + 3 = (x - 3)(x - 1) = 0$ giving $x = 1$
or 3. The region is shown in the figure.

$A = -\int_{-1}^{1}[(-x^2 + 4x - 3)\,dx + \int_{1}^{2}(-x^2 + 4x - 3)\,dx$

$= \tfrac{1}{3}x^3 - 2x^2 + 3x\Big|_{-1}^{1} + (-\tfrac{1}{3}x^3 + 2x^2 - 3x)\Big|_{1}^{2}$

$= (\tfrac{1}{3} - 2 + 3) - (-\tfrac{1}{3} - 2 - 3)$

$+ (-\tfrac{8}{3} + 8 - 6) - (-\tfrac{1}{3} + 2 - 3) = \tfrac{22}{3}$ sq units.

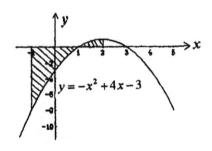

31. The region is shown in the figure at the right.

$A = \int_{0}^{1}(x^3 - 4x^2 + 3x)\,dx - \int_{1}^{2}(x^3 - 4x^2 + 3x)\,dx$

$= (\tfrac{1}{4}x^4 - \tfrac{4}{3}x^3 + \tfrac{3}{2}x^2)\Big|_{0}^{1}$

$-(\tfrac{1}{4}x^4 - \tfrac{4}{3}x^3 + \tfrac{3}{2}x^2)\Big|_{1}^{2} = \tfrac{3}{2}$ sq units.

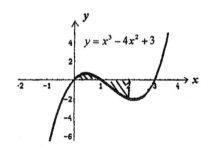

33. The region is shown in the figure at the right.

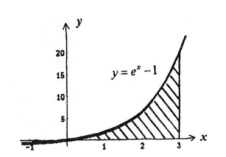

$$A = -\int_{-1}^{0} (e^x - 1)\, dx + \int_{0}^{3} (e^x - 1)\, dx$$

$$= (-e^x + x)\Big|_{-1}^{0} + (e^x - x)\Big|_{0}^{3}$$

$$= -1 - (-e^{-1} - 1) + (e^3 - 3) - 1$$

$$= e^3 - 4 + \tfrac{1}{e} \approx 16.5 \quad \text{sq units.}$$

35. To find the points of intersection of the two curves, we solve the equation

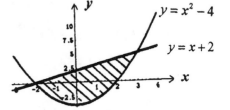

$$x^2 - 4 = x + 2$$
$$x^2 - x - 6 = (x - 3)(x + 2) = 0, \quad \text{obtaining}$$
$x = -2$ or $x = 3$. The region
is shown in the figure at the right.

$$A = \int_{-2}^{3} [(x+2) - (x^2 - 4)]\, dx = \int_{-2}^{3} (-x^2 + x + 6)\, dx = (-\tfrac{1}{3}x^3 + \tfrac{1}{2}x^2 + 6x)\Big|_{-2}^{3}$$
$$= (-9 + \tfrac{9}{2} + 18) - (\tfrac{8}{3} + 2 - 12) = \tfrac{125}{6} \text{ sq units.}$$

37. To find the points of intersection of the two curves, we solve the equation $x^3 = x^2$
or $x^3 - x^2 = x^2(x - 1) = 0$ giving $x = 0$ or 1.
The region is shown in the figure.

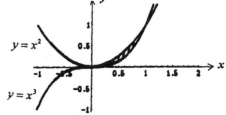

$$A = -\int_{0}^{1} (x^2 - x^3)\, dx$$

$$= (\tfrac{1}{3}x^3 - \tfrac{1}{4}x^4)\Big|_{0}^{1} = \tfrac{1}{3} - \tfrac{1}{4} = \tfrac{1}{12} \text{ sq units}$$

39. To find the points of intersection of the two curves, we solve the equation
$$x^3 - 6x^2 + 9x = x^2 - 3x,$$
or $x^3 - 7x^2 + 12x = x(x-4)(x-3) = 0$ obtaining $x = 0$, 3, or 4.

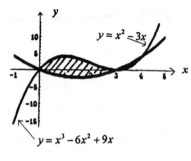

$$A = \int_0^3 [(x^3 - 6x^2 + 9x) - (x^2 + 3x)]dx$$

$$+ \int_3^4 [(x^2 - 3x) - (x^3 - 6x^2 + 9x)]dx$$

$$= \int_0^3 (x^3 - 7x^2 + 12x)dx - \int_3^4 (x^3 - 7x^2 + 12x)dx$$

$$= (\tfrac{1}{4}x^4 - \tfrac{7}{3}x^3 + 6x^2)\Big|_0^3 - (\tfrac{1}{4}x^4 - \tfrac{7}{3}x^3 + 6x^2)\Big|_3^4$$

$$= (\tfrac{81}{4} - 63 + 54) - (64 - \tfrac{448}{3} + 96) + (\tfrac{81}{4} - 63 + 54) = \tfrac{71}{6}.$$

41. By symmetry, $A = 2\int_0^3 x(9 - x^2)^{1/2}\, dx$. We integrate by substitution with

$u = 9 - x^2$, $du = -2x\, dx$. If $x = 0$, $u = 9$, and if $x = 3, u = 0$. So

$$A = 2\int_9^0 -\tfrac{1}{2}u^{1/2}\, du = -\int_9^0 u^{1/2}\, du = -\tfrac{2}{3}u^{3/2}\Big|_9^0 = \tfrac{2}{3}(9)^{3/2} = 18 \text{ sq units.}$$

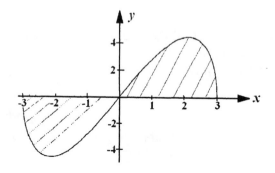

43. S gives the additional revenue that the company would realize if it used a different advertising agency. $S = \int_0^b [g(x) - f(x)]dx.$

45. Shortfall $= \int_{2010}^{2050} [f(t) - g(t)]dt$

47. a. $\int_{T_1}^{T}[g(t)-f(t)]dt - \int_{0}^{T_1}[f(t)-g(t)]dt = A_2 - A_1.$

   b. The number $A_2 - A_1$ gives the distance car 2 is ahead of car 1 after t seconds.

49. The turbo-charged model is moving at
$$A = \int_{0}^{10}[(4+1.2t+0.03t^2)-(4+0.8t)]dt$$
$$= \int_{0}^{10}(0.4t+0.03t^2)dt = (0.2t^2 + 0.1t^3)\Big|_{0}^{10}$$
$$= 20 + 10, \text{ or } 30 \text{ ft/sec faster than the standard model.}$$

51. The additional number of cars will be given by
$$\int_{0}^{5}(5e^{0.3t} - 5 - 0.5t^{3/2})dt = \frac{5}{0.3}e^{0.3t} - 5t - 0.2t^{5/2}\Big|_{0}^{5}$$
$$= \frac{5}{0.3}e^{1.5} - 25 - 0.2(5)^{5/2} - \frac{5}{0.3} = 74.695 - 25 - 0.2(5)^{5/2} - \frac{50}{3}$$

   ≈ 21.85, or 21,850 cars. (Remember t is measured in thousands.)

53. True. If $f(x) \geq g(x)$ on $[a, b]$, then the area of the said region is
$$\int_{a}^{b}[f(x)-g(x)]dx = \int_{a}^{b}|f(x)-g(x)|dx$$
   If $f(x) \leq g(x)$ on $[a, b]$, then the area of the region is
$$\int_{a}^{b}[g(x)-f(x)]dx = \int_{a}^{b}-[f(x)-g(x)]dx = \int_{a}^{b}|f(x)-g(x)|dx$$

55. The area of R' is
$$A = \int_{a}^{b}\{[f(x)+C]-[g(x)+C]\}dx = \int_{a}^{b}[f(x)+C-g(x)-C]dx$$
$$= \int_{a}^{b}[f(x)-g(x)]dx$$

1. a.

b. 1074.2857 sq units

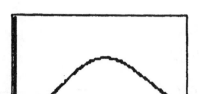

3. a.

5. a.

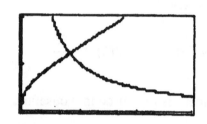

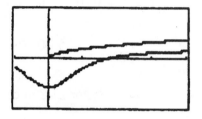

b. 0.9961 sq units

b. 5.4603 sq units

7. a.

9. a.

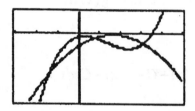

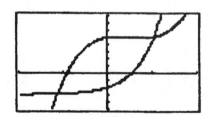

b. 25.8549 sq units

b. 10.5144 sq units

11. a.

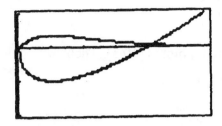

   b. 3.5799 sq units

13. 207.43 sq units

6.7 CONCEPT QUESTIONS, page 474

1. a. See definition in text on page 465.   b. See definition in text on page 465.
3. See definition in text on page 470.

EXERCISES 6.7, page 478

1. When $p = 4$, $-0.01x^2 - 0.1x + 6 = 4$ or $x^2 + 10x - 200 = 0$, $(x - 10)(x + 20) = 0$
    and $x = 10$ or -20. We reject the root $x = -20$. The consumers' surplus is

$$CS = \int_0^{10} (-0.01x^2 - 0.1x + 6)\,dx - (4)(10)$$

$$= -\frac{0.01}{3}x^3 - 0.05x^2 + 6x \Big|_0^{10} - 40 \approx 11.667, \text{ or } \$11{,}667.$$

3. Setting $p = 10$, we have $\sqrt{225 - 5x} = 10$, $225 - 5x = 100$, or $x = 25$.
    Then $CS = \int_0^{25} \sqrt{225 - 5x}\,dx - (10)(25) = \int_0^{25} (225 - 5x)^{1/2}\,dx - 250$.
    To evaluate the integral, let $u = 225 - 5x$ so that $du = -5\,dx$ or
    $dx = -\frac{1}{5}\,du$. If $x = 0$, $u = 225$ and if $x = 25$, $u = 100$. So

$$CS = -\frac{1}{5}\int_{225}^{100} u^{1/2}\,du - 250 = -\frac{2}{15} u^{3/2} \Big|_{225}^{100} - 250$$

$$= -\frac{2}{15}(1000 - 3375) - 250 = 66.667, \text{ or } \$6{,}667.$$

5. To find the equilibrium point, we solve
$$0.01x^2 + 0.1x + 3 = -0.01x^2 - 0.2x + 8, \quad \text{or} \quad 0.02x^2 + 0.3x - 5 = 0,$$
$$2x^2 + 30x - 500 = (2x - 20)(x + 25) = 0$$

obtaining $x = -25$ or 10. So the equilibrium point is (10,5). Then
$$PS = (5)(10) - \int_0^{10} (0.01x^2 + 0.1x + 3)\,dx$$
$$= 50 - (\frac{0.01}{3}x^3 + 0.05x^2 + 3x)\Big|_0^{10} = 50 - \frac{10}{3} - 5 - 30 = \frac{35}{3},$$
or approximately \$11,667.

7. To find the market equilibrium, we solve
$$-0.2x^2 + 80 = 0.1x^2 + x + 40, \quad 0.3x^2 + x - 40 = 0,$$
$$3x^2 + 10x - 400 = 0, \quad (3x + 40)(x - 10) = 0$$
giving $x = -\frac{40}{3}$ or $x = 10$. We reject the negative root. The corresponding

equilibrium price is \$60. The consumers' surplus is
$$CS = \int_0^{10} (-0.2x^2 + 80)\,dx - (60)(10) = -\frac{0.2}{3}x^3 + 80x\Big|_0^{10} - 600 = 133\tfrac{1}{3},$$
or \$13,333. The producers' surplus is
$$PS = 600 - \int_0^{10} (0.1x^2 + x + 40)\,dx = 600 - [\tfrac{0.1}{3}x^3 + \tfrac{1}{2}x^2 + 40x]\Big|_0^{10}$$
$$= 116\tfrac{2}{3}, \text{ or } \$11,667.$$

9. Here $P = 200,000$, $r = 0.08$, and $T = 5$. So
$$PV = \int_0^5 200,000e^{-0.08t}\,dt = -\frac{200,000}{0.08}e^{-0.08t}\Big|_0^5 = -2,500,000(e^{-0.4} - 1)$$
$$\approx 824,199.85, \text{ or } \$824,200.$$

11. Here $P = 250$, $m = 12$, $T = 20$, and $r = 0.08$. So
$$A = \frac{mP}{r}(e^{rT} - 1) = \frac{12(250)}{0.08}(e^{1.6} - 1) \approx 148,238.70$$
or approximately \$148,239.

13. Here $P = 150$, $m = 12$, $T = 15$, and $r = 0.08$. So
$$A = \frac{12(150)}{0.08}(e^{1.2} - 1) \approx 52{,}202.60, \text{ or approximately } \$52{,}203.$$

15. Here $P = 2000$, $m = 1$, $T = 15.75$, and $r = 0.1$. So
$$A = \frac{1(2000)}{0.1}(e^{1.575} - 1) \approx 76{,}615, \text{ or approximately } \$76{,}615.$$

17. Here $P = 1200$, $m = 12$, $T = 15$, and $r = 0.1$. So
$$PV = \frac{12(1200)}{0.1}(1 - e^{-1.5}) \approx 111{,}869, \text{ or approximately } \$111{,}869.$$

19. We want the present value of an annuity with $P = 300$, $m = 12$, $T = 10$, and $r = 0.12$. So
$$PV = \frac{12(300)}{0.12}(1 - e^{-1.2}) \approx 20{,}964, \text{ or approximately } \$20{,}964.$$

21. a.

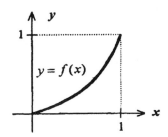

b.
$f(0.4) = \frac{15}{16}(0.4)^2 + \frac{1}{16}(0.4) \approx 0.175$; $f(0.9) = \frac{15}{16}(0.9)^2 + \frac{1}{16}(0.9) \approx 0.816$.

So, the lowest 40 percent of the people receive 17.5 percent of the total income and the lowest 90 percent of the people receive 81.6 percent of the income.

23. a.

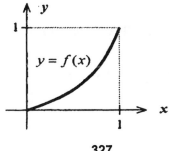

6 Integration

b. $f(0.3) = \frac{14}{15}(0.03)^2 + \frac{1}{15}(0.3) = 0.104$;

$f(0.7) = \frac{14}{15}(0.7)^2 + \frac{1}{15}(0.7) \approx 0.504$.

USING TECHNOLOGY EXERCISES 6.7, page 476

1. Consumer's surplus: $18,000,000; producer's surplus: $11,700,000.

3. Consumer's surplus: $33,120; producer's surplus: $2,880.

5. Investment

CHAPTER 6, CONCEPT REVIEW, page 479

1. a. $F'(x) = f(x)$ b. $F(x) + C$

3. a. Unknown   b. Function

5. a. $\int_a^b f(x)\,dx$ b. Minus

7. a. $\dfrac{1}{b-a} \int_a^b f(x)\,dx$ b. Area; area

9. $\int_0^{\bar{x}} D(x)\,dx - \bar{p}\,\bar{x}$ b. $\bar{p}\,\bar{x} - \int_0^{\bar{x}} S(x)\,dx$

11. $\dfrac{mP}{r}(e^{rT} - 1)$

CHAPTER 6 REVIEW EXERCISES, page 479

1. $\int (x^3 + 2x^2 - x)\,dx = \frac{1}{4}x^4 + \frac{2}{3}x^3 - \frac{1}{2}x^2 + C$.

3. $\displaystyle \int \left(x^4 - 2x^3 + \frac{1}{x^2} \right) dx = \frac{x^5}{5} - \frac{1}{2}x^4 - \frac{1}{x} + C$

5. $\displaystyle \int x(2x^2 + x^{1/2}) dx = \int (2x^3 + x^{3/2}) dx = \frac{1}{2}x^4 + \frac{2}{5}x^{5/2} + C.$

7. $\displaystyle \int (x^2 - x + \frac{2}{x} + 5) dx = \int x^2 \, dx - \int x \, dx + 2\int \frac{dx}{x} + 5\int dx$
$$= \frac{1}{3}x^3 - \frac{1}{2}x^2 + 2\ln|x| + 5x + C.$$

9. Let $u = 3x^2 - 2x + 1$ so that $du = (6x - 2)\, dx = 2(3x - 1)\, dx$ or $(3x - 1)\, dx = \frac{1}{2}\, du.$

   So $\displaystyle \int (3x - 1)(3x^2 - 2x + 1)^{1/3} \, dx = \frac{1}{2}\int u^{1/3} \, du = \frac{3}{8}u^{4/3} + C = \frac{3}{8}(3x^2 - 2x + 1)^{4/3} + C.$

11. Let $u = x^2 - 2x + 5$ so that $du = 2(x - 1)\, dx$ or $(x - 1)\, dx = \frac{1}{2}\, du.$

   $$\int \frac{x-1}{x^2 - 2x + 5} dx = \frac{1}{2}\int \frac{du}{u} = \frac{1}{2}\ln|u| + C = \frac{1}{2}\ln(x^2 - 2x + 5) + C.$$

13. Put $u = x^2 + x + 1$ so that $du = (2x + 1)\, dx = 2(x + \frac{1}{2})\, dx$ and $(x + \frac{1}{2})dx = \frac{1}{2}\, du.$

   $$\int (x + \tfrac{1}{2})e^{x^2 + x + 1} dx = \frac{1}{2}\int e^u \, du = \frac{1}{2}e^u + C = \frac{1}{2}e^{x^2 + x + 1} + C.$$

15. Let $u = \ln x$ so that $du = \frac{1}{x}\, dx$. Then

   $$\int \frac{(\ln x)^5}{x} dx = \int u^5 \, du = \frac{1}{6}u^6 + C = \frac{1}{6}(\ln x)^6 + C.$$

17. Let $u = x^2 + 1$ so that $du = 2x\, dx$ or $x\, dx = \frac{1}{2}\, du$. Then

   $$\int x^3 (x^2 + 1)^{10} \, dx = \frac{1}{2}\int (u - 1)u^{10} \, du \qquad \text{(}x^2 = u - 1\text{)}$$
   $$= \frac{1}{2}\int (u^{11} - u^{10}) \, du = \frac{1}{2}\left(\frac{1}{12}u^{12} - \frac{1}{11}u^{11}\right) + C$$
   $$= \frac{1}{264}u^{11}(11u - 12) + C = \frac{1}{264}(x^2 + 1)^{11}(11x^2 - 1) + C.$$

6 Integration

19. Put $u = x - 2$ so that $du = dx$. Then $x = u + 2$ and

$$\int \frac{x}{\sqrt{x-2}}\,dx = \int \frac{u+2}{\sqrt{u}}\,du = \int (u^{1/2} + 2u^{-1/2})\,du = \int u^{1/2}\,du + 2\int u^{-1/2}\,du$$

$$= \tfrac{2}{3}u^{3/2} + 4u^{1/2} + C = \tfrac{2}{3}u^{1/2}(u+6) + C = \tfrac{2}{3}\sqrt{x-2}\,(x-2+6) + C$$

$$= \tfrac{2}{3}(x+4)\sqrt{x-2} + C.$$

21. $\displaystyle\int_0^1 (2x^3 - 3x^2 + 1)\,dx = \tfrac{1}{2}x^4 - x^3 + x\Big|_0^1 = \tfrac{1}{2} - 1 + 1 = \tfrac{1}{2}.$

23. $\displaystyle\int_1^4 (x^{1/2} + x^{-3/2})\,dx = \tfrac{2}{3}x^{3/2} - 2x^{-1/2}\Big|_1^4 = \tfrac{2}{3}x^{3/2} - \frac{2}{\sqrt{x}}\Big|_1^4 = (\tfrac{16}{3} - 1) - (\tfrac{2}{3} - 2) = \tfrac{17}{3}.$

25. Put $u = x^3 - 3x^2 + 1$ so that $du = (3x^2 - 6x)\,dx = 3(x^2 - 2x)\,dx$ or $(x^2 - 2x)\,dx = \tfrac{1}{3}\,du$. Then if $x = -1$, $u = -3$, and if $x = 0$, $u = 1$,

$$\int_{-1}^0 12(x^2 - 2x)(x^3 - 3x^2 + 1)^3\,dx = (12)(\tfrac{1}{3})\int_{-3}^1 u^3\,du = 4(\tfrac{1}{4})u^4\Big|_{-3}^1$$

$$= 1 - 81 = -80.$$

27. Let $u = x^2 + 1$ so that $du = 2x\,dx$ or $x\,dx = \tfrac{1}{2}\,du$. Then, if $x = 0$, $u = 1$, and if $x = 2$, $u = 5$, so

$$\int_0^2 \frac{x}{x^2 + 1}\,dx = \frac{1}{2}\int_1^5 \frac{du}{u} = \frac{1}{2}\ln u\Big|_1^5 = \frac{1}{2}\ln 5.$$

29. Let $u = 1 + 2x^2$ so that $du = 4x\,dx$ or $x\,dx = \tfrac{1}{4}\,du$. If $x = 0$, then $u = 1$ and if $x = 2$, then $u = 9$.

$$\int_0^2 \frac{4x}{\sqrt{1 + 2x^2}}\,dx = \int_1^9 \frac{du}{u^{1/2}} = 2u^{1/2}\Big|_1^9 = 2(3 - 1) = 4.$$

31. Let $u = 1 + e^{-x}$ so that $du = -e^{-x}\,dx$ and $e^{-x}\,dx = -\,du$. Then

$$\int_{-1}^0 \frac{e^{-x}}{(1 + e^{-x})^2}\,dx = -\int_{1+e}^2 \frac{du}{u^2} = \frac{1}{u}\Big|_{1+e}^2 = \frac{1}{2} - \frac{1}{1+e} = \frac{e-1}{2(1+e)}.$$

33. $f(x) = \int f'(x)\,dx = \int (3x^2 - 4x + 1)\,dx = 3\int x^2\,dx - 4\int x\,dx + \int dx$

    $= x^3 - 2x^2 + x + C.$

    The given condition implies that $f(1) = 1$ or $1 - 2 + 1 + C = 1$, and $C = 1$.
    Therefore, the required function is $f(x) = x^3 - 2x^2 + x + 1$.

35. $f(x) = \int f'(x)\,dx = \int (1 - e^{-x})\,dx = x + e^{-x} + C,\ f(0) = 2$ implies $0 + 1 + C = 2$
    or $C = 1$. So $f(x) = x + e^{-x} + 1$.

37. $\Delta x = \frac{2-1}{5} = \frac{1}{5};\ x_1 = \frac{6}{5},\ x_2 = \frac{7}{5},\ x_3 = \frac{8}{5},\ x_4 = \frac{9}{5},\ x_5 = \frac{10}{5}$. The Riemann sum is
    $f(x_1)\Delta x + \cdots + f(x_5)\Delta x = \left\{\left[-2\left(\frac{6}{5}\right)^2 + 1\right] + \left[-2\left(\frac{7}{5}\right)^2 + 1\right] + \cdots + \left[-2\left(\frac{10}{5}\right)^2 + 1\right]\right\}\left(\frac{1}{5}\right)$
    $= \frac{1}{5}(-1.88 - 2.92 - 4.12 - 5.48 - 7) = -4.28.$

39. a. $R(x) = \int R'(x)\,dx = \int (-0.03x + 60)\,dx = -0.015x^2 + 60x + C.$

    $R(0) = 0$ implies that $C = 0$. So, $R(x) = -0.015x^2 + 60x.$
    b. From $R(x) = px$, we have $-0.015x^2 + 60x = px$ or $p = -0.015x + 60.$

41. The total number of systems that Vista may expect to sell t months from the time
    they are put on the market is given by $f(t) = 3000t - 50{,}000(1 - e^{-0.04t}).$

    The number is $\int_0^{12} (3000 - 2000e^{-0.04t})\,dt = \left(3000t - \frac{2000}{-0.04}e^{-0.04t} \right)\Big|_0^{12}$

    $= 3000(12) + 50{,}000e^{-0.48} - 50{,}000 = 16{,}939.$

43. The number of speakers sold at the end of 5 years is
    $f(t) = \int f'(t)\,dt = \int_0^5 2000(3 - 2e^{-t})\,dt = 2000[3(5) - 2e^{-5}] - 2000[3 - 2(1)]$
    $= 26{,}027.$

45. The number will be
    $\int (0.00933t^3 + 0.019t^2 - 0.10833t + 1.3467)\,dt$

$$= 0.0023325t^4 + 0.0063333t^3 - 0.054165t^2 + 1.3467\,t\Big|_0^{10} = 37.7,$$

or approximately 37.7 million Americans.

47. $A = \int_{-1}^{2} (3x^2 + 2x + 1)\,dx = x^3 + x^2 + x\big|_{-1}^{2} = [2^3 + 2^2 + 2] - [(-1)^3 + 1 - 1]$
   $= 14 - (-1) = 15.$

49. $A = \int_{1}^{3} \dfrac{1}{x^2}\,dx = \int_{1}^{3} x^{-2}\,dx = -\dfrac{1}{x}\Big|_{1}^{3} = -\dfrac{1}{3} + 1 = \dfrac{2}{3}.$

51.

$A = \int_a^b [f(x) - g(x)]\,dx$

$= \int_0^2 (e^x - x)\,dx$

$= \left(e^x - \dfrac{1}{2}x^2 \right)\Big|_0^2$

$= (e^2 - 2) - (1 - 0) = e^2 - 3.$

$= \tfrac{3}{10}$ sq units.

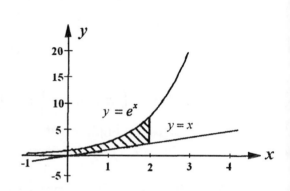

53. $A = \int_0^1 (x^3 - 3x^2 + 2x)\,dx - \int_1^2 (x^3 - 3x^2 + 2x)\,dx$

$= \dfrac{x^4}{4} - x^3 + x^2\Big|_0^1 - \left(\dfrac{x^4}{4} - x^3 + x^2 \right)\Big|_1^2$

$= \tfrac{1}{4} - 1 + 1 - [(4 - 8 + 4) - (\tfrac{1}{4} - 1 + 1)]$

$= \tfrac{1}{4} + \tfrac{1}{4} = \tfrac{1}{2},$

or ½ sq units.

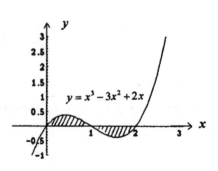

55.
$$A = \frac{1}{3}\int_0^3 \frac{x}{\sqrt{x^2+16}}\,dx = \frac{1}{3}\cdot\frac{1}{2}\cdot 2(x^2+16)^{1/2}\Big|_0^3$$

$$= \frac{1}{3}(x^2+16)^{1/2}\Big|_0^3 = \frac{1}{3}(5-4) = \frac{1}{3}\text{ sq units.}$$

57. Setting $p = 8$, we have $\quad -0.01x^2 - 0.2x + 23 = 8,\ -0.01x^2 - 0.2x + 15 = 0,$ or

$x^2 + 20x - 1500 = (x-30)(x+50) = 0$, giving $x = -50$ or 30.

$$CS = \int_0^{30}(-0.01x^2 - 0.2x + 23)\,dx - 8(30) = -\frac{0.01}{3}x^3 - 0.1x^2 + 23x\Big|_0^{30} - 240$$

$$= -\frac{0.01(30)^3}{3} - 0.1(900) + 23(30) - 240 = 270,\ \text{ or } \$270{,}000.$$

59. Use Equation (17) with $P = 4000$, $r = 0.08$, $T = 20$, and $m = 1$, obtaining
$$A = \frac{(1)(4000)}{0.08}(e^{1.6} - 1) \approx 197{,}651.62$$
that is, Chi-Tai will have approximately \$197,652 in his account after 20 years.

61. Here $P = 80{,}000$, $m = 1$, $T = 10$, and $r = 0.1$, so
$$PV = \frac{(1)(80{,}000)}{0.1}(1 - e^{-1}) \approx 505{,}696, \text{ or approximately } \$505{,}696.$$

63. The average population will be
$$\tfrac{1}{5}\int 80{,}000e^{-0.05t}\,dt = \frac{80{,}000}{5}\cdot\left(-\frac{1}{0.05}\right)e^{-0.05t}\Big|_0^5 = -320{,}000(e^{-0.25} - 1) \approx 70{,}784.$$

1. $\int (2x^3 + \sqrt{x} + \dfrac{2}{x} - \dfrac{2}{\sqrt{x}})dx = 2\int x^3\,dx + \int x^{1/2}\,dx + 2\int \dfrac{1}{x}dx - 2\int x^{-1/2}\,dx$

$$= \tfrac{1}{2}x^4 + \tfrac{2}{3}x^{3/2} + 2\ln|x| - 4x^{1/2} + C$$

2. $f(x) = \int f'(x)\,dx = \int (e^x + x)\,dx = e^x + \tfrac{1}{2}x^2 + C$

   $f(0) = 2$ implies $f(0) = e^0 + 0 + C = 2$ or $C = 1$.

   Therefore, $f(x) = e^x + \tfrac{1}{2}x^2 + 1$.

3. Let $u = x^2 + 1$ so that $du = 2x\,dx$ or $x\,dx = \tfrac{1}{2}du$. Then

$$\int \dfrac{x}{\sqrt{x^2 + 1}}dx = \dfrac{1}{2}\int \dfrac{du}{\sqrt{u}} = \dfrac{1}{2}\int u^{-1/2}\,du = \dfrac{1}{2}\left(2u^{1/2}\right) + C = \sqrt{u} + C = \sqrt{x^2 + 1} + C.$$

4. Let $u = 2 - x^2$. Then $du = -2x\,dx$ or $x\,dx = -\tfrac{1}{2}du$. If $x = 0$, then $u = 2$ and if $x = 1$, then $u = 1$. Therefore,

$$\int_0^1 x\sqrt{2 - x^2}\,dx = -\tfrac{1}{2}\int_2^1 u^{1/2}\,du = -\tfrac{1}{2}\cdot\tfrac{2}{3}u^{3/2}\Big|_2^1 = -\tfrac{1}{3}u^{3/2}\Big|_2^1$$

$$= -\tfrac{1}{3}(1 - 2^{3/2}) = \tfrac{1}{3}(2\sqrt{2} - 1).$$

5. To find the points of intersection, we solve

$$x^2 - 1 = 1 - x,\ x^2 + x - 2 = 0,$$
$$(x + 2)(x - 1) = 0$$

Giving $x = -2$ or $x = 1$. The points of intersection are (-2,3) and (1,0). The required area is

$$A = \int_{-2}^1 [(1 - x) - (x^2 - 1)]\,dx = \int_{-2}^1 (2 - x - x^2)\,dx = (2x - \tfrac{1}{2}x^2 - \tfrac{1}{3}x^3)\Big|_{-2}^1$$

$$= (2 - \tfrac{1}{2} - \tfrac{1}{3}) - (4 - 2 + \tfrac{8}{3}) = \tfrac{9}{2},\ \text{or } \tfrac{9}{2}\ \text{sq units.}$$

CHAPTER 7

7.1 Problem Solving Tips

1. When you use integration by parts, remember to choose u and dv so that du is simpler than u and dv is easy to integrate.

2. When you use the method of integration of parts, it is helpful follow the pattern used in the examples in this section.

$$u = \underline{\hspace{1.5cm}} \qquad dv = \underline{\hspace{1.5cm}}$$

and $\qquad du = \underline{\hspace{1.5cm}} \qquad v = \underline{\hspace{1.5cm}}$

Then $\qquad \int u\,dv = uv - \int v\,du$

7.1 CONCEPT QUESTIONS, page 488

1. $\int u\,dv = uv - \int vdu$

EXERCISES 7.1, page 489

1. $I = \int xe^{2x}\,dx$. Let $u = x$ and $dv = e^{2x}\,dx$. Then $du = dx$ and $v = \frac{1}{2}e^{2x}$. Therefore,

$$I = uv - \int v\,du = \frac{1}{2}xe^{2x} - \int \frac{1}{2}e^{2x}\,dx = \frac{1}{2}xe^{2x} - \frac{1}{4}e^{2x} = \frac{1}{4}e^{2x}(2x-1) + C.$$

3. $I = \int xe^{x/4}dx$. Let $u = x$ and $dv = e^{x/4}\,dx$. Then $du = dx$ and $v = 4e^{x/4}$.

$$\int xe^{x/4}\,dx = uv - \int v\,du = 4xe^{x/4} - 4\int e^{x/4}\,dx = 4xe^{x/4} - 16e^{x/4} + C$$

$$= 4(x-4)e^{x/4} + C.$$

5. $\int (e^x - x)^2\, dx = \int (e^{2x} - 2xe^x + x^2)\, dx = \int e^{2x}\, dx - 2\int xe^x\, dx + \int x^2\, dx.$

Using the result $\int xe^x\, dx = (x-1)e^x + k$, from Example 1, we see that

$\int (e^x - x)^2\, dx = \frac{1}{2}e^{2x} - 2(x-1)e^x + \frac{1}{3}x^3 + C.$

7. $I = \int (x+1)e^x\, dx.$ Let $u = x + 1$, $dv = e^x\, dx$. Then $du = dx$ and $v = e^x$. Therefore,

$I = (x+1)e^x - \int e^x\, dx = (x+1)e^x - e^x + C = xe^x + C.$

9.   Let $u = x$ and $dv = (x+1)^{-3/2}\, dx$. Then $du = dx$ and $v = -2(x+1)^{-1/2}$.

$$\int x(x+1)^{-3/2}\, dx = uv - \int v\, du = -2x(x+1)^{-1/2} + 2\int (x+1)^{-1/2}\, dx$$

$$= -2x(x+1)^{-1/2} + 4(x+1)^{1/2} + C$$

$$= 2(x+1)^{-1/2}[-x+2(x+1)] + C = \frac{2(x+2)}{\sqrt{x+1}} + C.$$

11. $I = \int x(x-5)^{1/2}\, dx.$ Let $u = x$ and $dv = (x-5)^{1/2}\, dx$. Then $du = dx$ and

$v = \frac{2}{3}(x-5)^{3/2}$. Therefore,

$$I = \frac{2}{3}x(x-5)^{3/2} - \int \frac{2}{3}(x-5)^{3/2}\, dx = \frac{2}{3}x(x-5)^{3/2} - \frac{2}{3}\cdot\frac{2}{5}(x-5)^{5/2} + C$$

$$= \frac{2}{3}(x-5)^{3/2}[x - \frac{2}{5}(x-5)] + C = \frac{2}{15}(x-5)^{3/2}(5x - 2x + 10) + C$$

$$= \frac{2}{15}(x-5)^{3/2}(3x+10) + C.$$

13. $I = \int x \ln 2x\, dx.$ Let $u = \ln 2x$ and $dv = x\, dx$. Then $du = \frac{1}{x}\, dx$ and $v = \frac{1}{2}x^2$.

Therefore, $I = \frac{1}{2}x^2 \ln 2x - \int \frac{1}{2}x\, dx = \frac{1}{2}x^2 \ln 2x - \frac{1}{4}x^2 + C = \frac{1}{4}x^2(2\ln 2x - 1) + C.$

15. Let $u = \ln x$ and $dv = x^3\,dx$, then $du = \frac{1}{x}\,dx$, and $v = \frac{1}{4}x^4$.

$$\int x^3 \ln x\,dx = \frac{1}{4}x^4 \ln x - \frac{1}{4}\int x^3\,dx = \frac{1}{4}x^4 \ln x - \frac{1}{16}x^4 + C$$
$$= \frac{1}{16}x^4(4 \ln x - 1) + C.$$

17. Let $u = \ln x^{1/2}$ and $dv = x^{1/2}\,dx$. Then $du = \frac{1}{2x}\,dx$ and $v = \frac{2}{3}x^{3/2}$,

and $\int \sqrt{x} \ln \sqrt{x}\,dx = uv - \int v\,du = \frac{2}{3}x^{3/2} \ln x^{1/2} - \frac{1}{3}\int x^{1/2}\,dx$
$$= \frac{2}{3}x^{3/2} \ln x^{1/2} - \frac{2}{9}x^{3/2} + C = \frac{2}{9}x\sqrt{x}(3 \ln \sqrt{x} - 1) + C.$$

19. Let $u = \ln x$ and $dv = x^{-2}\,dx$. Then $du = \frac{1}{x}\,dx$ and $v = -x^{-1}$,

$$\int \frac{\ln x}{x^2}\,dx = uv - \int v\,du = -\frac{\ln x}{x} + \int x^{-2}\,dx = -\frac{\ln x}{x} - \frac{1}{x} + C$$
$$= -\frac{1}{x}(\ln x + 1) + C.$$

21. Let $u = \ln x$ and $dv = dx$. Then $du = \frac{1}{x}\,dx$ and $v = x$ and

$$\int \ln x\,dx = uv - \int v\,du = x \ln x - \int dx = x \ln x - x + C = x(\ln x - 1) + C.$$

23. Let $u = x^2$ and $dv = e^{-x}\,dx$. Then $du = 2x\,dx$ and $v = -e^{-x}$, and

$$\int x^2 e^{-x}\,dx = uv - \int v\,du = -x^2 e^{-x} + 2\int xe^{-x}\,dx.$$

We can integrate by parts again, or, using the result of Problem 2, we find

$$\int x^2 e^{-x}\,dx = -x^2 e^{-x} + 2[-(x+1)e^{-x}] + C = -x^2 e^{-x} - 2(x+1)e^{-x} + C$$
$$= -(x^2 + 2x + 2)e^{-x} + C.$$

25. $I = \int x(\ln x)^2\,dx$. Let $u = (\ln x)^2$ and $dv = x\,dx$, so that

$$du = 2(\ln x)\left(\frac{1}{x}\right) = \frac{2 \ln x}{x} \text{ and } v = \frac{1}{2}x^2. \text{ Then } I = \frac{1}{2}x^2(\ln x)^2 - \int x \ln x\,dx.$$

Next, we evaluate $\int x \ln x\,dx$, by letting $u = \ln x$ and $dv = x\,dx$, so that $du = \frac{1}{x}\,dx$

7 *Additional Topics in Integration*

and $v = \frac{1}{2}x^2$. Then $\int x \ln x\, dx = \frac{1}{2}x^2 (\ln x) - \frac{1}{2}\int x\, dx = \frac{1}{2}x^2 \ln x - \frac{1}{4}x^2 + C.$

Therefore, $\int x(\ln x)^2\, dx = \frac{1}{2}x^2 (\ln x)^2 - \frac{1}{2}x^2 \ln x + \frac{1}{4}x^2 + C$

$$= \frac{1}{4}x^2 [2(\ln x)^2 - 2\ln x + 1] + C.$$

27. $\displaystyle\int_0^{\ln 2} xe^x\, dx = (x-1)e^x\Big|_0^{\ln 2}$ (Using the results of Example 1.)

$$= (\ln 2 - 1)e^{\ln 2} - (-e^0) = 2(\ln 2 - 1) + 1 \text{ (Recall } e^{\ln 2} = 2.) = 2\ln 2 - 1.$$

29. We first integrate $I = \int \ln x\, dx$. Integrating by parts with $u = \ln x$ and $dv = dx$ so

that $du = \frac{1}{x}\, dx$ and $v = x$, we find

$$I = x \ln x - \int dx = x \ln x - x + C = x(\ln x - 1) + C.$$

Therefore, $\displaystyle\int_1^4 \ln x\, dx = x(\ln x - 1)\Big|_1^4 = 4(\ln 4 - 1) - 1(\ln 1 - 1) = 4\ln 4 - 3.$

31. Let $u = x$ and $dv = e^{2x}\, dx$. Then $u = dx$ and $v = \frac{1}{2}e^{2x}$ and

$$\int_0^2 xe^{2x}\, dx = \frac{1}{2}xe^{2x}\Big|_0^2 - \frac{1}{2}\int_0^2 e^{2x}\, dx = e^4 - \frac{1}{4}e^{2x}\Big|_0^2$$
$$= e^4 - \frac{1}{4}e^4 + \frac{1}{4} = \frac{1}{4}(3e^4 + 1).$$

33. Let $u = x$ and $dv = e^{-2x}\, dx$, so that $du = dx$ and $v = -\frac{1}{2}e^{-2x}$.

$$f(x) = \int xe^{-2x}\, dx = -\frac{1}{2}xe^{-2x} - \frac{1}{4}e^{-2x} + C \;\; ; \; f(0) = -\frac{1}{4} + C = 3 \text{ and } C = \frac{13}{4}.$$

Therefore, $y = -\frac{1}{2}xe^{-2x} - \frac{1}{4}e^{-2x} + \frac{13}{4}.$

35. The required area is given by $\displaystyle\int_1^5 \ln x\, dx$. We first find $\int \ln x\, dx$. Using the

technique of integration by parts with $u = \ln x$ and $dv = dx$ so that $du = \frac{1}{x}\, dx$ and

$v = x$, we have

$$\int \ln x\, dx = x \ln x - \int dx = x \ln x - x = x(\ln x - 1) + C.$$

Therefore, $\int_1^5 \ln x \, dx = x(\ln x - 1)\Big|_1^5 = 5(\ln 5 - 1) - 1(\ln 1 - 1) = 5 \ \ln 5 - 4$

and the required area is (5 ln 5 - 4) sq units.

37. The distance covered is given by $\int_0^{10} 100te^{-0.2t} \, dt = 100\int_0^{10} te^{-0.2t} \, dt.$

We integrate by parts, letting $u = t$ and $dv = e^{-0.2t} \, dt$ so that $du = dt$ and

$v = -\dfrac{1}{0.2}e^{-0.2t} = -5e^{-0.2t}$. Therefore,

$$100\int_0^{10} te^{-0.2t} \, dt = 100\left[-5te^{-0.2t}\Big|_0^{10}\right] + 5\int_0^{10} e^{-0.2t} \, dt$$

$$= 100[-5te^{-0.2t} - 25e^{-0.2t}]\Big|_0^{10} = -500e^{-0.2t}(t + 5)\Big|_0^{10}$$

$$= -500e^{-2}(15) + 500(5) = 1485, \text{ or } 1485 \text{ feet.}$$

39. The average concentration is $C = \dfrac{1}{12}\int_0^{12} 3te^{-t/3} \, dt = \dfrac{1}{4}\int_0^{12} te^{-t/3} \, dt.$

Let $u = t$ and $dv = e^{-t/3} \, dt$. So $du = dt$ and $v = -3e^{-t/3}$. Then

$$C = \frac{1}{4}\left[-3te^{-t/3}\Big|_0^{12} + 3\int_0^{12} e^{-t/3} \, dt\right] = \frac{1}{4}\left\{-36e^{-4} - \left[9e^{-t/3}\Big|_0^{12}\right]\right\}$$

$$= \tfrac{1}{4}(-36e^{-4} - 9e^{-4} + 9) \approx 2.04 \text{ mg/ml.}$$

41. $N = 2\int te^{-0.1t} \, dt.$ Let $u = t$ and $dv = e^{-0.1t}$, so that $du = dt$ and $v = -10e^{-0.1t}$. Then

$v = -10e^{-0.1t}$. Then

$$N(t) = 2[-10te^{-0.1t} + 10\int e - 0.1t \, dt] = 2(-10te^{-0.1t} - 100e^{-0.1t}) + C$$

$$= -20e^{-0.1t}(t + 10) + 200. \qquad\qquad [N(0) = 0]$$

43. $PV = \int_0^5 (30{,}000 + 800t)e^{-0.08t} \, dt = 30{,}000\int_0^5 e^{-0.08t}dt + 800\int_0^5 te^{-0.08t} \, dt.$

Let $I = \int te^{-0.08t}dt.$ To evaluate I by parts, let $u = t,\ dv = e^{-0.08t}dt$

and $du = dt,\ v = -\dfrac{1}{0.08}e^{-0.08t} = -12.5e^{-0.08t}.$

Therefore, $I = -12.5te^{-0.08t} + 12.5\int e^{-0.08t} \, dt = -12.5te^{-0.08t} - 156.25e^{-0.08t} + C.$

$$PV = \left[-\frac{30{,}000}{0.08}e^{-0.08t} - 800(12.5)te^{-0.08t} - 800(156.25)e^{-0.08t} \right]_0^5$$

$$= -375{,}000\,e^{-0.4} + 375{,}000 - 50{,}000e^{-0.4} - 125{,}000e^{-0.4} + 125{,}000$$

$$= 500{,}000 - 550{,}000e^{-0.4} = 131{,}323.97, \text{ or approximately } \$131{,}324.$$

45. The membership will be

$$N(5) = N(0) + \int_0^5 9\sqrt{t+1}\,\ln\sqrt{t+1}\,dt = 50 + 9\int_0^5 \sqrt{t+1}\,\ln\sqrt{t+1}\,dt$$

To evaluate the integral, let $u = t + 1$ so that $du = dt$. Also, if $t = 0$, then $u = 1$ and if $t = 5$, then $u = 6$. So $9\int_0^5 \sqrt{t+1}\,\ln\sqrt{t+1}\,dt = 9\int_1^6 \sqrt{u}\,\ln\sqrt{u}\,du$.

Using the results of Problem 17, we find $9\int_1^6 \sqrt{u}\,\ln\sqrt{u}\,du = 2u\sqrt{u}(3\ln\sqrt{u} - 1)\Big|_1^6$.

Therefore, $N = 50 + 51.606 \approx 101.606$ or 101,606 people.

47. The average concentration from $r = r_1$ to $r = r_2$ is

$$A = \frac{1}{r^2 - r_1}\int_{r_1}^{r_2} c(r)\,dr = \frac{1}{r^2 - r_1}\int_{r_1}^{r_2}\left[\left(\frac{c_1 - c_2}{\ln r_1 - \ln r_2}\right)(\ln r - \ln r_2) + c_2\right]dr$$

$$= \frac{1}{r_2 - r_1}\left(\frac{c_1 - c_2}{\ln r_1 - \ln r_2}\right)\left[\int_{r_1}^{r_2}\ln r\,dr - \int_{r_1}^{r_2}\ln r_2\,dr\right] + \frac{1}{r_2 - r_1}\int_{r_1}^{r_2} c_2\,dr$$

We integrate $\int \ln r\,dr$ by parts, letting $u = \ln r$, $dv = dr$, or $du = \dfrac{dr}{r}$, and $v = r$, so

that $\int \ln r\,dr = r\ln r - \int r\,\dfrac{dr}{r} = r\ln r - r = r(\ln r - 1)$.

Therefore,

$$A = \frac{1}{r_2 - r_1}\left(\frac{c_1 - c_2}{\ln r_1 - \ln r_2}\right)\left[r(\ln r - 1)\Big|_{r_1}^{r_2} - (\ln r_2)\Big|_{r_1}^{r_2}\right] + c_2$$

$$= \frac{1}{r_2 - r_1}\left(\frac{c_1 - c_2}{\ln r_1 - \ln r_2}\right)\{r_2(\ln r_2 - 1) - r_1(\ln r_1 - 1) - (r_2 - r_1)\ln r_2\} + c_2$$

$$= \frac{1}{r_2 - r_1}\left(\frac{c_1 - c_2}{\ln r_1 - \ln r_2}\right)[r_1(\ln r_2 - \ln r_1) - (r_2 - r_1)] + c_2]$$

$$= (c_2 - c_1)\left[\frac{r_1}{r_2 - r_1} + \frac{1}{\ln r_1 - \ln r_2}\right] + c_2.$$

49. True. This is just the integration by parts formula.

7.2 Problem Solving Tips

1. The integrals that follow in the exercise set may not be exactly in the same form as those in the table of integrals. Sometimes you may need to rewrite the integral (as in Example 2, page 494 of the text) or you may need to apply one rule more than once (as in Example 5, page 495 in the text).

7.2 CONCEPT QUESTIONS, page 496

1. a. We would chose Formula 19.
   b. Put $a = \sqrt{2}$ and $x = u$. Then, using Formula 19, we find

$$\int \frac{\sqrt{2-x^2}}{x}\,dx = \int \frac{\sqrt{(\sqrt{2})^2 - x^2}}{x}\,dx = \sqrt{2-x^2} - \sqrt{2}\ln\left|\frac{\sqrt{a} + \sqrt{2-x^2}}{x}\right| + C$$

EXERCISES 7.2, page 497

1. First we note that

$$\int \frac{2x}{2+3x}\,dx = 2\int \frac{x}{2+3x}\,dx.$$

Next, we use Formula 1 with $a = 2$, $b = 3$, and $u = x$. Then

$$\int \frac{2x}{2+3x}\,dx = \frac{2}{9}[2+3x - 2\ln|2+3x|] + C.$$

3. $\displaystyle\int \frac{3x^2}{2+4x}\,dx = \frac{3}{2}\int \frac{x^2}{1+2x}\,dx.$
   Use Formula 2 with $a = 1$ and $b = 2$ obtaining

$$\int \frac{3x^2}{2+4x}\,dx = \frac{3}{32}[(1+2x)^2 - 4(1+2x) + 2\ln|1+2x|] + C.$$

5. $\displaystyle\int x^2\sqrt{9+4x^2}\,dx = \int x^2\sqrt{4(\tfrac{9}{4}) + x^2)}\,dx = 2\int x^2\sqrt{(\tfrac{3}{2})^2 + x^2}\,dx$.

   Use Formula 8 with $a = 3/2$, we find that

$$\int x^2\sqrt{9+4x^2}\,dx == 2[\tfrac{x}{8}(\tfrac{9}{4} + 2x^2)\sqrt{\tfrac{9}{4} + x^2} - \tfrac{81}{128}\ln\left|x + \sqrt{\tfrac{9}{4} + x^2}\right| + C.$$

7. Use Formula 6 with $a = 1$, $b = 4$, and $u = x$, then

$$\int \frac{dx}{x\sqrt{1+4x}} = \ln\left|\frac{\sqrt{1+4x}-1}{\sqrt{1+4x}+1}\right| + C.$$

9. Use Formula 9 with $a = 3$ and $u = 2x$. Then $du = 2\,dx$ and

$$\int_0^2 \frac{dx}{\sqrt{9+4x^2}} = \frac{1}{2}\int_0^4 \frac{du}{\sqrt{3^2 + u^2}} = \frac{1}{2}\ln\left|u + \sqrt{9+u^2}\right|\Big|_0^4$$

$$= \frac{1}{2}(\ln 9 - \ln 3) = \frac{1}{2}\ln 3.$$

   Note that the limits of integration have been changed from $x = 0$ to $x = 2$ and from $u = 0$ to $u = 4$.

11. Using Formula 22 with $a = 3$, we see that $\displaystyle\int \frac{dx}{(9-x^2)^{3/2}} = \frac{x}{9\sqrt{9-x^2}} + C.$

13. $\displaystyle\int x^2\sqrt{x^2 - 4}\,dx.$

   Use Formula 14 with $a = 2$ and $u = x$, obtaining

$$\int x^2\sqrt{x^2-4}\,dx = \tfrac{x}{8}(2x^2 - 4)\sqrt{x^2-4} - 2\ln\left|x + \sqrt{x^2-4}\right| + C.$$

15. Using Formula 19 with $a = 2$ and $u = x$, we have

$$\int \frac{\sqrt{4-x^2}}{x}\,dx = \sqrt{4-x^2} - 2\ln\left|\frac{2+\sqrt{4-x^2}}{x}\right| + C.$$

17. $\int xe^{2x}\,dx.$

Use Formula 23 with $u = x$ and $a = 2$, obtaining

$$\int xe^{2x}\,dx = \frac{1}{4}(2x-1)e^{2x} + C.$$

19. $\int \dfrac{dx}{(x+1)\ln(x+1)}.$

Let $u = x + 1$ so that $du = dx$. Then $\displaystyle\int \frac{dx}{(x+1)\ln(x+1)} = \int \frac{du}{u\ln u}.$

Use Formula 28 with $u = x$, obtaining $\displaystyle\int \frac{du}{u\ln u} = \ln|\ln u| + C.$

Therefore, $\displaystyle\int \frac{dx}{(x+1)\ln(x+1)} = \ln|\ln(x+1)| + C$

21. $\int \dfrac{e^{2x}}{(1+3e^x)^2}\,dx.$

Put $u = e^x$ then $du = e^x dx$. Then we use Formula 3 with $a = 1, b = 3$. Then

$$I = \int \frac{u}{(1+3u)^2}\,du = \frac{1}{9}\left[\frac{1}{1+3u} + \ln|1+3u|\right] + C = \frac{1}{9}\left[\frac{1}{1+3e^x} + \ln(1+3e^x)\right] + C$$

23. $\int \dfrac{3e^x}{1+e^{x/2}}\,dx = 3\int \dfrac{e^{x/2}}{e^{-x/2}+1}\,dx.$

Let $v = e^{x/2}$ so that $dv = \frac{1}{2}e^{x/2}dx$ or $e^{x/2}\,dx = 2\,dv$. Then

$$\int \frac{3e^x}{1+e^{x/2}}\,dx = 6\int \frac{dv}{\frac{1}{v}+1} = 6\int \frac{v}{v+1}\,dv.$$

Use Formula 1 with $a = 1$, $b = 1$, and $u = v$, obtaining

$$6\int \frac{v}{v+1}\,dv = 6[1+v-\ln|1+v|] + C. \text{ So } \int \frac{3e^x}{1+e^{x/2}}\,dx = 6[1+e^{x/2}-\ln(1+e^{x/2})] + C.$$

This answer may be written in the form $6[e^{x/2}-\ln(1+e^{x/2})] + C$ since C is an arbitrary constant.

25. $\int \dfrac{\ln x}{x(2+3\ln x)}\,dx.$ Let $v = \ln x$ so that $dv = \dfrac{1}{x}\,dx.$ Then

$$\int \frac{\ln x}{x(2+3\ln x)}\,dx = \int \frac{v}{2+3v}\,dv.$$

Use Formula 1 with $a = 2$, $b = 3$, and $u = v$ to obtain

$$\int \frac{v}{2+3v}\,dv = \tfrac{1}{9}[2+3\ln x - 2\ln|2+3\ln x|]+C\ .\ \text{ So}$$

$$\int \frac{\ln x}{x(2+3\ln x)}\,dx = \tfrac{1}{9}[2+3\ln x - 2\ln|2+3\ln x|+C.$$

27. Using Formula 24 with $a = 1$, $n = 2$, and $u = x$. Then

$$\int_0^1 x^2 e^x\,dx = x^2 e^x\Big|_0^1 - 2\int_0^1 xe^x\,dx = x^2 e^x - 2(xe^x - e^x)\Big|_0^1$$

$$= x^2 e^x - 2xe^x + 2e^x\Big|_0^1 = e - 2e + 2e - 2 = e - 2.$$

29. $\int x^2 \ln x\,dx.$ Use Formula 27 with $n = 2$ and $u = x$, obtaining

$$\int x^2 \ln x\,dx = \frac{x^3}{9}(3\ln x - 1)+C.$$

31. $\int (\ln x)^3 dx.$ Use Formula 29 with $n = 3$ to write

$$\int (\ln x)^3\,dx = x(\ln x)^3 - 3\int (\ln x)^2\,dx.\qquad \text{Using Formula 29 again with } n = 2, \text{ we}$$

obtain

$$\int (\ln x)^3\,dx = x(\ln x)^3 - 3[x(\ln x)^2 - 2\int \ln x\,dx].$$

Using Formula 29 one more time with $n = 1$ gives

$$\int (\ln x)^3\,dx = x(\ln x)^3 - 3x(\ln x)^2 + 6(x\ln x - x)+C$$

$$= x(\ln x)^3 - 3x(\ln x)^2 + 6x\ln x - 6x + C.$$

33. Letting $p = 50$ gives $50 = \dfrac{250}{\sqrt{16+x^2}}$, from which we deduce that

$\sqrt{16+x^2} = 5$, $16+x^2 = 25$, and $x = 3$. Using Formula 9 with $u = 3$, we see that

$$CS = \int_0^3 \frac{250}{\sqrt{16+x^2}}\,dx = 50(3) = 250\int_0^3 \frac{1}{\sqrt{16+x^2}}\,dx - 150$$

$$= 250\ln\left|x + \sqrt{16+x^2}\right|\Big|_0^3 - 150 = 250[\ln 8 - \ln 4] - 150$$

$$= 23.286795, \text{ or approximately } \$2,329.$$

35. The number of visitors admitted to the amusement park by noon is found by evaluating the integral

$$\int_0^3 \frac{60}{(2+t^2)^{3/2}}\,dt = 60\int_0^3 \frac{dt}{(2+t)^{3/2}}.$$

Using Formula 12 with $a = \sqrt{2}$ and $u = t$, we find

$$60\int_0^3 \frac{dt}{(2+t^2)^{3/2}} = 60\left[\frac{t}{2\sqrt{2+t^3}}\right]\Big|_0^3 = 60\left[\frac{3}{2\sqrt{11}-0}\right] = \frac{90}{\sqrt{11}} = 27.136, \text{ or } 27,136.$$

37. In the first 10 days

$$\frac{1}{10}\int_0^{10} \frac{1000}{1+24e^{-0.02t}}\,dt = 100\int_0^{10} \frac{1}{1+24e^{-0.02t}}\,dt = 100\left[t + \frac{1}{0.02}\ln(1+24e^{-0.02t})\right]\Big|_0^{10}$$

(Use Formula 25 with $a = 0.02$ and $b = 24$.)

$$= 100[10 + 50\ln 20.64953807 - 50\ln 25] = 44.0856,$$

or approximately 44 fruitflies. In the first 20 days:

$$\frac{1}{20}\int_0^{20} \frac{1000}{1+24e^{-0.02t}}\,dt = 50\int_0^{10} \frac{1}{1+24e^{-0.02t}}\,dt$$

$$= 500[t + \ln(1+24e^{-0.02t})]\Big|_0^{20}$$

$$= 50[20 + 50\ln 17.0876822 - 50\ln 25] = 48.71$$

or approximately 49 fruitflies.

39. $\dfrac{1}{5}\displaystyle\int_0^5 \frac{100,000}{2(1+1.5e^{-0.2t})}\,dt = 10,000\int_0^5 \frac{1}{1+1.5e^{-0.2t}}\,dt$

$$= 10,000[t + 5\ln(1 + 1.5e^{-0.2t})] \Big|_0^5$$

(Use Formula 25 with a = -0.2 and b = 1.5.)

$$= 10,000[5 + 5\ln 1.551819162 - 5\ln 2.5] \approx 26157,$$

or approximately 26,157 people.

41. $\displaystyle\int_0^5 20,000te^{0.15t}\,dt = 20,000\int_0^5 te^{0.15t}\,dt = 20,000\left[\frac{1}{(0.15)^2}(0.15t-1)e^{0.15t}\right]\Big|_0^5$

(Use Formula 23 with a = 0.15.)

$$= 888,888.8889[-0.25e^{0.75} + 1] = \$418,444.$$

7.3 CONCEPT QUESTIONS, page 509

1. In the trapezoidal rule, each region under the graph of f (or over the graph of f) is approximated by the area of a trapezoid whose base consists of two consecutive points in the partition. Therefore, n can be odd or even. In Simpson's rule, the area of each subregion is approximated by part of a parabola passing through those points. Therefore, there are two subintervals involved in the approximations. This implies that n must be even.

3. If we use the trapezoidal rule and f is a linear function, then $f''(x) = 0$, and therefore, $M = 0$, and consequently the maximum error is 0. If we use Simpson's rule, then $f^4(x) = 0$ and therefore, $M = 0$ and consequently the maximum error is 0.

EXERCISES 7.3, page 510

1. $\Delta x = \frac{2}{6} = \frac{1}{3}, x_0 = 0, x_1 = \frac{1}{3}, x_2 = \frac{2}{3}, x_3 = 1, x_4 = \frac{4}{3}, x_5 = \frac{5}{3}, x_6 = 2.$

   Trapezoidal Rule:

   $\displaystyle\int_0^2 x^2\,dx \approx \frac{1}{6}\left[0 + 2(\tfrac{1}{3})^2 + 2(\tfrac{2}{3})^2 + 2(1)^2 + 2(\tfrac{4}{3})^2 + 2(\tfrac{5}{3})^2 + 2^2\right]$

   $\approx \frac{1}{6}\,(0.22222 + 0.88889 + 2 + 3.55556 + 5.55556 + 4) \approx 2.7037.$

   Simpson's Rule:

   $\displaystyle\int x^2\,dx = \frac{1}{9}\left[0 + 4(\tfrac{1}{3})^2 + 2(\tfrac{2}{3})^2 + 4(1)^2 + 2(\tfrac{4}{3})^2 + 4(\tfrac{5}{3})^2 + 2^2\right]$

   $\approx \frac{1}{9}\,(0.44444 + 0.88889 + 4 + 3.55556 + 11.11111 + 4) \approx 2.6667.$

Exact Value: $\int_0^2 x^2\,dx = \frac{1}{3}x^3\Big|_0^2 = \frac{8}{3} = 2\frac{2}{3}$.

3. $\Delta x = \frac{b-a}{n} = \frac{1-0}{4} = \frac{1}{4}; x_0 = 0, x_1 = \frac{1}{4}, x_2 = \frac{1}{2}, x_3 = \frac{3}{4}, x_4 = 1.$

Trapezoidal Rule:

$$\int_0^1 x^3\,dx \approx \frac{1}{8}\left[0 + 2(\tfrac{1}{4})^3 + 2(\tfrac{1}{2})^3 + 2(\tfrac{3}{4})^3 + 1^3\right] \approx \frac{1}{8}(0 + 0.3125 + 0.25 + 0.8)$$
$$\approx 0.265625.$$

Simpson's Rule:

$$\int_0^1 x^3\,dx \approx \frac{1}{12}\left[0 + 4(\tfrac{1}{4})^3 + 2(\tfrac{1}{2})^3 + 4(\tfrac{3}{4})^3 + 1\right] \approx \frac{1}{12}[0 + 0.625 + 0.25 + 1.6875 + 1]$$
$$\approx 0.25.$$

Exact Value: $\int_0^1 x^3\,dx = \frac{1}{4}x^4\Big|_0^1 = \frac{1}{4} - 0 = \frac{1}{4}$.

5. a. Here $a = 1$, $b = 2$, and $n = 4$; so $\Delta x = \frac{2-1}{4} = \frac{1}{4} = 0.25$, and $x_0 = 1$, $x_1 = 1.25$, $x_2 = 1.5$, $x_3 = 1.75$, $x_4 = 2$.

Trapezoidal Rule:

$$\int_1^2 \frac{1}{x}\,dx \approx \frac{0.25}{2}\left[1 + 2\left(\frac{1}{1.25}\right) + 2\left(\frac{1}{1.5}\right) + 2\left(\frac{1}{1.75}\right) + \frac{1}{2}\right] \approx 0.697.$$

Simpson's Rule:

$$\int_1^2 \frac{1}{x}\,dx \approx \frac{0.25}{3}\left[1 + 4\left(\frac{1}{1.25}\right) + 2\left(\frac{1}{1.5}\right) + 4\left(\frac{1}{1.75}\right) + \frac{1}{2}\right] \approx 0.6933.$$

$$\int_1^2 \frac{1}{x}\,dx = \ln x\Big|_1^2 = \ln 2 - \ln 1 \approx 0.6931.$$

7. $\Delta x = \frac{1}{4}$, $x_0 = 1$, $x_1 = \frac{5}{4}$, $x_2 = \frac{3}{2}$, $x_3 = \frac{7}{4}$, $x_4 = 2$.

Trapezoidal Rule:

$$\int_1^2 \frac{1}{x^2}\,dx \approx \frac{1}{8}\left[1 + 2(\tfrac{4}{5})^2 + 2(\tfrac{2}{3})^2 + 2(\tfrac{4}{7})^2 + (\tfrac{1}{2})^2\right] \approx 0.5090.$$

Simpson's Rule:

$$\int_1^2 \frac{1}{x^2}\,dx \approx \frac{1}{12}\left[1 + 4(\tfrac{4}{5})^2 + 2(\tfrac{2}{3})^2 + 4(\tfrac{4}{7})^2 + (\tfrac{1}{2})^2\right] \approx 0.5004.$$

Exact Value: $\displaystyle\int_1^2 \frac{1}{x^2}\,dx = -\frac{1}{x}\Big|_1^2 = -\frac{1}{2}+1 = \frac{1}{2}$.

9. $\Delta x = \frac{b-a}{n} = \frac{4-0}{8} = \frac{1}{2}; x_0 = 0, x_1 = \frac{1}{2}, x_2 = \frac{2}{2}, x_3 = \frac{3}{2}, \ldots, x_8 = \frac{8}{2}$.

Trapezoidal Rule:

$\displaystyle\int_0^4 \sqrt{x}\,dx \approx \frac{\frac{1}{2}}{2}\left[0 + 2\sqrt{0.5} + 2\sqrt{1} + 2\sqrt{1.5} + \cdots + 2\sqrt{3.5} + \sqrt{4}\right] \approx 5.26504$.

Simpson's Rule:

$\displaystyle\int_0^4 \sqrt{x}\,dx \approx \frac{\frac{1}{2}}{3}\left[0 + 4\sqrt{0.5} + 2\sqrt{1} + 4\sqrt{1.5} + \cdots + 4\sqrt{3.5} + \sqrt{4}\right] \approx 5.30463$.

The actual value is $\displaystyle\int_0^4 \sqrt{x}\,dx \approx \frac{2}{3}x^{3/2}\Big|_0^4 = \frac{2}{3}(8) = \frac{16}{3} \approx 5.333333$.

11. $\Delta x = \frac{1-0}{6} = \frac{1}{6}; x_0 = 0, x_1 = \frac{1}{6}, x_2 = \frac{2}{6}, \ldots, x_6 = \frac{6}{6}$.

Trapezoidal Rule:

$\displaystyle\int_0^1 e^{-x}\,dx \approx \frac{\frac{1}{6}}{2}[1 + 2e^{-1/6} + 2e^{-2/6} + \cdots + 2e^{-5/6} + e^{-1}] \approx 0.633583$.

Simpson's Rule:

$\displaystyle\int_0^1 e^{-x}\,dx \approx \frac{\frac{1}{6}}{3}[1 + 4e^{-1/6} + 2e^{-2/6} + \cdots + 4e^{-5/6} + e^{-1}] \approx 0.632123$.

The actual value is $\displaystyle\int_0^1 e^{-x}\,dx = -e^{-x}\Big|_0^1 = -e^{-1} + 1 \approx 0.632121$.

13. $\Delta x = \frac{1}{4}; x_0 = 0, x_1 = \frac{5}{4}, x_2 = \frac{3}{2}, x_3 = \frac{7}{4}, x_4 = 2$.

Trapezoidal Rule:

$\displaystyle\int_1^2 \ln x\,dx \approx \frac{1}{8}[\ln 1 + 2\ln\frac{5}{4} + 2\ln\frac{3}{2} + 2\ln\frac{7}{4} + \ln 2] \approx 0.38370$.

Simpson's Rule:

$\displaystyle\int_1^2 \ln x\,dx \approx \frac{1}{12}[\ln 1 + 4\ln\frac{5}{4} + 2\ln\frac{3}{2} + 4\ln\frac{7}{4} + \ln 2] \approx 0.38626$.

Exact Value: $\displaystyle\int_1^2 \ln x\,dx \approx x(\ln x - 1)\Big|_1^2 = 2(\ln 2 - 1) + 1 = 2\ln 2 - 1 \approx 0.3863$.

15. $\Delta x = \frac{1-0}{4} = \frac{1}{4}; x_0 = 0, x_1 = \frac{1}{4}, x_2 = \frac{2}{4}, x_3 = \frac{3}{4}, x_4 = \frac{4}{4}$.

Trapezoidal Rule:

$$\int_0^1 \sqrt{1+x^3}\,dx \approx \frac{1}{4}\left[\sqrt{1}+2\sqrt{1+(\tfrac{1}{4})^3}+\cdots+2\sqrt{1+(\tfrac{3}{4})^3}+\sqrt{2}\right] \approx 1.1170.$$

Simpson's Rule:

$$\int_0^1 \sqrt{1+x^3}\,dx \approx \frac{1}{3}\left[\sqrt{1}+4\sqrt{1+(\tfrac{1}{4})^3}+2\sqrt{1+(\tfrac{2}{4})^3}\cdots+4\sqrt{1+(\tfrac{3}{4})^3}+\sqrt{2}\right] \approx 1.1114.$$

17. $\Delta x = \frac{2-0}{4} = \frac{1}{2}; x_0 = 0, x_1 = \frac{1}{2}, x_2 = \frac{2}{2}, x_3 = \frac{3}{2}, x_4 = \frac{4}{2}.$

Trapezoidal Rule:

$$\int_0^2 \frac{1}{\sqrt{x^3+1}}\,dx = \frac{\frac{1}{2}}{2}\left[1+\frac{2}{\sqrt{(\tfrac{1}{2})^3+1}}+\frac{2}{\sqrt{(1)^3+1}}+\frac{2}{\sqrt{(\tfrac{3}{2})^3+1}}+\frac{1}{\sqrt{(2)^3+1}}\right]$$
$$\approx 1.3973$$

Simpson's Rule:

$$\int_0^2 \frac{1}{\sqrt{x^3+1}}\,dx = \frac{\frac{1}{2}}{3}\left[1+\frac{4}{\sqrt{(\tfrac{1}{2})^3+1}}+\frac{2}{\sqrt{(1)^3+1}}+\frac{4}{\sqrt{(\tfrac{3}{2})^3+1}}+\frac{1}{\sqrt{(2)^3+1}}\right]$$
$$\approx 1.4052$$

19. $\Delta x = \frac{2}{4} = \frac{1}{2}; x_0 = 0, x_1 = \frac{1}{2}, x_2 = 1, x_3 = \frac{3}{2}, x_4 = 2.$

Trapezoidal Rule:

$$\int_0^2 e^{-x^2}\,dx = \frac{1}{4}[e^{-0}+2e^{-(1/2)^2}+2e^{-1}+2e^{-(3/2)^2}+e^{-4}] \approx 0.8806.$$

Simpson's Rule:

$$\int_0^2 e^{-x^2}\,dx = \frac{1}{6}[e^{-0}+4e^{-(1/2)^2}+2e^{-1}+4e^{-(3/2)^2}+e^{-4}] \approx 0.8818.$$

21. $\Delta x = \frac{2-1}{4} = \frac{1}{4}; x_0 = 1, x_1 = \frac{5}{4}, x_2 = \frac{6}{4}, x_3 = \frac{7}{4}, x_4 = \frac{8}{4}.$

Trapezoidal Rule:

$$\int_1^2 x^{-1/2}e^x\,dx = \frac{\frac{1}{4}}{2}\left[e+\frac{2e^{5/4}}{\sqrt{\frac{5}{4}}}+\cdots+\frac{2e^{7/4}}{\sqrt{\frac{7}{4}}}+\frac{e^2}{\sqrt{2}}\right] \approx 3.7757.$$

Simpson's Rule:

$$\int_1^2 x^{-1/2} e^x\, dx = \tfrac{1}{3}\left[e + \frac{4e^{5/4}}{\sqrt{\frac{5}{4}}} + \cdots + \frac{4e^{7/4}}{\sqrt{\frac{7}{4}}} + \frac{e^2}{\sqrt{2}}\right] \approx 3.7625.$$

23. a. Here $a = -1$, $b = 2$, $n = 10$, and $f(x) = x^5$. $f'(x) = 5x^4$ and $f''(x) = 20x^3$.
Because $f'''(x) = 60x^2 > 0$ on $(-1,0) \cup (0,2)$, we see that $f''(x)$ is increasing on
$(-1,0) \cup (0,2)$. So, we take $M = f''(2) = 20(2^3) = 160$.
Using (7), we see that the maximum error incurred is
$$\frac{M(b-a)^3}{12n^2} = \frac{160[2-(-1)]^3}{12(100)} = 3.6.$$
b. We compute $f''' = 60x^2$ and $f^{(iv)}(x) = 120x$. $f^{(iv)}(x)$ is clearly increasing on
$(-1,2)$, so we can take $M = f^{(iv)}(2) = 240$. Therefore, using (8), we see that an
error bound is $\dfrac{M(b-a)^3}{180n^4} = \dfrac{240(3)^5}{180(10^4)} \approx 0.0324$.

25. a. Here $a = 1$, $b = 3$, $n = 10$, and $f(x) = \dfrac{1}{x}$. We find $f'(x) = -\dfrac{1}{x^2}$, $f'''(x) = \dfrac{2}{x^3}$.
Since $f'''(x) = -\dfrac{6}{x^4} < 0$ on $(1,3)$, we see that $f''(x)$ is decreasing there. We
may take $M = f''(1) = 2$. Using (7), we find an error bound is
$$\frac{M(b-a)^3}{12n^2} = \frac{2(3-1)^3}{12(100)} \approx 0.013.$$
b. $f'''(x) = -\dfrac{6}{x^4}$ and $f^{(iv)}(x) = \dfrac{24}{x^5}$. $f^{(iv)}(x)$ is decreasing on $(1,3)$, so we can
take $M = f^{(iv)}(1) = 24$. Using (8), we find an error bound is $\dfrac{24(3-1)^5}{180(10^4)} \approx 0.00043$.

27. a. Here $a = 0$, $b = 2$, $n = 8$, and $f(x) = (1+x)^{-1/2}$. We find
$$f'(x) = -\tfrac{1}{2}(1+x)^{-3/2},\ f''(x) = \tfrac{3}{4}(1+x)^{-5/2}.$$
Since f'' is positive and decreasing on $(0,2)$, we see that $|f''(x)| \le \tfrac{3}{4}$.
So the maximum error is $\dfrac{\tfrac{3}{4}(2-0)^3}{12(8)^2} = 0.0078125$.

b. $f''' = -\frac{15x}{8}(1+x)^{-7/2}$ and $f^{(4)}(x) = \frac{105}{16}(1+x)^{-9/2}$. Since $f^{(4)}$ is positive

and decreasing on $(0,2)$, we find $\left|f^{(4)}(x)\right| \le \frac{105}{16}$.

Therefore, the maximum error is $\dfrac{\frac{105}{16}(2-0)^5}{180(8)^4} = 0.000285$.

29. The distance covered is given by

$d = \int_0^2 V(t)\,dt = \frac{\frac{1}{4}}{2}\left[V(0) + 2V(\frac{1}{4}) + \cdots + 2V(\frac{7}{4}) + V(2)\right]$

$= \frac{1}{8}[19.5 + 2(24.3) + 2(34.2) + 2(40.5) + 2(38.4) + 2(26.2)$

$\quad + 2(18) + 2(16) + 8] \approx 52.84,$ or 52.84 miles.

31. $\dfrac{1}{13}\int_0^{13} f(t)\,dt = (\frac{1}{13})(\frac{1}{2})\{13.2 + 2[14.8 + 16.6 + 17.2 + 18.7 + 19.3 + 22.6 + 24.2 + 25$

$\quad\quad\quad + 24.6 + 25.6 + 26.4 + 26.6] + 26.6\} \approx 21.65,$ or 21.65 mpg.

33. The average daily consumption of oil is

$A = \dfrac{1}{b-1}\int_a^b f(t)\,dt$

where $f(t)$ has the values shown in the table where $t = 0$ corresponds to 1980.
Using Simpson's Rule with $n = 10$ and $\Delta t = 2$

$A = \dfrac{1}{20-0}\int_0^{20} f(t)\,dt \approx \dfrac{1}{20}\cdot\dfrac{2}{3}[f(0) + 4f(2) + 2f(4) + 4f(6) + \cdots + 4f(18) + f(20)]$

$= \dfrac{1}{30}[17.1 + 4(15.3) + 2(15.7) + 4(16.3) + 2(17.3) + 4(17) + 2(17) + 4(17.7)$

$\quad + 2(18.3) + 4(18.9) + 19.7]$

$= 17.14,$ \quad\quad or 17.14 million barrels.

35. The required rate of flow is

$R = $ (area of cross section of the river) $\times$ rate of flow

$\quad = (4)$(area of cross section)

$\quad = 4\int_0^{78} y(x)\,dx$

Approximating the integral using the trapezoidal rule,

$$R \approx (4)(\tfrac{6}{2})[0.8 + 2(2.6) + 2(5.8) + 2(6.2) + 2(7.6) + 2(6.4) + 2(5.2) + 2(3.9)$$
$$+ 2(8.2) + 2(10.1) + 2(10.8) + 2(9.8) + 2(2.4) + 1.4]$$
$$= 1922.4$$

or 1922.4 cu ft/sec.

37. We solve the equation $8 = \sqrt{0.01x^2 + 0.11x + 38}$.
$64 = 0.01x^2 + 0.11x + 38$, $0.01x^2 + 0.11x - 26 = 0$, $x^2 + 11x - 2600 = 0$,

and $x = \dfrac{-11 \pm \sqrt{121 + 10{,}400}}{2} \approx 45.786$. Therefore

$$PS = (8)(45.786) - \int_0^{45.786} \sqrt{0.01x^2 + 0.11x + 38}\, dx.$$

a. $\Delta x = \dfrac{45.786}{8} = 5.72$; $x_0 = 0$, $x_1 = 5.72$, $x_2 = 11.44$, ..., $x_8 = 45.79$

$$PS = 366.288 - \frac{5.72}{2}\left[\sqrt{38} + 2\sqrt{0.01(5.72)^2 + 0.11(5.72) + 38} + \cdots\right.$$
$$\left. + \sqrt{0.01(45.79)^2 + 0.11(45.79) + 38}\right] \qquad \approx 51{,}558, \text{ or } \$51{,}558.$$

$$PS = 366.288 - \frac{5.72}{2}\left[\sqrt{38} + 4\sqrt{0.01(5.72)^2 + 0.11(5.72) + 38} + \cdots\right.$$
$$\left. + \sqrt{0.01(45.79)^2 + 0.11(45.79) + 38}\right] \qquad \approx 51{,}708, \text{ or } \$51{,}708.$$

39. The average petroleum reserves from 1981 through 1990 were
$$A = \frac{1}{9-0}\int_0^9 S(t)\, dt = \frac{1}{9}\int_0^9 \frac{613.7t^2 + 1449.1}{t^2 + 6.3}\, dt$$
Using the trapezoidal rule with $a = 0$, $b = 9$, and $n = 9$ so that $\Delta t = (9-0)/9 = 1$, we
have $t_0 = 0$, $t = 1, ..., t_9 = 9$ so that

$$A = \frac{1}{9}\int_0^9 S(t)\, dt = \left(\frac{1}{9}\right)\left(\frac{1}{2}\right)[S(0) + 2S(1) + 2f(x) + 2f(x) + \cdots$$

$$\approx \frac{1}{18}[130.02 + 2(282.58) + 2(379.02) + 2(455.71) + 2(505.30) + 2(536.47)$$

$$+ 2(556.56) + 2(569.99) + 2(579.32) + 586.01]$$

$$\approx 474.77$$

or approximately 474.77 million barrels.

41. $\Delta x = \dfrac{40{,}000 - 30{,}000}{10} = 1000;\ x_0 = 30{,}000,\ x_1 = 31{,}000,\ x_2, \ldots, x_{10} = 40{,}000.$

$P = \dfrac{100}{2000\sqrt{2\pi}} \displaystyle\int_{30{,}000}^{40{,}000} e^{-0.5[x-40{,}000)/[2000]^2}\, dx$

$P = \dfrac{100(1000)}{2000\sqrt{2\pi}} \left[e^{-0.5[30{,}000-40{,}000)/[2000]^2} + 4e^{-0.5[(31{,}000-40{,}000)/2000]^2} + \cdots + 1] \right]$

≈ 50, or 50 percent.

43. $R = \dfrac{60D}{\displaystyle\int_0^T C(t)\,dt} = \dfrac{480}{\displaystyle\int_0^{24} C(t)\,dt}$. Now,

$\displaystyle\int_0^{24} C(t)\,dt \approx \tfrac{24}{12}\cdot\tfrac{1}{3}[0 + 4(0) + 2(2.8) + 4(6.1) + 2(9.7) + 4(7.6) + 2(4.8)$

$\qquad\qquad + 4(3.7) + 2(1.9) + 4(0.8) + 2(0.3) + 4(0.1) + 0] \approx 74.8$

and $R = \dfrac{480}{74.8} \approx 6.42$, or 6.42 liters/min.

45. False. The number n can be odd or even.

47. True.

49. Taking the limit and recalling the definition of the Riemann sum, we find

$$\lim_{\Delta t \to 0}[c(t_1)R\Delta t + c(t_2)R\Delta t + \cdots + c(t_n)R\Delta t]/60 = D$$

$$\frac{R}{60}\lim_{\Delta t \to 0}[c(t_1)\Delta t + c(t_2)\Delta t + \cdots + c(t_n)\Delta t] = D$$

$$\frac{R}{60}\int_0^T c(t)\,dt = D, \text{ or } R = \frac{60D}{\displaystyle\int_0^T c(t)\,dt}.$$

7.4 Problem Solving Tips

1. The improper integral on the left-hand side of the equation

$$\int_{-\infty}^{\infty} f(x)\,dx = \int_{-\infty}^{c} f(x)\,dx + \int_{c}^{\infty} f(x)\,dx \text{ is only convergent if both integrals on the}$$

right-hand side of the equation converge.

2. It is often convenient to choose $c = 0$ in the formula given in (1).

7.4 CONCEPT QUESTIONS, page 520

1. a. $\displaystyle\int_{a}^{\infty} f(x)\,dx = \lim_{b\to\infty} \int_{a}^{b} f(x)\,dx$

   b. $\displaystyle\int_{-\infty}^{b} f(x)\,dx = \lim_{a\to-\infty} \int_{a}^{b} f(x)\,dx$

   c. $\displaystyle\int_{-\infty}^{\infty} f(x)\,dx = \int_{-\infty}^{c} f(x)\,dx + \int_{c}^{\infty} f(x)\,dx$ where c is any real number.

EXERCISES 7.4, page 520

1. The required area is given by

$$\int_{3}^{\infty} \frac{2}{x^2}\,dx = \lim_{b\to\infty} \int_{3}^{b} \frac{2}{x^2}\,dx = \lim_{b\to\infty} \left(-\frac{2}{x} \right)\Big|_{3}^{b} = \lim_{b\to\infty} \left(-\frac{2}{b} + \frac{2}{3} \right) = \frac{2}{3} \text{ or } \frac{2}{3} \text{ sq units.}$$

3. $A = \displaystyle\int_{3}^{\infty} \frac{1}{(x-2)^2}\,dx = \lim_{b\to\infty} \int_{3}^{b} (x-2)^{-2}\,dx = \lim_{b\to\infty} -\frac{1}{x-2}\Big|_{3}^{b} = \lim_{b\to\infty} \left(-\frac{1}{b-2} + 1 \right)$

   $= 1$ sq unit.

5. $A = \displaystyle\int_{1}^{\infty} \frac{1}{x^{3/2}}\,dx = \lim_{b\to\infty} \int_{1}^{b} x^{-3/2}\,dx = \lim_{b\to\infty} -\frac{2}{\sqrt{x}}\Big|_{1}^{b} = \lim_{b\to\infty} \left(-\frac{2}{\sqrt{b}} + 2 \right) = 2$ sq units.

7. $A = \displaystyle\int_{0}^{\infty} \frac{1}{(x+1)^{5/2}}\,dx = \lim_{b\to\infty} \int_{1}^{b} (x+1)^{-5/2}\,dx = \lim_{b\to\infty} -\frac{2}{3}(x+1)^{-3/2}\Big|_{0}^{b}$

$$= \lim_{b\to\infty}\left[-\frac{2}{3(b+1)^{3/2}}+\frac{2}{3}\right]=\frac{2}{3}\text{ sq units.}$$

9. $A=\displaystyle\int_{-\infty}^{2}e^{2x}\,dx=\lim_{a\to-\infty}\int_{a}^{2}e^{2x}\,dx=\lim_{a\to-\infty}\tfrac{1}{2}e^{2x}\Big|_{a}^{2}=\lim_{a\to-\infty}\left(\tfrac{1}{2}e^{4}-\tfrac{1}{2}e^{2a}\right)=\tfrac{1}{2}e^{4}\text{ sq units.}$

11. Using symmetry, the required area is given by

$$2\int_{0}^{\infty}\frac{x}{(1+x^2)^2}\,dx=2\lim_{b\to\infty}\int_{0}^{\infty}\frac{x}{(1+x^2)^2}\,dx.$$

To evaluate the indefinite integral $\displaystyle\int\frac{x}{(1+x^2)^2}\,dx$, put $u=1+x^2$ so that

$du=2x\,dx$ or $x\,dx=\tfrac{1}{2}\,du$.

Then $\displaystyle\int\frac{x}{(1+x^2)^2}\,dx=\frac{1}{2}\int\frac{du}{u^2}=-\frac{1}{2u}+C=-\frac{1}{2(1+x^2)}+C.$

Therefore, $\displaystyle 2\lim_{b\to\infty}\int_{0}^{b}\frac{x}{(1+x^2)}\,dx=\lim_{b\to\infty}-\frac{1}{(1+x^2)^2}\Big|_{0}^{b}=\lim_{b\to\infty}\left[-\frac{1}{(1+b^2)}+1\right]=1,$

or 1 sq unit.

13. a. $I(b)=\displaystyle\int_{0}^{b}\sqrt{x}\,dx=\tfrac{2}{3}x^{3/2}\Big|_{0}^{b}=\tfrac{2}{3}b^{3/2}.$ b. $\displaystyle\lim_{b\to\infty}I(b)=\lim_{b\to\infty}\tfrac{2}{3}b^{3/2}=\infty.$

15. $\displaystyle\int_{1}^{\infty}\frac{3}{x^4}\,dx=\lim_{b\to\infty}\int_{1}^{b}3x^{-4}\,dx=\lim_{b\to\infty}\left(-\frac{1}{x^3}\right)\Big|_{1}^{b}=\lim_{b\to\infty}\left(-\frac{1}{b^3}+1\right)=1.$

17. $A=\displaystyle\int_{4}^{\infty}\frac{2}{x^{3/2}}\,dx=\lim_{b\to\infty}\int_{4}^{b}2x^{-3/2}\,dx=\lim_{b\to\infty}-4x^{-1/2}\Big|_{4}^{b}=\lim_{b\to\infty}\left(-\frac{4}{\sqrt{b}}+2\right)=2.$

19. $\displaystyle\int_{1}^{\infty}\frac{4}{x}\,dx=\lim_{b\to\infty}\int_{1}^{b}\frac{4}{x}\,dx=\lim_{b\to\infty}4\ln x\Big|_{1}^{b}=\lim_{b\to\infty}(4\ln b)=\infty.$

21. $\displaystyle\int_{-\infty}^{0}(x-2)^{-3}\,dx=\lim_{a\to-\infty}\int_{a}^{0}(x-2)^{-3}\,dx=\lim_{a\to-\infty}-\frac{1}{2(x-2)^2}\Big|_{a}^{0}=-\frac{1}{8}.$

23. $\displaystyle\int_1^\infty \frac{1}{(2x-1)^{3/2}}\,dx = \lim_{b\to\infty}\int_1^b (2x-1)^{-3/2}\,dx = \lim_{b\to\infty}-\frac{1}{(2x-1)^{1/2}}\,\bigg|_1^b$

$$= \lim_{b\to\infty}\left(-\frac{1}{\sqrt{2b-1}}+1\right) = 1.$$

25. $\displaystyle\int_0^\infty e^{-x}\,dx = \lim_{b\to\infty}\int_0^b e^{-x}\,dx = \lim_{b\to\infty}-e^{-x}\,\big|_0^b = \lim_{b\to\infty}(-e^{-b}+1) = 1.$

27. $\displaystyle\int_{-\infty}^0 e^{2x}\,dx = \lim_{a\to-\infty}\tfrac{1}{2}e^{2x}\,\big|_a^0 = \lim_{a\to-\infty}\left(\tfrac{1}{2}-\tfrac{1}{2}e^{2a}\right) = \tfrac{1}{2}.$

29. $\displaystyle\int_1^\infty \frac{e^{\sqrt{x}}}{\sqrt{x}}\,dx = \lim_{b\to\infty}\int_1^b \frac{e^{\sqrt{x}}}{\sqrt{x}}\,dx = \lim_{b\to\infty}2e^{\sqrt{x}}\,\big|_1^b$ (Integrate by substitution: $u = \sqrt{x}$.)

$$= \lim_{b\to\infty}(2e^{\sqrt{b}}-2e) = \infty, \text{ and so it diverges.}$$

31. $\displaystyle\int_{-\infty}^0 xe^x\,dx = \lim_{a\to-\infty}\int_a^0 xe^x\,dx = \lim_{a\to-\infty}(x-1)e^x\,\big|_a^0 = \lim_{a\to-\infty}[-1+(a-1)e^a] = -1.$

Note: We have used integration by parts to evaluate the integral.

33. $\displaystyle\int_{-\infty}^\infty x\,dx = \lim_{a\to-\infty}\tfrac{1}{2}x^2\,\big|_a^0 + \lim_{b\to\infty}\tfrac{1}{2}x^2\,\big|_0^b$ both of which diverge and so the integral diverges.

35. $\displaystyle\int_{-\infty}^\infty x^3(1+x^4)^{-2}\,dx = \int_{-\infty}^0 x^3(1+x^4)^{-2}\,dx + \int_0^\infty x^3(1+x^4)^{-2}\,dx$

$$= \lim_{a\to-\infty}\int_a^0 x^3(1+x^4)^{-2}\,dx + \lim_{b\to\infty}\int_0^b x^3(1+x^4)^{-2}\,dx$$

$$= \lim_{a\to-\infty}\left[-\frac{1}{4}(1+x^4)^{-1}\,\bigg|_a^0\right] + \lim_{b\to\infty}\left[-\frac{1}{4}(1+x^4)^{-1}\,\bigg|_0^b\right]$$

$$= \lim_{a\to-\infty}\left[-\frac{1}{4}+\frac{1}{4(1+a^4)}\right] + \lim_{b\to\infty}\left[-\frac{1}{4(1+b^4)}+\frac{1}{4}\right]$$

$$= -\tfrac{1}{4}+\tfrac{1}{4} = 0.$$

37. $\displaystyle\int_{-\infty}^{\infty} xe^{1-x^2}\,dx = \lim_{a\to-\infty}\int_{a}^{0} xe^{1-x^2}\,dx + \lim_{b\to\infty}\int_{0}^{b} xe^{1-x^2}\,dx$

$\displaystyle = \lim_{a\to-\infty} -\tfrac{1}{2}e^{1-x^2}\Big|_{a}^{0} + \lim_{b\to\infty} -\tfrac{1}{2}e^{1-x^2}\Big|_{0}^{b}$

$\displaystyle = \lim_{a\to-\infty}\left(-\tfrac{1}{2}e + \tfrac{1}{2}e^{1-a^2}\right) + \lim_{b\to\infty}\left(-\tfrac{1}{2}e^{1-b^2} + \tfrac{1}{2}e\right) = 0.$

39. $\displaystyle\int_{-\infty}^{\infty} \frac{e^{-x}}{1+e^{-x}}\,dx = \lim_{a\to-\infty} -\ln(1+e^{-x})\Big|_{a}^{0} + \lim_{b\to\infty} -\ln(1+e^{-x})\Big|_{0}^{b} = \infty$, and it is divergent.

41. First, we find the indefinite integral $I = \displaystyle\int \frac{1}{x\ln^3 x}\,dx$. Let $u = \ln x$ so that

$du = \dfrac{1}{x}\,dx$. Therefore, $I = \displaystyle\int \frac{du}{u^3} = -\frac{1}{2u^2+C} = -\frac{1}{2\ln^2 x}+C.$ So

$\displaystyle\int_{e}^{\infty} \frac{1}{x\ln^3 x}\,dx = \lim_{b\to\infty}\int_{e}^{b} \frac{1}{x\ln^3 x}\,dx$

$\displaystyle = \lim_{b\to\infty}\left[-\frac{1}{2\ln^2 x}\Big|_{e}^{b}\right] = \lim_{b\to\infty}\left(-\frac{1}{2(\ln b)^2}+\frac{1}{2}\right)=\frac{1}{2}$

and so the given integral is convergent.

43. We want the present value PV of a perpetuity with $m = 1$, $P = 1500$, and $r = 0.08$.

We find $\quad PV = \dfrac{(1)(1500)}{0.08} = 18{,}750,\ $ or $\$18{,}750.$

45. $PV = \displaystyle\int_{0}^{\infty}(10{,}000+4000t)e^{-rt}\,dt = 10{,}000\int_{0}^{\infty}e^{-rt}\,dt + 4000\int_{0}^{\infty}te^{-rt}\,dt$

$\displaystyle = \lim_{b\to\infty}\left(-\frac{10{,}000}{r}e^{-rt}\Big|_{0}^{b}\right) + 4000\left(\frac{1}{r^2}\right)\left(-rt-1)e^{-rt}\Big|_{0}^{b}\right)$

(Integrating by parts.)

$\displaystyle = \frac{10{,}000}{r} + \frac{4000}{r^2} = \frac{10{,}000r+4000}{r^2}\ $ dollars.

47. True. $\displaystyle\int_{a}^{\infty} f(x)\,dx = \int_{a}^{b} f(x)\,dx + \int_{b}^{\infty} f(x)\,dx.$ So if $\displaystyle\int_{a}^{\infty} f(x)\,dx$ exists then

$$\int_b^\infty f(x)\,dx = \int_a^\infty f(x)\,dx - \int_a^b f(x)\,dx.$$

49. False. Let $f(x) = \begin{cases} e^{2x} & \text{if } -\infty < x \le 0 \\ e^{-x} & \text{if } 0 < x < \infty \end{cases}$. Then

$\int_{-\infty}^\infty f(x)\,dx = \int_{-\infty}^0 e^{2x}\,dx + \int_0^\infty e^{-x}\,dx = \dfrac{1}{2} + 1 = \dfrac{3}{2}$. But $2\int_0^\infty f(x)\,dx = 2\int_0^\infty e^{-x}\,dx = 2$.

51. a. $CV \approx \int_0^\infty Re^{-it}\,dt = \lim\limits_{b\to\infty} \int_0^b Re^{-it}\,dt = \lim\limits_{b\to\infty} -\dfrac{R}{i}e^{-it}\Big|_0^b = \lim\limits_{b\to\infty}\left(-\dfrac{R}{i}e^{-ib} + \dfrac{R}{i}\right) = \dfrac{R}{i}.$

  b. $CV \approx \dfrac{10{,}000}{0.12} \approx 83{,}333$, or \$83,333.

53. $\int_{-\infty}^b e^{px}\,dx = \lim\limits_{a\to-\infty} \int_a^b e^{px}\,dx = \lim\limits_{a\to-\infty}\left[\dfrac{1}{p}e^{px}\right]_a^b = \lim\limits_{a\to-\infty}\left(\dfrac{1}{p}e^{pb} - \dfrac{1}{p}e^{pa}\right)$

$= -\dfrac{1}{p}e^{pa}$ if $p > 0$ and is divergent if $p < 0$.

7.5 Problem Solving Tips

1. To show that a function f is a probability density function of a random variable over a given interval, first show that the function is nonnegative over that interval, and then show that the area of the region under the graph of f over that interval is equal to 1.

7.5 CONCEPT QUESTIONS, page 530

1. See the definition given on page 523 of the test. For example, $f(x) = \dfrac{3x^2}{125}$ on the interval [0, 5].

1. $f(x) \geq 0$ on $[2,6]$. Next $\int_2^6 \frac{2}{32} x \, dx = \frac{1}{32} x^2 \Big|_2^6 = \frac{1}{32}(36-4) = 1$,

    and so f is a probability density function on $[2,6]$.

3. $f(x) = \frac{3}{8} x^2$ is nonnegative on $[0,2]$. Next, we compute

    $$\int_0^2 \frac{3}{8} x^2 \, dx = \frac{1}{8} x^3 \Big|_0^2 = 1$$

    and so f is a probability density function.

5. $\int_0^1 20(x^3 - x^4) dx = 20(\frac{1}{4} x^4 - \frac{1}{5} x^5) \Big|_0^1 = 20(\frac{1}{4} - \frac{1}{5}) = 20(\frac{1}{20}) = 1$.

    Furthermore, $f(x) = 20(x^3 - x^4) = 20x^3(1 - x) \geq 0$ on $[0,1]$.

    Therefore, f is a density function on $[0,1]$ as asserted.

7. Clearly $f(x) \geq 0$ on $[1,4]$. Next,

    $$\int_1^4 f(x) \, dx = \frac{3}{14} \int_1^4 x^{1/2} \, dx = (\frac{3}{14})(\frac{2}{3}) x^{3/2} \Big|_1^4 = \frac{1}{7}(8-1) = 1,$$

    and so f is a probability density function on $[1,4]$.

9. First, $f(x) \geq 0$ on $[0, \infty)$. Next, we compute

    $$I = \int_0^{\infty} \frac{x}{(x^2 + 1)^{3/2}} \, dx = \lim_{b \to \infty} \int_0^b x(x^2 + 1)^{-3/2} \, dx.$$

    Letting $u = x^2 + 1$, so that $du = 2x \, dx$, we find

    $$I = \lim_{b \to \infty} \int_1^{b^2+1} u^{-3/2} \, du = \frac{1}{2} \lim_{b \to \infty} -2u^{-1/2} \Big|_1^{b^2+1} = \lim_{b \to \infty} \left(-\frac{1}{\sqrt{b^2+1}} + 1 \right) = 1.$$

    So the given function is a probability density function on $[0, \infty)$.

11. a. $\int_0^4 k(4 - x) \, dx = k \int_0^4 (4 - x) \, dx = k(4x - \frac{1}{2} x^2) \Big|_0^4 = k(16-8) = 8k = 1$

    implies that $k = 1/8$.

    b. $P(1 \leq x \leq 3) = \frac{1}{8} \int_1^3 (4 - x) \, dx = \frac{1}{8}(4x - \frac{1}{2} x^2) \Big|_1^3 = \frac{1}{8}[(12 - \frac{9}{2}) - (4 - \frac{1}{2})] = \frac{1}{2}$.

13. a. $\displaystyle\int_0^4 f(x)\,dx = \int_0^4 2ke^{-kx}\,dx = (2k)\left(-\frac{1}{k}\right)e^{-kx}\Big|_0^4$

$$= -2e^{-kx}\Big|_0^4 = -2e^{-4k} + 2 = 2(1 - e^{-4k}) = 1$$

gives

$$1 - e^{-4k} = \frac{1}{2}$$

$$e^{-4k} = \frac{1}{2}$$

$$-4k = \ln\frac{1}{2} = \ln 1 - \ln 2 = -\ln 2$$

So
$$k = \frac{\ln 2}{4}$$

b. $\displaystyle P(1 \le x \le 2) = \frac{2\ln 2}{4}\int_1^2 e^{-\left(\frac{\ln 2}{4}\right)x}\,dx = \frac{2\ln 2}{4}\cdot\left(-\frac{4}{\ln 2}\right)e^{-\left(\frac{\ln 2}{4}\right)x}\Big|_1^2$

$$= -2e^{-\left(\frac{\ln 4}{2}\right)2} + 2e^{-\frac{\ln 4}{2}} \approx 0.2676.$$

15. a. Here $k = \frac{1}{15}$ and so $f(x) = \frac{1}{15}e^{(-1/15)x}$.

b. The probability is
$$\int_{10}^{12} \frac{1}{15}e^{(-1/15)x}\,dx = -e^{(-1/15)x}\Big|_{10}^{12} = -e^{-12/15} + e^{-10/15} \approx 0.06.$$

c. The probability is
$$\int_{15}^{\infty} \frac{1}{15}e^{(-1/15)x}\,dx = \lim_{b\to\infty}\int_{15}^{b}\frac{1}{15}e^{(-1/15)x}\,dx$$

$$= \lim_{b\to\infty} -e^{(-1/15)x}\Big|_{15}^{b} = \lim_{b\to\infty} -e^{(-1/15)b} + e^{-1} \approx 0.37.$$

17. $\displaystyle \mu = \int_0^5 t\cdot\frac{2}{25}t\,dt = \frac{2}{25}\int_0^5 t^2\,dt = \frac{2}{75}t^3\Big|_0^5 = \frac{2}{75}(125) = 3\frac{1}{3}.$

So a shopper is expected to spend $3\frac{1}{3}$ minutes in the magazine section.

19. $\displaystyle \mu = \int_0^5 x\cdot\frac{6}{125}x(5-x)\,dx = \frac{6}{125}\int_0^5 (5x^2 - x^3)\,dx = \frac{6}{125}\left(\frac{5}{3}x^3 - \frac{1}{4}x^4\right)\Big|_0^5$

$$= \frac{6}{125}\left(\frac{625}{3} - \frac{625}{4}\right) = 2.5.$$

So the expected demand is 2500 lb.

21. The required probability is given by
$$P(0 \le x \le \tfrac{1}{2}) = \int_0^{1/2} 12x^2(1-x)\,dx = 12\int_0^{1/2}(x^2 - x^3)\,dx$$
$$= 12\left(\tfrac{1}{3}x^3 - \tfrac{1}{4}x^4\right)\Big|_0^{1/2} = 12x^3\left[\tfrac{1}{3} - \tfrac{1}{4}(x)\right]\Big|_0^{1/2}$$
$$= 12\left(\tfrac{1}{2}\right)^3\left[\tfrac{1}{3} - \tfrac{1}{4}\left(\tfrac{1}{2}\right)\right] - 0 = \tfrac{5}{16}.$$

23. $\mu = \displaystyle\int_0^{\infty} t \cdot 9(9+t^2)^{-3/2}\,dt = \lim_{b\to\infty}\int_0^b 9t(9+t^2)^{-3/2}\,dt$
$$= \lim_{b\to\infty}\left[(9)(\tfrac{1}{2})(-2)(9+t^2)^{-1/2}\right]\Big|_0^b = \lim_{b\to\infty}\left[-\frac{9}{\sqrt{9+t^2}} + 3\right] = 3.$$
So the tubes are expected to last 3 years.

25. The probability function is $f(x) = \tfrac{1}{8}e^{-x/8}$. The required probability is
$$P(x \ge 8) = \tfrac{1}{8}\int_8^{\infty} e^{-x/8}\,dx = \lim_{b\to\infty}-e^{-x/8}\Big|_8^b = \lim_{b\to\infty}(-e^{-b/8} + e^{-1}) = e^{-1} \approx 0.37.$$

27. The probability function is $f(x) = 0.00001e^{-0.00001x}$. The required probability is
$$P(x \le 20{,}000) = 0.00001\int_0^{20{,}000} e^{-0.00001x}\,dx = -e^{-0.00001x}\Big|_0^{20{,}000} = -e^{0.2} + 1 \approx 0.18.$$

29. False. f must be nonnegative on $[a, b]$ as well.

31. False. Let $f(x) = 1$ for all x in the interval $[0, 1]$. Then f is a probability density function on $[0, 1]$, but f is not a probability density function on $\left[\tfrac{1}{2}, \tfrac{3}{4}\right]$ since
$$\int_{1/2}^{3/4} 1\,dx = \tfrac{1}{4} \ne 1.$$

CHAPTER 7 CONCEPT REVIEW, page 533

1. Product; $uv - \displaystyle\int v\,du$; u; easy to integrate

3. $\dfrac{\Delta x}{2}[f(x_0) + 2f(x_1) + 2f(x_2) + \cdots 2f(x_{n-1}) + f(x_n)]$; even; $\dfrac{M(b-a)^3}{12n^2}$

7 Additional Topics in Integration

5. $\lim\limits_{a\to-\infty}\int_a^b f(x)\,dx$; $\lim\limits_{b\to\infty}\int_a^b f(x)\,dx$; $\int_{-\infty}^c f(x)\,dx + \int_c^\infty f(x)\,dx$

CHAPTER 7 REVIEW EXERCISES, page 534

1.   Let $u = 2x$ and $dv = e^{-x}\,dx$ so that $du = 2\,dx$ and $v = -e^{-x}$. Then

$$\int 2xe^{-x}\,dx = uv - \int v\,du = -2xe^{-x} + 2\int e^{-x}\,dx$$

$$= -2xe^{-x} - 2e^{-x} + C = -2(1+x)e^{-x} + C.$$

3.   Let $u = \ln 5x$ and $dv = dx$, so that $du = \frac{1}{x}\,dx$ and $v = x$. Then

$$\int \ln 5x\,dx = x\ln 5x\,dx - \int dx = x\ln 5x - x + C = x(\ln 5x - 1) + C.$$

5.   Let $u = x$ and $dv = e^{-2x}\,dx$ so that $du = dx$ and $v = -\frac{1}{2}e^{-2x}$. Then

$$\int_0^1 xe^{-2x}\,dx = -\frac{1}{2}xe^{-2x}\Big|_0^1 + \frac{1}{2}\int_0^1 e^{-2x}\,dx = -\frac{1}{2}e^{-2} - \frac{1}{4}e^{-2x}\Big|_0^1$$

$$= -\frac{1}{2}e^{-2} - \frac{1}{4}e^{-2} + \frac{1}{4} = \frac{1}{4}(1 - 3e^{-2}).$$

7. $f(x) = \int f'(x)\,dx = \int \dfrac{\ln x}{\sqrt{x}}\,dx$. To evaluate the integral, we integrate by parts

with $u = \ln x$, $dv = x^{-1/2}\,dx$, $du = \frac{1}{x}\,dx$ and $v = 2x^{1/2}\,dx$. Then

$$\int \frac{\ln x}{x^{1/2}}\,dx = 2x^{1/2}\ln x - \int 2x^{-1/2}\,dx = 2x^{1/2}\ln x - 4x^{1/2} + C$$

$$= 2x^{1/2}(\ln x - 2) + C = 2\sqrt{x}(\ln x - 2) + C.$$

But $f(1) = -2$ and this gives $2\sqrt{1}(\ln 1 - 2) + C = -2$, or $C = 2$. Therefore, $f(x) = 2\sqrt{x}(\ln x - 2) + 2.$

9. Using Formula 4 with $a = 3$ and $b = 2$, we obtain

$$\int \frac{x^2}{(3+2x)^2}\,dx = \frac{1}{8}\left[3 + 2x - \frac{9}{3+2x} - 6\ln|3+2x|\right] + C.$$

11. Use Formula 24 with $a = 4$ and $n = 2$, obtaining $\int x^2 e^{4x}\, dx = \frac{1}{4} x^2 e^{4x} - \frac{1}{2}\int x e^{4x}\, dx$.

Use Formula 23 to obtain

$$\int x^2 e^{4x}\, dx = \frac{1}{4} x^2 e^{4x} - \frac{1}{2}\left[\frac{1}{16}(4x-1)e^{4x}\right] + C$$
$$= \frac{1}{32}(8x^2 - 4x + 1)e^{4x} + C.$$

13. Use Formula 17 with $a = 2$ obtaining $\int \dfrac{dx}{x^2\sqrt{x^2-4}} = \dfrac{\sqrt{x^2-4}}{4x} + C.$

15. $\displaystyle\int_0^\infty e^{-2x}\, dx = \lim_{b\to\infty}\int_0^b e^{-2x}\, dx = \lim_{b\to\infty}\left(-\tfrac{1}{2}e^{-2x}\right)\Big|_0^b = \lim_{b\to\infty}\left(-\tfrac{1}{2}e^{-2b} + \tfrac{1}{2}\right) = \tfrac{1}{2}.$

17. $\displaystyle\int_3^\infty \frac{2}{x}\, dx = \lim_{b\to\infty}\int_3^b \frac{2}{x}\, dx = \lim_{b\to\infty} 2\ln x\Big|_3^b = \lim_{b\to\infty}(2\ln b - 2\ln 3) = \infty.$

19. $\displaystyle\int_2^\infty \frac{dx}{(1+2x)^2} = \lim_{b\to\infty}\int_2^b (1+2x)^{-2}\, dx = \lim_{b\to\infty}\left(\tfrac{1}{2}\right)(-1)(1+2x)^{-1}\Big|_2^b$

$$= \lim_{b\to\infty}\left(-\frac{1}{2(1+2b)} + \frac{1}{2(5)}\right) = \frac{1}{10}.$$

21. $\Delta x = \frac{b-a}{n} = \frac{3-1}{4} = \frac{1}{2};\ x_0 = 1,\ x_1 = \frac{3}{2},\ x_2 = 2,\ x_3 = \frac{5}{2},\ x_4 = 3.$

Trapezoidal Rule:

$$\int_1^3 \frac{dx}{1+\sqrt{x}} \approx \frac{\frac{1}{2}}{2}\left[\frac{1}{2} + \frac{2}{1+\sqrt{1.5}} + \frac{2}{1+\sqrt{2}} + \frac{2}{1+\sqrt{2.5}} + \frac{1}{1+\sqrt{3}}\right] \approx 0.8421.$$

Simpson's Rule

$$\int_1^3 \frac{dx}{1+\sqrt{x}} \approx \frac{\frac{1}{2}}{3}\left[\frac{1}{2} + \frac{4}{1+\sqrt{1.5}} + \frac{2}{1+\sqrt{2}} + \frac{4}{1+\sqrt{2.5}} + \frac{1}{1+\sqrt{3}}\right] \approx 0.8404.$$

23. $\Delta x = \frac{1-(-1)}{4} = \frac{1}{2};\ x_0 = -1,\ x_1 = -\frac{1}{2},\ x_2 = 0,\ x_3 = \frac{1}{2},\ x_4 = 1.$

Trapezoidal Rule:

$$\int_{-1}^1 \sqrt{1+x^4}\, dx \approx \frac{0.5}{2}\left[\sqrt{2} + 2\sqrt{1+(-0.5)^4} + 2 + 2\sqrt{1+(0.5)^4} + \sqrt{2}\right]$$
$$\approx 2.2379.$$

7 *Additional Topics in Integration*

Simpson's Rule:

$$\int_{-1}^{1} \sqrt{1+x^4}\, dx \approx \frac{0.5}{3}\left[\sqrt{2} + 4\sqrt{1+(-0.5)^4} + 2 + 4\sqrt{1+(0.5)^4} + \sqrt{2}\right]$$

$$\approx 2.1791.$$

25. a. Here $a = 0, b = 1$, and $f(x) = \dfrac{1}{x+1}$. We have

$$f'(x) = -\frac{1}{(x+1)^2} \quad \text{and} \quad f''(x) = \frac{2}{(x+1)^3}$$

Since f'' is positive and decreasing on $(0,1)$, it attains its maximum value of 2 at $x = 0$. So we take $M = 2$. Using (7), we see that the maximum error incurred is

$$\frac{M(b-a)^3}{12n^2} = \frac{2(1^3)}{12(8^2)} = \frac{1}{384} \approx 0.002604$$

b. We compute

$$f'''(x) = -\frac{6}{(x+1)^4} \quad \text{and} \quad f^{(iv)}(x) = \frac{24}{(x+1)^5}$$

Since $f^{(4)}(x)$ is positive and decreasing on $(0,1)$, we take $M = 24$. The maximum

error is $\dfrac{24(1^5)}{180(8^4)} = \dfrac{1}{30720} \approx 0.000033$.

27. $f(x) \geq 0$ on $[0,3]$. Next,

$$\frac{1}{9}\int_0^3 x\sqrt{9-x^2}\, dx = \frac{1}{9}\int_0^3 x(9-x^2)^{1/2}\, dx = \left(\frac{1}{9}\right)\left(-\frac{1}{2}\right)\left(\frac{2}{3}\right)(9-x^2)^{3/2}\Big|_0^3$$

$$= -\frac{1}{27}(9-x^2)^{3/2}\Big|_0^3 = 0 + \frac{1}{27}(9)^{3/2} = 1.$$

29. a. $\displaystyle\int_1^4 \frac{k}{\sqrt{x}}\, dx = k\int_1^4 x^{-1/2}\, dx = 2k\sqrt{x}\Big|_1^4 = 2k(2-1) = 2k = 1$, or $k = 1/2$.

b. $\displaystyle\int_2^3 \frac{\frac{1}{2}}{\sqrt{x}}\, dx = \frac{1}{2}\int_2^3 x^{-1/2}\, dx = 2\left(\frac{1}{2}\right)\sqrt{x}\Big|_2^3 = \sqrt{3} - \sqrt{2} = 0.3178,$

31. a. The probability that a woman entering the maternity wing stays in the hospital more than 6 days is given by

$$\int_6^\infty \tfrac{1}{4} e^{-0.25t}\, dt = \lim_{b \to \infty} \int_6^b \tfrac{1}{4} e^{-0.25t}\, dt = \lim_{b \to \infty} -e^{-0.25t}\Big|_6^b = \lim_{b \to \infty}(-e^{-0.25b} + e^{-1.5}) = 0.22$$

b. The probability that a woman entering the maternity wing at the hospital stays there less than 2 days is given by $\int_0^2 \tfrac{1}{4} e^{-0.25t}\, dt = -e^{-0.25t}\Big|_0^2 = -e^{-0.5} + 1 = 0.39$.

c. $E = \dfrac{1}{k} - \dfrac{1}{\frac{1}{4}} = 4$, or 4 days.

33. Let $u = t$ and $dv = e^{-0.05t}$ so that $du = 1$ and $v = -20e^{-0.05t}$, and integrate by parts obtaining $S(t) = -20te^{-0.05t} + \displaystyle\int 20e^{-0.05t}\, dt = -20te^{-0.05t} - 400e^{-0.05t} + C$

$$= -20te^{-0.05t} - 400e^{-0.05t} + C = -20e^{-0.05t}(t + 20) + C.$$

The initial condition implies $S(0) = 0$ giving $-20(20) + C = 0$, or $C = 400$.
Therefore, $S(t) = -20e^{-0.05t}(t + 20) + 400$. By the end of the first year, the number of units sold is given by $S(12) = -20e^{-0.6}(32) + 400 = 48.761$, or 48,761 cartridges.

35. Trapezoidal Rule:
$A = \dfrac{100}{2}[0 + 480 + 520 + 600 + 680 + 680 + 800 + 680 + 600 + 440 + 0]$
$= 274{,}000$, or 274,000 sq ft.
Simpson's Rule:
$A = \dfrac{100}{3}[0 + 960 + 520 + 1200 + 680 + 1360 + 800 + 1360 + 600 + 880 + 0]$
$= 278{,}667$, or 278,667 sq ft.

37. We want the present value of a perpetuity with $m = 1$, $P = 10{,}000$, and $r = 0.09$.
We find $PV = \dfrac{(1)(10{,}000)}{0.09} \approx 111{,}111$ or approximately \$111,111.

CHAPTER 7 BEFORE MOVING ON, page 535

1. Let $u = \ln x$ and $dv = x^2\, dx$. Then $du = \dfrac{1}{x}\, dx$ and $v = \dfrac{1}{3}x^3$.

$$\int x^2 \ln x\, dx = \dfrac{1}{3}x^3 \ln x - \int \dfrac{1}{3}x^2\, dx = \dfrac{1}{3}x^3 \ln x - \dfrac{1}{9}x^3 + C$$

$$= \dfrac{1}{9}x^3(3\ln x - 1) + C$$

2. $I = \int \dfrac{du}{x^2\sqrt{x+2x^2}}$. Let $u = \sqrt{2}x$. Then $du = \sqrt{2}\,dx$ or $dx = \dfrac{du}{\sqrt{2}} = \dfrac{\sqrt{2}}{2}\,du$.

$$I = \frac{\sqrt{2}}{2}\int \frac{du}{\dfrac{u^2}{2}\sqrt{8+u^2}} = \sqrt{2}\int \frac{du}{u^2\sqrt{(2\sqrt{2})^2 + u^2}}$$

With $a = 2\sqrt{2}$ and $x = u$,

$$I = \sqrt{2}\int \frac{du}{u^2\sqrt{(2\sqrt{2})^2 + u^2}} = \sqrt{2}\left[-\frac{\sqrt{8+u^2}}{8u}\right] + C = -\frac{\sqrt{8+2x^2}}{8x} + C$$

3. $n = 5$, $\Delta x = \dfrac{4-2}{5} = 0.4$; $x_0 = 2$, $x_1 = 2.4$, $x_2 = 2.8$, $x_3 = 3.2$, $x_4 = 3.6$, $x_5 = 4$

$$\int_2^4 \sqrt{x^2+1}\,dx \approx \frac{0.4}{2}[f(2) + 2f(2.4) + 2f(2.8) + 2f(3.2) + 2f(3.2) + 2f(3.6) + f(4)]$$

$$= 0.2[2.23607 + 2(2.6) + 2(2.97321) + 2(3.35261) + 2(3.73631) + 4.12311]$$

$$\approx 6.3367.$$

4. $n = 6$, $\Delta x = \dfrac{3-1}{6} = \dfrac{1}{3}$; $x_0 = 1$, $x_1 = \frac{4}{3}$, $x_2 = \frac{5}{3}$, $x_3 = 2$, $x_4 = \frac{7}{3}$, $x_5 = \frac{8}{3}$, $x_6 = 3$

$$\int_1^3 e^{0.2x}\,dx \approx \frac{\frac{1}{3}}{3}[f(1) + 4f(\tfrac{4}{3}) + 2f(\tfrac{5}{3}) + 4f(2) + 2f(\tfrac{7}{3}) + 4f(\tfrac{8}{3}) + f(3)]$$

$$\approx \frac{1}{9}[1.2214 + 4(1.30561) + 2(1.39561) + 4(1.49182) + 2(1.59467)$$

$$+ 4(1.7046) + 1.82212]$$

$$\approx 3.0036$$

5. $\displaystyle\int_1^\infty e^{-2x}\,dx = \lim_{b\to\infty}\int_1^b e^{-2x}\,dx = \lim_{b\to\infty}\left[-\tfrac{1}{2}e^{-2x}\Big|_1^b\right] = \lim_{b\to\infty}\left(-\frac{1}{2}e^{-2b} + \frac{1}{2}e^{-2}\right) = \frac{1}{2}e^{-2} = \frac{1}{2e^2}$

6. a. $f(x) \geq 0$ on $[0,8]$ and $\displaystyle\int_0^8 \tfrac{5}{96}x^{2/3}\,dx = \tfrac{5}{96}\left[\tfrac{3}{5}x^{5/3}\Big|_0^8\right] = \tfrac{5}{96}(\tfrac{3}{5})(8^{5/3}) = 1$

Therefore f is a probability density function on $[0, 8]$.

b. $P(1 \leq x \leq 8) = \displaystyle\int_1^8 \tfrac{5}{96}x^{2/3}\,dx = \tfrac{5}{96}(\tfrac{3}{5})x^{5/3}\Big|_1^8 = \tfrac{1}{32}(32-1) = \tfrac{31}{32}$.

CHAPTER 8

8.1 CONCEPT QUESTIONS, page 544

1. A function of two variables is a rule that assigns to each point (x, y) in a subset of the plane, a unique number $f(x, y)$. For example, $f(x, y) = x^2 + 2y^2$ has the whole xy-plane as its domain.

3. a. The graph of $f(x, y)$ is the set $S = \{(x, y, z) \mid z = f(x, y), (x, y) \in \text{Domain of } f\}$

   b. The level curve of f is the projection onto the xy-plane of the trace of $f(x, y)$ in the plane $z = k$, where k is a constant in the range of f.

EXERCISES 8.1, page 544

1. $f(0, 0) = 2(0) + 3(0) - 4 = -4.$ $\qquad$ $f(1, 0) = 2(1) + 3(0) - 4 = -2.$
   $f(0, 1) = 2(0) + 3(1) - 4 = -1.$ $\qquad$ $f(1, 2) = 2(1) + 3(2) - 4 = 4.$
   $f(2, -1) = 2(2) + 3(-1) - 4 = -3.$

3. $f(1, 2) = 1^2 + 2(1)(2) - 1 + 3 = 7;\ f(2, 1) = 2^2 + 2(2)(1) - 2 + 3 = 9$
   $f(-1, 2) = (-1)^2 + 2(-1)(2) - (-1) + 3 = 1;\ f(2, -1) = 2^2 + 2(2)(-1) - 2 + 3 = 1.$

5. $g(s, t) = 3s\sqrt{t} + t\sqrt{s} + 2;\ g(1, 2) = 3(1)\sqrt{2} + 2\sqrt{1} + 2 = 4 + 3\sqrt{2}$
   $g(2, 1) = 3(2)\sqrt{1} + \sqrt{2} + 2 = 8 + \sqrt{2};$
   $g(0, 4) = 0 + 0 + 2 = 2,\ g(4, 9) = 3(4)\sqrt{9} + 9\sqrt{4} + 2 = 56.$

7. $h(1, e) = \ln e - e \ln 1 = \ln e = 1;\ h(e, 1) = e \ln 1 - \ln e = -1;$
   $h(e, e) = e \ln e - e \ln e = 0.$

9. $g(r, s, t) = re^{s/t};\ g(1, 1, 1) = e,\ g(1, 0, 1) = 1,\ g(-1, -1, -1) = -e^{-1/(-1)} = -e.$

11. The domain of f is the set of all ordered pairs (x, y) where x and y are real numbers.

13. All real values of u and v except those satisfying the equation $u = v$.

15. The domain of g is the set of all ordered pairs (r,s) satisfying $rs \geq 0$, that is the set of all ordered pairs where both $r \geq 0$ and $s \geq 0$, or in which both $r \leq 0$ and $s \leq 0$.

17. The domain of h is the set of all ordered pairs (x, y) such that $x + y > 5$.

19. The level curves of
$z = f(x, y) = 2x + 3y$
for $z = -2, -1, 0, 1, 2$, follow.

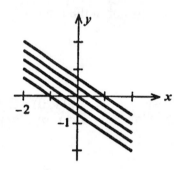

21. The level curves of
$f(x,y) = 2x^2 + y$ for
$z = -2, -1, 0, 1, 2$, are shown below.

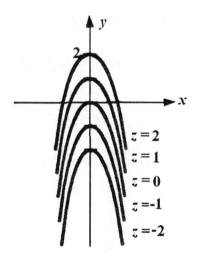

23. The level curves of $f(x,y) = \sqrt{16 - x^2 - y^2}$
for $z = 0, 1, 2, 3, 4$ follow.

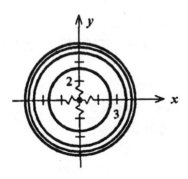

25. $V = f(1.5,4) = \pi(1.5)^2(4) = 9\pi$, or 9π cu ft

27. a. $M = \dfrac{80}{(1.8)^2} = 24.69$.

b. $\dfrac{w}{(1.8)^2} < 25$; that is, $w < 25(1.8)^2 = 81$, that is less than 81 kg.

29. a. $R(x,y) = xp + yq = x(200 - \frac{1}{5}x - \frac{1}{10}y) + y(160 - \frac{1}{10}x - \frac{1}{4}y)$
$= -\frac{1}{5}x^2 - \frac{1}{4}y^2 - \frac{1}{5}xy + 200x + 160y$.

b. The domain of R is the set of all points (x,y) satisfying
$200 - \frac{1}{5}x - \frac{1}{10}y \geq 0$, $160 - \frac{1}{10}x - \frac{1}{4}y \geq 0$

31. a. $R(x,y) = xp + yq = 20x - 0.005x^2 - 0.001xy + 15y - 0.001xy - 0.003y^2$
$= -0.005x^2 - 0.003y^2 - 0.002xy + 20x + 15y$.

b. Since p and q must both be nonnegative, the domain of R is the set of all
ordered pairs (x, y) for which
$20 - 0.005x - 0.001y \geq 0$
and $15 - 0.001x - 0.003y \geq 0$.

33. a. The domain of V is the set of all ordered pairs (P, T) where P and T are positive real numbers.

   b. $V = \dfrac{30.9(273)}{760} = 11.10$ liters.

35. The output is $f(32, 243) = 100(32^{3/5})(243)^{2/5} = 100(8)(9) = 7200$ or $7,200$ billlion.

37. The number of suspicious fires is

   $$N(100, 20) = \frac{100[1000 + 0.03(100^2)(20)]^{1/2}}{[5 + 0.2(20)]^2} = 103.29, \text{ or approximately } 103.$$

39. a. $P = f(100,000, 0.08, 30) = \dfrac{100,000(0.08)}{12\left[1 - \left(1 + \dfrac{0.08}{12}\right)^{-360}\right]} \approx 733.76$, or $733.76.

   $P = f(100,000, 0.1, 30) = \dfrac{100,000(0.1)}{12\left[1 - \left(1 + \dfrac{0.1}{12}\right)^{-360}\right]} \approx 877.57$, or $877.57.

   b. $P = f(100,000, 0.08, 20) = \dfrac{100,000(0,08)}{12\left[1 - \left(1 + \dfrac{0.08}{12}\right)^{-240}\right]} \approx 836.44$, or $836.44.

41. $f(M, 600, 10) = \dfrac{\pi^2 (360,000) M (10)}{900} \approx 39,478.42 M$

   or $\dfrac{39,478.42}{980} \approx 40.28$ times gravity.

43. False. Let $h(x, y) = xy$. Then there are no functions f and g such that $h(x, y) = f(x) + g(y)$.

45. False. Since $x^2 - y^2 = (x + y)(x - y)$, we see that $x^2 - y^2 = 0$ if $y = \pm x$. Therefore, the domain of f is $\{(x, y) \mid y \neq \pm x\}$.

8.2 Problem Solving Tips

1. The notation f_{xy} and f_{yx} is used to denote the second partial derivative of the function $f(x, y)$. Note that when this notation is used the differentiation is carried out in the order (left to right) in which x and y appear.

   The notation $\dfrac{\partial^2 f}{\partial y \partial x}$ and $\dfrac{\partial^2 f}{\partial x \partial y}$ is also used to denote the second partial derivatives of the function $f(x, y)$, but in this case the differentiation is carried out in the reverse order (right to left) in which x and y appear.

8.2 CONCEPT QUESTIONS, page 556

1. a. $\dfrac{\partial f}{\partial x}(a, b) = \dfrac{\partial f}{\partial x}(x, y)\bigg|_{(a,b)} = \left[\lim_{h \to 0} \dfrac{f(x+h, y) - f(x, y)}{h} \right]\bigg|_{(a,b)}$

   b. See page 547-548 of the text.

3. f_x, f_y, f_{xx}, f_{xy}, and f_{yy}.

EXERCISES 8.2, page 557

1. $f_x = 2$, $f_y = 3$

3. $g_x = 4x$, $g_y = 4$

5. $f_x = -\dfrac{4y}{x^3}$; $f_y = \dfrac{2}{x^2}$.

7. $g(u, v) = \dfrac{u - v}{u + v}$; $\dfrac{\partial g}{\partial u} = \dfrac{(u + v)(1) - (u - v)(1)}{(u + v)^2} = \dfrac{2v}{(u + v)^2}$.

   $\dfrac{\partial g}{\partial v} = \dfrac{(u + v)(-1) - (u - v)(1)}{(u + v)^2} = -\dfrac{2u}{(u + v)^2}$.

9. $f(s,t) = (s^2 - st + t^2)^3$; $f_s = 3(s^2 - st + t^2)^2(2s - t)$ and $f_t = 3(s^2 - st + t^2)^2(2t - s)$

11. $f(x,y) = (x^2 + y^2)^{2/3}$; $f_x = \frac{2}{3}(x^2 + y^2)^{-1/3}(2x) = \frac{4}{3}x(x^2 + y^2)^{-1/3}$. Similarly, $f_y = \frac{4}{3}y(x^2 + y^2)^{-1/3}$.

13. $f(x,y) = e^{xy+1}$; $f_x = ye^{xy+1}$, $f_y = xe^{xy+1}$.

15. $f(x,y) = x \ln y + y \ln x$; $f_x = \ln y + \dfrac{y}{x}$, $f_y = \dfrac{x}{y} + \ln x$.

17. $g(u,v) = e^u \ln v$. $g_u = e^u \ln v$, $g_v = \dfrac{e^u}{v}$.

19. $f(x,y,z) = xyz + xy^2 + yz^2 + zx^2$; $f_x = yz + y^2 + 2xz$, $f_y = xz + 2xy + z^2$, $f_z = xy + 2yz + x^2$.

21. $h(r,s,t) = e^{rst}$; $h_r = ste^{rst}$, $h_s = rte^{rst}$, $h_t = rse^{rst}$.

23. $f(x,y) = x^2y + xy^2$; $f_x(1,2) = 2xy + y^2\big|_{(1,2)} = 8$; $f_y(1,2) = x^2 + 2xy\big|_{(1,2)} = 5$.

25. $f(x,y) = x\sqrt{y} + y^2 = xy^{1/2} + y^2$; $f_x(2,1) = \sqrt{y}\,\big|_{(2,1)} = 1$,

$f_y(2,1) = \dfrac{x}{2\sqrt{y}} + 2y\,\big|_{(2,1)} = 3$.

27. $f(x,y) = \dfrac{x}{y}$; $f_x(1,2) = \dfrac{1}{y}\bigg|_{(1,2)} = \dfrac{1}{2}$, $f_y(1,2) = -\dfrac{x}{y^2}\bigg|_{(1,2)} = -\dfrac{1}{4}$.

29. $f(x,y) = e^{xy}$. $f_x(1,1) = ye^{xy}\big|_{(1,1)} = e$, $f_y(1,1) = xe^{xy}\big|_{(1,1)} = e$.

31. $f(x,y,z) = x^2 yz^3$; $f_x(1,0,2) = 2xyz^3\big|_{(1,0,2)} = 0$; $f_y(1,0,2) = x^2 z^3\big|_{(1,0,2)} = 8$.

$f_z(1,0,2) = 3x^2 yz^2\big|_{(1,0,2)} = 0$.

33. $f(x,y) = x^2 y + xy^3$; $f_x = 2xy + y^3$, $f_y = x^2 + 3xy^2$.

Therefore, $f_{xx} = 2y$, $f_{xy} = 2x + 3y^2 = f_{yx}$, $f_{yy} = 6xy$.

35. $f(x,y) = x^2 - 2xy + 2y^2 + x - 2y$; $f_x = 2x - 2y + 1$, $f_y = -2x + 4y - 2$; $f_{xx} = 2$,

$f_{xy} = -2$, $f_{yx} = -2$, $f_{yy} = 4$.

37. $f(x,y) = (x^2 + y^2)^{1/2}$; $f_x = \tfrac{1}{2}(x^2 + y^2)^{-1/2}(2x) = x(x^2 + y^2)^{-1/2}$;

$f_y = y(x^2 + y^2)^{-1/2}$.

$f_{xx} = (x^2 + y^2)^{-1/2} + x(-\tfrac{1}{2})(x^2 + y^2)^{-3/2}(2x) = (x^2 + y^2)^{-1/2} - x^2(x^2 + y^2)^{-3/2}$

$\quad = (x^2 + y^2)^{-3/2}(x^2 + y^2 - x^2) = \dfrac{y^2}{(x^2 + y^2)^{3/2}}$.

$f_{xy} = x(-\tfrac{1}{2})(x^2 + y^2)^{-3/2}(2y) = -\dfrac{xy}{(x^2 + y^2)^{3/2}} = f_{yx}$.

$f_{yy} = (x^2 + y^2)^{-1/2} + y(-\tfrac{1}{2})(x^2 + y^2)^{-3/2}(2y) = (x^2 + y^2)^{-1/2} - y^2(x^2 + y^2)^{-3/2}$

$\quad = (x^2 + y^2)^{-3/2}(x^2 + y^2 - y^2) = \dfrac{x^2}{(x^2 + y^2)^{3/2}}$.

39. $f(x,y) = e^{-x/y}$; $f_x = -\dfrac{1}{y}e^{-x/y}$; $f_y = \dfrac{x}{y^2}e^{-x/y}$; $f_{xx} = \dfrac{1}{y^2}e^{-x/y}$;

$f_{xy} = -\dfrac{x}{y^3}e^{-x/y} + \dfrac{1}{y^2}e^{-x/y} = \left(\dfrac{-x+y}{y^3}\right)e^{-x/y} = f_{yx}$.

$f_{yy} = -\dfrac{2x}{y^3}e^{-x/y} + \dfrac{x^2}{y^4}e^{-x/y} = \dfrac{x}{y^3}\left(\dfrac{x}{y} - 2\right)e^{-x/y}$.

41. a. $f(x,y) = 20x^{3/4}y^{1/4}$. $f_x(256,16) = 15\left(\dfrac{y}{x}\right)^{1/4}\bigg|_{(256,16)}$

373

$$= 15\left(\frac{16}{256}\right)^{1/4} = 15\left(\frac{2}{4}\right) = 7.5.$$

$$f_y(256,16) = 5\left(\frac{x}{y}\right)^{3/4}\Bigg|_{(256,16)} = 5\left(\frac{256}{16}\right)^{3/4} = 5(80) = 40.$$

b.   Yes.

43. $p(x,y) = 200 - 10(x-\tfrac{1}{2})^2 - 15(y-1)^2.$ $\dfrac{\partial p}{\partial x}(0,1) = -20(x-\tfrac{1}{2})\big|_{(0,1)} = 10;$

At the location (0,1) in the figure, the price of land is changing at the rate of $10 per sq ft per mile change to the right.

$$\frac{\partial p}{\partial y}(0,1) = -30(y-1)\big|_{(0,1)} = 0;$$

At the location (0,1) in the figure, the price of land is constant per mile change upward.

45. $f(p,q) = 10,000 - 10p - e^{0.5q};$ $g(p,q) = 50,000 - 4000q - 10p.$

$$\frac{\partial f}{\partial q} = -0.5e^{0.5q} < 0 \text{ and } \frac{\partial g}{\partial p} = -10 < 0$$

and so the two commodities are complementary commodities.

47. $R(x,y) = -0.2x^2 - 0.25y^2 - 0.2xy + 200x + 160y.$

$$\frac{\partial R}{\partial x}(300,250) = -0.4x - 0.2y + 200\big|_{(300,250)}$$

$$= -0.4(300) - 0.2(250) + 200 = 30$$

and this says that at a sales level of 300 finished and 250 unfinished units the revenue is increasing at the rate of $30 per week per unit increase in the finished units.

$$\frac{\partial R}{\partial y}(300,250) = -0.5y - 0.2x + 160\big|_{(300,250)}$$

$$= -0.5(250) - 0.2(300) + 160 = -25$$

and this says that at a level of 300 finished and 250 unfinished units the revenue is decreasing at the rate of $25 per week per increase in the unfinished units.

49. a. $T = f(32,20) = 35.74 + 0.6215(32) - 35.75(20^{0.16}) + 0.4275(32)(20^{0.16})$

≈ 19.99, or approximately $20°F$.

b. $\dfrac{\partial T}{\partial s} = -35.75(0.16S^{-0.84}) + 0.4275t(0.16S^{-0.84})$

$= 0.16(-35.75 + 0.4275t)s^{-0.84}$

$\dfrac{\partial T}{\partial s}\bigg|_{(32,20)} = 0.16[-35.75 + 0.4275(32)]20^{-0.84} \approx -0.285$;

that is, the wind chill will drop by 0.3 degrees for each 1 mph increase in wind speed.

51. $V = \dfrac{30.9T}{P}$. $\dfrac{\partial V}{\partial T} = \dfrac{30.9}{P}$ and $\dfrac{\partial V}{\partial P} = -\dfrac{30.9T}{P^2}$.

Therefore, $\dfrac{\partial V}{\partial T}\bigg|_{T=300, P=800} = \dfrac{30.9}{800} = 0.039$, or 0.039 liters/degree.

$\dfrac{\partial V}{\partial P}\bigg|_{T=300, P=800} = -\dfrac{(30.9)(300)}{800^2} = -0.015$

or -0.015 liters/mm of mercury.

53. $V = \dfrac{kT}{P}$ and $\dfrac{\partial V}{\partial T} = \dfrac{k}{P}$; $T = \dfrac{VP}{k}$ and $\dfrac{\partial T}{\partial P} = \dfrac{V}{k} = \dfrac{T}{P}$; and

$P = \dfrac{kT}{V}$ and $\dfrac{\partial P}{\partial V} = -\dfrac{kT}{V^2} = -kT \cdot \dfrac{P^2}{(kT)^2} = -\dfrac{P^2}{kT}$

Therefore $\dfrac{\partial V}{\partial T} \cdot \dfrac{\partial T}{\partial P} \cdot \dfrac{\partial P}{\partial V} = \dfrac{k}{P} \cdot \dfrac{T}{P} \cdot -\dfrac{P^2}{kT} = -1$.

55. False. Let $f(x, y) = xy^{1/2}$. Then $f_x = y^{1/2}$ is defined at $(0, 0)$. But

$f_y = \dfrac{1}{2}xy^{-1/2} = \dfrac{x}{2y^{1/2}}$ is not defined at $(0, 0)$.

57. True. See Section 8.2.

1. 1.3124; 0.4038   3. −1.8889; 0.7778   5. −0.3863; −0.8497

8.3 Problem Solving Tips

1. To find the relative extrema of a function of several variables, first find the critical

   points of $f(x,y)$ by solving the simultaneous equations $f_x = 0$ and $f_y = 0$, and

   then use the second derivative test to classify those points.

2. To use the second derivative test, first evaluate the function $D(x,y) = f_{xx}f_{yy} - f_{xy}^2$ for

   each critical point found in (1). If $D(a,b) > 0$ and $f_{xx}(a,b) < 0$ then $f(x,y)$ has a

   relative maximum at the point (a,b). If $D(a,b) > 0$ and $f_{xx}(a,b) > 0$, then $f(x,y)$

   has a relative minimum at the point (a,b). If $D(a,b) < 0$ then $f(x,y)$ has neither a

   relative maximum nor a relative minimum at the point (a,b). If $D(a,b) = 0$, then the

   test is inconclusive.

8.3 CONCEPT QUESTIONS, page 569

1. a. A function $f(x,y)$ has a relative maximum at (a,b) if $f(a,b)$ is the largest
    number compared to $f(x,y)$ for all (x,y) near (a,b).
    b. $f(a,y)$ has an absolute maximum at (a,b) if $f(a,b)$ is the largest number
    compared to $f(x,y)$ for all (x,y) in the domain of f.
3. See the procedure given in the text on page 564.

1. $f(x,y) = 1 - 2x^2 - 3y^2$. To find the critical point(s) of f, we solve the system
$$\begin{cases} f_x = -4x = 0 \\ f_y = -6y = 0 \end{cases}$$
obtaining $(0,0)$ as the only critical point of f. Next,
$$f_{xx} = -4, f_{xy} = 0, \text{ and } f_{yy} = -6.$$
In particular, $f_{xx}(0,0) = -4, f_{xy}(0,0) = 0$, and $f_{yy}(0,0) = -6$, giving
$$D(0,0) = (-4)(-6) - 0^2 = 24 > 0.$$
Since $f_{xx}(0,0) < 0$, the Second Derivative Test implies that $(0,0)$ gives rise to a relative maximum of f. Finally, the relative maximum of f is $f(0,0) = 1$.

3. To find the critical points of f, we solve the system
$$\begin{cases} f_x = 2x - 2 = 0 \\ f_y = -2y + 4 = 0 \end{cases}$$
obtaining $x = 1$ and $y = 2$ so that $(1,2)$ is the only critical point.
$$f_{xx} = 2, f_{xy} = 0, \text{ and } f_{yy} = -2.$$
So $D(x,y) = f_{xx}f_{yy} - f_{xy}^2 = -4$. In particular, $D(1,2) = -4 < 0$ and so $(1,2)$ affords a saddle point of f and $f(1,2) = 4$.

5. $f(x,y) = x^2 + 2xy + 2y^2 - 4x + 8y - 1$. To find the critical point(s) of f, we solve
the system
$$\begin{cases} f_x = 2x + 2y - 4 = 0 \\ f_y = 2x + 4y + 8 = 0 \end{cases}$$
obtaining $(8,-6)$ as the critical point of f. Next, $f_{xx} = 2, f_{xy} = 2, f_{yy} = 4$. In particular, $f_{xx}(8,-6) = 2, f_{xy}(8,-6) = 2, f_{yy}(8,-6) = 4$, giving $D = 2(4) - 4 = 4 > 0$. Since $f_{xx}(8,-6) > 0$, $(8,-6)$ gives rise to a relative minimum of f. Finally, the relative minimum value of f is $f(8,-6) = -41$.

7. $f(x,y) = 2x^3 + y^2 - 9x^2 - 4y + 12x - 2.$. To find the critical points of f, we solve
the system
$$\begin{cases} f_x = 6x^2 - 18x + 12 = 0 \\ f_y = 2y - 4 = 0 \end{cases}$$
The first equation is equivalent to $x^2 - 3x + 2 = 0$, or $(x - 2)(x - 1) = 0$ which gives $x = 1$ or 2. The second equation of the system gives $y = 2$. Therefore, there are two critical points, $(1,2)$ and $(2,2)$. Next, we compute

$f_{xx} = 12x - 18 = 6(2x - 3), f_{xy} = 0, f_{yy} = 2.$

At the point $(1,2)$:

$f_{xx}(1,2) = 6(2 - 3) = -6, f_{xy}(1,2) = 0,$ and $f_{yy}(1,2) = 2.$

Therefore, $D = (-6)(2) - 0 = -12 < 0$ and we conclude that $(1,2)$ gives rise to a saddle point of f. At the point $(2,2)$:

$f_{xx}(2,2) = 6(4 - 3) = 6, f_{xy}(2,2) = 0,$ and $f_{yy}(2,2) = 2.$

Therefore, $D = (6)(2) - 0 = 12 > 0$. Since $f_{xx}(2,2) > 0$, we see that $(2,2)$ gives rise to a relative minimum with value $f(2,2) = -2$.

9. To find the critical points of f, we solve the system

$$\begin{cases} f_x = 3x^2 - 2y + 7 = 0 \\ f_y = 2y - 2x - 8 = 0 \end{cases}$$

Adding the two equations gives $3x^2 - 2x - 1 = 0$, or $(3x + 1)(x - 1) = 0$.

Therefore, $x = -1/3$ or 1. Substituting each of these values of x into the second equation gives $y = 8/3$ and $y = 5$, respectively. Therefore, $(-\frac{1}{3}, \frac{11}{3})$ and $(1,5)$ are critical points of f.

Next, $f_{xx} = 6x, f_{xy} = -2,$ and $f_{yy} = 2$. So $D(x,y) = 12x - 4 = 4(3x - 1)$. Then

$$D(-\tfrac{1}{3}, \tfrac{11}{3}) = 4(-1 - 1) = -8 < 0$$

and so $(-\frac{1}{3}, \frac{11}{3})$ gives a saddle point. Next, $D(1,5) = 4(3 - 1) = 8 > 0$ and since $f_{xx}(1,5) = 6 > 0$, we see that $(1,5)$ gives rise to a relative minimum.

11. To find the critical points of f, we solve the system

$$\begin{cases} f_x = 3x^2 - 3y = 0 \\ f_y = -3x + 3y^2 = 0 \end{cases}$$

The first equation gives $y = x^2$ which when substituted into the second equation gives $-3x + 3x^4 = 3x(x^3 - 1) = 0$. Therefore, $x = 0$ or 1. Substituting these values of x into the first equation gives $y = 0$ and $y = 1$, respectively. Therefore, $(0,0)$ and $(1,1)$ are critical points of f. Next, we find $f_{xx} = 6x, f_{xy} = -3,$ and $f_{yy} = 6y$. So $D = f_{xx}f_{yy} - f_{xy}^2 = 36xy - 9$. Since $D(0,0) = -9 < 0$, we see that $(0,0)$ gives a saddle point of f. Next, $D(1,1) = 36 - 9 = 27 > 0$ and since $f_{xx}(1,1) = 6 > 0$, we see that $f(1,1) = -3$ is a relative minimum value of f.

13. Solving the system of equations

$$\begin{cases} f_x = y - \dfrac{4}{x^2} = 0 \\ f_y = x - \dfrac{2}{y^2} = 0 \end{cases}$$

we obtain $y = \frac{4}{x^2}$. Therefore, $x - 2\left(\frac{x^4}{16}\right) = 0$ and $8x - x^4 = x(8 - x^3) = 0$, and $x = 0$, or $x = 2$. Since $x = 0$ is not in the domain of f, $(2,1)$ is the only critical point of f. Next, $f_{xx} = \frac{8}{x^3}$, $f_{xy} = 1$, and $f_{yy} = \frac{4}{y^3}$. Therefore,

$$D(2,1) = \left.\frac{32}{x^3 y^3} - 1\right|_{(2,1)} = 4 - 1 = 3 > 0 \text{ and } f_{xx}(2,1) = 1 > 0. \text{ Therefore, the relative}$$

minimum value of f is $f(2,1) = 2 + 4/2 + 2/1 = 6$.

15. Solving the system of equations $f_x = 2x = 0$ and $f_y = -2ye^{y^2} = 0$, we obtain $x = 0$ and $y = 0$. Therefore, $(0,0)$ is the only critical point of f. Next,
$$f_{xx} = 2, f_{xy} = 0, f_{yy} = -2e^{y^2} - 4y^2 e^{y^2}.$$
Therefore, $D(0,0) = \left.-4e^{y^2}(1 + 2y^2)\right|_{(0,0)} = -4(1) < 0$, and we conclude that $(0,0)$

gives rise to a saddle point.

17. $f(x,y) = e^{x^2 + y^2}$
Solving the system

$$\begin{cases} f_x = 2xe^{x^2 + y^2} = 0 \\ f_y = 2ye^{x^2 + y^2} = 0 \end{cases}$$

we see that $x = 0$ and $y = 0$ (recall that $e^{x^2 + y^2} \neq 0$). Therefore, $(0,0)$ is the only critical point of f. Next, we compute
$$f_{xx} = 2e^{x^2 + y^2} + 2x(2x)e^{x^2 + y^2} = 2(1 + 2x^2)e^{x^2 + y^2}$$
$$f_{xy} = 2x(2y)e^{x^2 + y^2} = 4xye^{x^2 + y^2}$$
$$f_{yy} = 2(1 + 2y^2)e^{x^2 + y^2}.$$

In particular, at the point $(0,0)$, $f_{xx}(0,0) = 2$, $f_{xy}(0,0) = 0$, and $f_{yy}(0,0) = 2$. Therefore, $D = (2)(2) - 0 = 4 > 0$. Furthermore, since $f_{xx}(0,0) > 0$, we conclude that $(0,0)$ gives rise to a relative minimum of f. The relative minimum value of f is $f(0,0) = 1$.

19. $f(x,y) = \ln(1+x^2+y^2)$. We solve the system of equations

$$f_x = \frac{2x}{1+x^2+y^2} = 0 \text{ and } f_y = \frac{2y}{1+x^2+y^2} = 0,$$

obtaining $x = 0$ and $y = 0$. Therefore, $(0,0)$ is the only critical point of f. Next,

$$f_{xx} = \frac{(1+x^2+y^2)2-(2x)(2x)}{(1+x^2+y^2)^2} = \frac{2+2y^2-2x^2}{(1+x^2+y^2)^2}$$

$$f_{yy} = \frac{(1+x^2+y^2)2-(2y)(2y)}{(1+x^2+y^2)^2} = \frac{2+2x^2-2y^2}{(1+x^2+y^2)^2}$$

$$f_{xy} = -2x(1+x^2+y^2)^{-2}(2y) = -\frac{4xy}{(1+x^2+y^2)^2}.$$

Therefore, $D(x,y) = \dfrac{(2+2y^2-2x^2)(2+2x^2-2y^2)}{(1+x^2+y^2)^4} - \dfrac{16x^2y^2}{(1+x^2+y^2)^4}.$

Since $D(0,0) = \frac{4}{1} > 0$ and $f_{xx}(0,0) = 2 > 0$, $f(0,0) = 0$ is a relative minimum value.

21. $P(x) = -0.2x^2 - 0.25y^2 - 0.2xy + 200x + 160y - 100x - 70y - 4000$
$= -0.2x^2 - 0.25y^2 - 0.2xy + 100x + 90y - 4000.$

Then $\begin{cases} P_x = -0.4x - 0.2y + 100 = 0 \\ P_y = -0.5y - 0.2x + 90 = 0 \end{cases}$

implies that $\begin{cases} 4x+2y = 1000 \\ 2x+5y = 900 \end{cases}$. Solving, we find $x = 200$ and $y = 100$.

Next, $P_{xx} = -0.4, P_{yy} = -0.5, P_{xy} = -0.2$, and

$D(200,100) = (-0.4)(-0.5)-(-0.2)^2 > 0$. Since $P_{xx}(200, 100) < 0$, we conclude that $(200,100)$ is a relative maximum of P. Thus, the company should manufacture 200 finished and 100 unfinished units per week. The maximum profit is

$$P(200,100) = -0.2(200)^2 - 0.25(100)^2 - 0.2(100)(200) + 100(200) + 90(100) - 4000$$
$$= 10{,}500, \text{ or } \$10{,}500.$$

23. $p(x,y) = 200 - 10(x - \frac{1}{2})^2 - 15(y - 1)^2$. Solving the system of equations

$$\begin{cases} p_x = -20(x - \tfrac{1}{2}) = 0 \\ p_y = -30(y - 1) = 0 \end{cases}$$

we obtain $x = 1/2$, $y = 1$. We conclude that the only critical point of f is $(\tfrac{1}{2}, 1)$.

Next, $\quad p_{xx} = -20,\ p_{xy} = 0,\ p_{yy} = -30$

so $\quad D(\tfrac{1}{2}, 1) = (-20)(-30) = 600 > 0$.

Since $p_{xx} = -20 < 0$, we conclude that $f(\tfrac{1}{2}, 1)$ gives a relative maximum. So we conclude that the price of land is highest at $(\tfrac{1}{2}, 1)$.

25. We want to minimize
$$f(x,y) = D^2 = (x - 5)^2 + (y - 2)^2 + (x + 4)^2 + (y - 4)^2 + (x + 1)^2 + (y + 3)^2.$$
Next, $\begin{cases} f_x = 2(x - 5) + 2(x + 4) + 2(x + 1) = 6x = 0, \\ f_y = 2(y - 2) + 2(y - 4) + 2(y + 3) = 6y - 6 = 0 \end{cases}$

and we conclude that $x = 0$ and $y = 1$. Also,
$$f_{xx} = 6, f_{xy} = 0, f_{yy} = 6 \text{ and } D(x,y) = (6)(6) = 36 > 0.$$
Since $f_{xx} > 0$, we conclude that the function is minimized at $(0,1)$ and so $(0,1)$ gives the desired location.

27. Refer to the figure in the text.
$$xy + 2xz + 2yz = 300; \quad z(2x + 2y) = 300 - xy; \text{ and } z = \frac{300 - xy}{2(x + y)}.$$

Then the volume is given by
$$V = xyz = xy\left[\frac{300 - xy}{2(x + y)}\right] = \frac{300xy - x^2 y^2}{2(x + y)}.$$

We find
$$\frac{\partial V}{\partial x} = \frac{1}{2} \frac{(x + y)(300y - 2xy^2) - (300xy - x^2 y^2)}{(x + y)^2}$$
$$= \frac{300xy - 2x^2 y^2 + 300y^2 - 2xy^3 - 300xy + x^2 y^2}{2(x + y)^2}$$
$$= \frac{300y^2 - 2xy^3 - x^2 y^2}{2(x + y)^2} = \frac{y^2(300 - 2xy - x^2)}{2(x + y)^2}.$$

Similarly, $\dfrac{\partial V}{\partial y} = \dfrac{x^2(300 - 2xy - y^2)}{2(x + y)^2}$. Setting both $\partial V / \partial x$ and $\partial V / \partial y$ equal to zero and observing that both $x > 0$ and $y > 0$, we have the system

$\begin{cases} 2yx + x^2 = 300 \\ 2yx + y^2 = 300 \end{cases}$. Subtracting, we find $y^2 - x^2 = 0$; that is $(y-x)(y+x) = 0$. So $y = x$ or $y = -x$. The latter is not possible since $x, y > 0$. Therefore, $y = x$. Substituting this value into the first equation in the system gives

$$2x^2 + x^2 = 300; \quad x^2 = 100; \quad \text{and} \quad x = 10.$$

Therefore, $y = 10$. Substituting this value into the expression for z

gives $z = \dfrac{300 - 10^2}{2(10+10)} = 5$. So the dimensions are $10" \times 10" \times 5"$. The volume is

500 cu in.

29. The heating cost is $C = 2xy + 8xz + 6yz$. But $xyz = 12{,}000$ or $z = 12{,}000/xy$. Therefore,

$$C = f(x,y) = 2xy + 8x\left(\frac{12{,}000}{xy}\right) + 6y\left(\frac{12{,}000}{xy}\right) = 2xy + \frac{96{,}000}{y} + \frac{72{,}000}{x}.$$

To find the minimum of f, we find the critical point of f by solving the system

$$\begin{cases} f_x = 2y - \dfrac{72{,}000}{x^2} = 0 \\ f_y = 2x - \dfrac{96{,}000}{y^2} = 0 \end{cases}.$$

The first equation gives $y = 36000/x^2$, which when substituted into the second equation yields

$$2x - 96{,}000\left(\frac{x^2}{36{,}000}\right)^2 = 0, \quad (36{,}000)^2 x - 48{,}000x^4 = 0$$

$$x(27{,}000 - x^3) = 0.$$

Solving this equation, we have $x = 0$, or $x = 30$. We reject the first root because $x \neq 0$. With $x = 30$, we find $y = 40$ and

$$f_{xx} = \frac{144{,}000}{x^3}, \quad f_{xy} = 2, \quad \text{and} \quad f_{yy} = \frac{192{,}000}{y^3}.$$

In particular, $f_{xx}(30,40) = 5.33, f_{xy} = (30,40) = 2$, and $f_{yy}(30,40) = 3$. So $D(30,40) = (5.33)(3) - 4 = 11.99 > 0$

and since $f_{xx}(30,40) > 0$, we see that $(30,40)$ gives a relative minimum. Physical considerations tell us that this is an absolute minimum. The minimal annual heating cost is

$$f(30,40) = 2(30)(40) + \frac{96,000}{40} + \frac{72,000}{30} = 7200, \text{ or } \$7,200.$$

31. False. Let $f(x,y) = xy$. Then $f_x(0,0) = 0$ and $f_y(0,0) = 0$. But $(0,0)$ does not afford a relative extremum of $(0,0)$. In fact, $f_{xx} = 0$, $f_{yy} = 0$, and $f_{xy} = 1$. Therefore, $D(x,y) = f_{xx}f_{yy} - f_{xy}^2 = -1$ and so $D(0,0) = -1$ which shows that $(0, 0, 0)$ is a saddle point.

8.4 Problem Solving Tips

1. You will find it helpful to organize the data in a least-squares problem in the form of a table as that given in Examples 2 and 3 in the text.

8.4 CONCEPT QUESTIONS, page 544

1. a. A scatter diagram is a graph showing the data points that describe the relationship between the two variables x and y.
   b. The least squares line is the straight line that best fits a set of data points when the points are scattered about a straight line.

EXERCISES 8.4, page 578

1. a. We first summarize the data:

x	y	x^2	xy
1	4	1	4
2	6	4	12
3	8	9	24
4	11	16	44
10	29	30	84

The normal equations are $4b + 10m = 29$
$$10b + 30m = 84.$$

Solving this system of equations, we obtain $m = 2.3$ and $b = 1.5$. So an equation is $y = 2.3x + 1.5$.

b. The scatter diagram and the least squares line for this data follow:

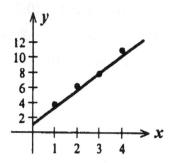

3. a. We first summarize the data:

x	y	x^2	xy
1	4.5	1	4.5
2	5	4	10
3	3	9	9
4	2	16	8
4	3.5	16	14
6	1	36	6
20	19	82	51.5

The normal equations are $6b + 20m = 19$

$$20b + 82m = 51.5.$$

The solutions are $m \approx -0.7717$ and $b \approx 5.7391$ and so a required equation is $y = -0.772x + 5.739$.

b. The scatter diagram and the least-squares line for these data follow.

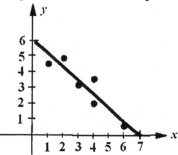

5. a. We first summarize the data:

x	y	x^2	xy
1	3	1	3
2	5	4	10
3	5	9	15
4	7	16	28
5	8	25	40
15	28	55	96

The normal equations are $55m + 15b = 96$
$$15m + 5b = 28.$$
Solving, we find $m = 1.2$ and $b = 2$, so that the required equation is $y = 1.2x + 2$.

b.   The scatter diagram and the least-squares line for the given data follow.

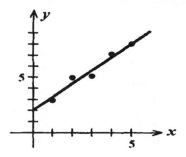

7. a. We first summarize the data:

x	y	x^2	xy
4	0.5	16	2
4.5	0.6	20.25	2.7
5	0.8	25	4
5.5	0.9	30.25	4.95
6	1.2	36	7.2
25	4	127.5	20.85

The normal equations are
$$5b + 25m = 4$$
$$25b + 127.5m = 20.85.$$

The solutions are $m = 0.34$ and $b = -0.9$, and so a required equation is $y = 0.34x - 0.9$.

b. The scatter diagram and the least-squares line for these data follow.

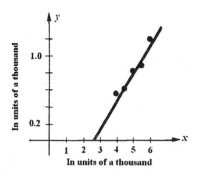

c. If $x = 6.4$, then $y = 0.34(6.4) - 0.9 = 1.276$ and so 1276 completed applications might be expected.

9. a. We first summarize the data:

x	y	x^2	xy
1	436	1	436
2	438	4	876
3	428	9	1284
4	430	16	1720
5	426	25	2130
15	2158	55	6446

The normal equations are
$$5b + 15m = 2158$$
$$15b + 55m = 6446.$$
Solving this system, we find $m = -2.8$ and $b = 440$.
Thus, the equation of the least-squares line is $y = -2.8x + 440$.

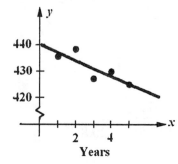

Years

b. The scatter diagram and the least-squares line for this data are shown in the figure that follows.

c. Two years from now, the average SAT verbal score in that area will be $y = -2.8(7) + 440 = 420.4$.

11. a. We first summarize the data:

x	y	x^2	xy
1	20	1	20
2	24	4	48
3	26	9	78
4	28	16	112
5	32	25	160
15	130	55	418

The normal equations are
$$5b + 15m = 130$$
$$15b + 55m = 418.$$

The solutions are $m = 2.8$ and $b = 17.6$, and so an equation of the line is
$$y = 2.8x + 17.6.$$

b. When $x = 8$, $y = 2.8(8) + 17.6 = 40$. Hence, the state subsidy is expected to be $40 million for the eighth year.

13. a.

t	y	t^2	ty
0	126	1	0
1	144	4	144
2	171	9	343
3	191	16	573
4	216	25	864
10	848	30	1923

The normal equations are
$$5b + 10m = 848$$
$$10b + 30m = 1923$$

The solutions are $m \approx 22.7$ and $b \approx 124.2$. Therefore, the required equation is $y = 22.7t + 124.2$.

b. $y = 22.7(6) + 124.2 = 260.4$, or $260.4 billion.

15. a.

x	y	x^2	xy
0	3.7	0	0
1	4.0	1	4
2	4.4	4	8.8
3	4.8	9	14.4
4	5.2	16	20.8
5	5.8	25	29.0
6	6.3	36	37.8
21	34.2	91	114.8

The normal equations are
$$7b + 21m = 34.2$$
$$21b + 91m = 114.8$$
The solutions are $m \approx 0.4357$ and $b \approx 3.5786$. Therefore, the required equation is $y = 0.4357x + 3.5786$.

b. The rate of change is given by the slope of the least-squares line, that is, approximately \$0.4357 billion/yr.

17. a.

x	y	x^2	xy
0	7.9	0	0
1	9.6	1	9.6
2	11.5	4	23
3	13.3	9	39.9
4	15.2	16	60.8
5	16	25	80
6	18.8	36	112.8
21	92.3	30	326.1

The normal equations are
$$7b + 21m = 92.3$$
$$21b + 91m = 326.1$$
The solutions are $m \approx 1.7571$ and $b \approx 7.9143$. Therefore, the required equation is $y = 1.7571x + 7.9143$

b. $y = 1.7571(5) + 7.9143 \approx 16.7$ or \$16.7 billion.

19. a.

x	y	x^2	xy
1	72.6	1	72.6
2	76.2	4	152.4
3	80.4	9	241.2
4	84.9	16	339.6
5	87	25	435
6	87.9	36	527.6
21	489	91	1768.2

The normal equations are

$$6b + 21m = 489$$

$$21b + 91m = 1768.2$$

The solutions are $m \approx 3.24$ and $b \approx 70.16$. Therefore the required equation is

$$y = 3.24x + 70.16$$

b. Then the FICA wage base for the year 2006 is given by

$$y = 3.24(10) + 70.16 = 102.560, \text{ or } \$102,560.$$

21. a. We first summarize the given data:

x	y	x^2	xy
0	15.9	0	0
10	16.8	100	168
20	17.6	400	352
30	18.5	900	555
40	19.3	1600	772
50	20.3	2500	1015
150	108.4	5500	2862

The normal equations are

$$6b + 150m = 108.4$$

$$150b + 5500m = 2862$$

The solutions are $b = 15.90$ and $m = 0.09$. Therefore, $y = 0.09x + 15.9$

b. The life expectancy at 65 of a male in 2040 is

$$y = 0.9(40) + 15.9 = 19.50 \quad \text{or} \quad 19.50 \text{ years}$$

The datum gives a life expectancy of 19.5 years.

c. The life expectancy at 65 of a male in 2030 is
$$y = 0.09(30) + 15.9 = 18.6 \quad \text{or} \quad 18.6 \text{ years.}$$

23. a. We first summarize the given data:

x	y	x^2	xy
0	90.4	0	0
1	100.0	1	100
2	110.4	4	220.8
3	120.4	9	361.2
4	130.8	16	523.2
5	140.4	25	702
6	150	36	900
21	842.4	91	2807.2

The normal equations are $\begin{cases} 7b + 21m = 842.4 \\ 21b + 91m = 2807.2 \end{cases}$.

The solutions are $m = 10$ and $b = 90.34$. Therefore, the required equation is $y = 10x + 90.34$.

b. If $x = 6$, then $y = 10(6) + 90.34 = 150.34$, or 150,340,000. This compares well with the actual data for that year-- 150,000,000 subscribers.

25. False. See Example 1, page 575.

27. True

USING TECHNOLOGY EXERCISES 8.4, page 582

1. $y = 2.3596x + 3.8639$ 3. $y = -1.1948x + 3.5525$

5. a. $y = 0.5471x + 1.1671$ b. $5.54 billion

7. a. $y = 13.321x + 72.57$ b. 192 million tons

9. a. $1.95x + 12.19$; $23.89 billion

8.5 Problem Solving Tips

1. The method of Lagrange Multipliers allows us to find the critical points of the function $f(x,y)$ subject to the constraint $g(x,y) = 0$ (if the extrema exist). However, it does not tell us whether those critical points lead to a relative extrema. Instead we rely on the geometric or physical nature of the problem to classify these points.

2. To use the method of Lagrange Multipliers, first form the function $F(x,y,\lambda) = f(x,y) + \lambda g(x,y)$ and then solve the system of equations $F_x = 0$, $F_y = 0$, and $F_\lambda = 0$ for all values of x, y and λ. The solutions are candidates for the extrema of f.

8.5 CONCEPT QUESTIONS, page 592

1. A constrained relative extremem of f is an extremum of f subject to a constraint of the form $g(x,y) = 0$.

EXERCISES 8.5, page 592

1. We form the Lagrangian function $F(x,y,\lambda) = x^2 + 3y^2 + \lambda(x + y - 1)$. We solve the system

$$\begin{cases} F_x = 2x + \lambda = 0 \\ F_y = 6y + \lambda = 0 \\ F_\lambda = x + y - 1 = 0. \end{cases}$$

Solving the first and the second equations for x and y in terms of λ we obtain $x = -\frac{\lambda}{2}$ and $y = -\frac{\lambda}{6}$ which, upon substitution into the third equation, yields

$-\frac{\lambda}{2}-\frac{\lambda}{6}-1=0$ or $\lambda=-\frac{3}{2}$. Therefore, $x=\frac{3}{4}$ and $y=\frac{1}{4}$ which gives the point $(\frac{3}{4},\frac{1}{4})$ as the sole critical point of F. Therefore, $(\frac{3}{4},\frac{1}{4})=\frac{3}{4}$ is a minimum of F.

3. We form the Lagrangian function $F(x,y,\lambda) = 2x + 3y - x^2 - y^2 + \lambda(x + 2y - 9)$. We then solve the system

$$\begin{cases} F_x = 2 - 2x + \lambda = 0 \\ F_y = 3 - 2y + 2\lambda = 0. \\ F_\lambda = x + 2y - 9 = 0 \end{cases}$$

Solving the first equation λ, we obtain $\lambda = 2x - 2$. Substituting into the second equation, we have $3 - 2y + 4x - 4 = 0$, or $4x - 2y - 1 = 0$.   Adding this equation to the third equation in the system, we have $5x - 10 = 0$, or $x = 2$. Therefore, $y = 7/2$ and $f(2,\frac{7}{2}) = -\frac{7}{4}$ is the maximum value of f.

5. Form the Lagrangian function $F(x,y,\lambda) = x^2 + 4y^2 + \lambda(xy - 1)$. We then solve the

system $\begin{cases} F_x = 2x + \lambda y = 0 \\ F_y = 8y + \lambda x = 0. \\ F_\lambda = xy - 1 = 0 \end{cases}$

Multiplying the first and second equations by x and y, respectively, and subtracting the resulting equations, we obtain $2x^2 - 8y^2 = 0$, or $x = \pm 2y$. Substituting this into the third equation gives $2y^2 - 1 = 0$ or $y = \pm\frac{\sqrt{2}}{2}$. We conclude that $f(-\sqrt{2}, -\frac{\sqrt{2}}{2}) = f(\sqrt{2}, \frac{\sqrt{2}}{2}) = 4$ is the minimum value of f.

7. We form the Lagrangian function
$$F(x,y,\lambda) = x + 5y - 2xy - x^2 - 2y^2 + \lambda(2x + y - 4).$$
Next, we solve the system
$$\begin{cases} F_x = 1 - 2y - 2x + 2\lambda = 0 \\ F_y = 5 - 2x - 4y + \lambda = 0 \\ F_\lambda = 2x + y - 4 = 0 \end{cases}$$
Solving the last two equations for x and y in terms of λ, we obtain
$$y = \tfrac{1}{3}(1 + \lambda) \text{ and } x = \tfrac{1}{6}(11 - \lambda)$$
which, upon substitution into the first equation, yields

$$1-\tfrac{2}{3}(1+\lambda)-\tfrac{1}{3}(11-\lambda)+2\lambda = 0$$

or $1-\tfrac{2}{3}-\tfrac{2}{3}\lambda-\tfrac{11}{3}+\tfrac{\lambda}{3}+2\lambda = 0$

or $\lambda = 2$. Therefore, $x = 3/2$ and $y = 1$. The maximum of f is
$$f(\tfrac{3}{2},1) = \tfrac{3}{2}+5-2(\tfrac{3}{2})-(\tfrac{3}{2})^2-2 = -\tfrac{3}{4}.$$

9. Form the Lagrangian $F(x,y,\lambda) = xy^2 + \lambda(9x^2 + y^2 - 9)$. We then solve
$$\begin{cases} F_x = y^2 + 18\lambda x = 0 \\ F_y = 2xy + 2\lambda y = 0 \\ F_\lambda = 9x^2 + y^2 - 9 = 0. \end{cases}$$

The first equation gives $\lambda = -\dfrac{y^2}{18x}$. Substituting into the second gives

$$2xy + 2y\left(-\dfrac{y^2}{18x}\right) = 0, \text{ or } 18x^2y - y^3 = y(18x^2 - y^2) = 0,$$

giving $y = 0$ or $y = \pm 3\sqrt{2}x$. If $y = 0$, then the third equation gives $9x^2 - 9 = 0$ or $x = \pm 1$. If $y = \pm 3\sqrt{3}/3$. Therefore, the points $(-1,0),(-\sqrt{3}/3,-\sqrt{6})$, $(-\sqrt{3}/3,\sqrt{6})$, $(\sqrt{3}/3,-\sqrt{6})$ and $(\sqrt{3}/3,\sqrt{6})$ give rise to extreme values of f subject to the given constraint. Evaluating $f(x,y)$ at each of these points, we see that $f(\sqrt{3}/3,-\sqrt{6}) = (\sqrt{3}/3,\sqrt{6}) = 2\sqrt{3}$ is the maximum value of f.

11. We form the Lagrangian function $F(x,y,\lambda) = xy + \lambda(x^2 + y^2 - 16)$. To find the critical points of F, we solve the system
$$\begin{cases} F_x = y + 2\lambda x = 0 \\ F_y = x + 2\lambda y = 0 \\ F_\lambda = x^2 + y^2 - 16 = 0 \end{cases}$$
Solving the first equation for λ and substituting this value into the second

equation yields $x - 2\left(\dfrac{y}{2x}\right)y = 0$, or $x^2 = y^2$. Substituting the last equation into

the third equation in the system, yields $x^2 + x^2 - 16 = 0$, or $x^2 = 8$, that is, $x = \pm 2\sqrt{2}$. The corresponding values of y are $y = \pm 2\sqrt{2}$. Therefore the critical points of F are $(-2\sqrt{2},-2\sqrt{2}), (-2\sqrt{2},2\sqrt{2}), (2\sqrt{2},-2\sqrt{2})(2\sqrt{2}, 2\sqrt{2})$. Evaluating f at each of these values, we find that $f(-2\sqrt{2},2\sqrt{2}) = -8$ and $f(2\sqrt{2},-2\sqrt{2}) = -8$ are relative minimum values and $f(-2\sqrt{2},-2\sqrt{2}) = 8$ and

$f(2\sqrt{2},2\sqrt{2}) = 8$, are relative maximum values.

13. We form the Lagrangian function $F(x,y,\lambda) = xy^2 + \lambda(x^2 + y^2 - 1)$. Next, we solve the system

$$\begin{cases} F_x = \quad y^2 + 2x\lambda = 0 \\ F_y = 2xy + 2y\lambda = 0 \\ F_\lambda = x^2 + y^2 - 1 = 0 \end{cases}.$$

We find that $x = \pm\sqrt{3}/3$ and $y = \pm\sqrt{6}/3$ and $x = \pm1$, $y = 0$. Evaluating f at each of the critical points $(-\frac{\sqrt{3}}{3},-\frac{\sqrt{6}}{3}), (-\frac{\sqrt{3}}{3},\frac{\sqrt{6}}{3})(\frac{\sqrt{3}}{3},-\frac{\sqrt{6}}{3})(\frac{\sqrt{3}}{3},\frac{\sqrt{6}}{3}),(-1,0)$, and $(1,0)$, we find that $f(-\frac{\sqrt{3}}{3},-\frac{\sqrt{6}}{3}) = -\frac{2\sqrt{3}}{9}$ and $f(-\frac{\sqrt{3}}{3},\frac{\sqrt{6}}{3}) = -\frac{2\sqrt{3}}{9}$ are relative minimum values and $f(\frac{\sqrt{3}}{3},-\frac{\sqrt{6}}{3}) = \frac{2\sqrt{3}}{9}$ and $f(\frac{\sqrt{3}}{3},\frac{\sqrt{6}}{3}) = \frac{2\sqrt{3}}{9}$ are relative maximum values.

15. Form the Lagrangian function $F(x,y,z,\lambda) = x^2 + y^2 + z^2 + \lambda(3x + 2y + z - 6)$. We solve the system

$$\begin{cases} F_x = 2x + 3\lambda = 0 \\ F_y = 2y + 2\lambda = 0 \\ F_z = 2x + \lambda = 0 \\ F_\lambda = 3x + 2y + z - 6 = 0. \end{cases}$$

The third equation give $\lambda = -2z$. Substituting into the first two equations gives

$$\begin{cases} 2x - 6z = 0 \\ 2y - 4z = 0. \end{cases}$$

So $x = 3z$ and $y = 2z$. Substituting into the third equation yields $9z + 4z + z - 6 = 0$, or $z = 3/7$. Therefore, $x = 9/7$ and $y = 6/7$. Therefore, $f(\frac{9}{7},\frac{6}{7},\frac{3}{7}) = \frac{18}{7}$ is the minimum value of F.

17. We want to maximize P subject to the constraint $x + y = 200$. The Lagrangian function is

$$F(x, y, \lambda) = -0.2x^2 - 0.25y^2 - 0.2xy + 100x + 90y - 4000 + \lambda(x + y - 200).$$

Next, we solve

$$\begin{cases} F_x = -0.4x - 0.2y + 100 + \lambda = 0 \\ F_y = -0.5y - 0.2x + 90 + \lambda = 0 \\ F_\lambda = x + y - 200 = 0. \end{cases}$$

Subtracting the first equation from the second yields
$$0.2x - 0.3y - 10 = 0, \text{ or } 2x - 3y - 100 = 0.$$

Multiplying the third equation in the system by 2 and subtracting the resulting equation from the last equation, we find $-5y + 300 = 0$ or $y = 60$. So $x = 140$ and the company should make 140 finished and 60 unfinished units.

19. Suppose each of the sides made of pine board is x feet long and those of steel are y feet long. Then $xy = 800$. The cost is $C = 12x + 3y$ and is to be minimized subject to the condition $xy = 800$. We form the Lagrangian function
$$F(x,y,\lambda) = 12x + 3y + \lambda(xy - 800).$$
We solve the system
$$\begin{cases} F_x = 12 + \lambda y = 0 \\ F_y = 3 + \lambda x = 0 \\ F_\lambda = xy - 800 = 0. \end{cases}$$
Multiplying the first equation by x and the second equation by y and subtracting the resulting equations, we obtain $12x - 3y = 0$, or $y = 4x$. Substituting this into the third equation of the system, we obtain
$$4x^2 - 800 = 0, \text{ or } x = \pm\sqrt{200} = \pm 10\sqrt{2}.$$
Since x must be positive, we take $x = 10\sqrt{2}$. So $y = 40\sqrt{2}$. So the dimensions are approximately 14.14 ft by 56.56 ft.

21. We want to minimize the function $C(r,h)$ subject to the constraint $\pi r^2 h - 64 = 0$. We form the Lagrangian function $F(r,h,\lambda) = 8\pi rh + 6\pi r^2 - \lambda(\pi r^2 h - 64)$. Then we solve the system
$$\begin{cases} F_r = 8\pi h + 12\pi r - 2\lambda\pi rh = 0 \\ F_h = 8\pi r - \lambda r^2 = 0 \\ F_\lambda = \pi r^2 h - 64 = 0 \end{cases}$$
Solving the second equation for λ yields $\lambda = 8/r$, which when substituted into the first equation yields
$$8\pi h + 12\pi r - 2\pi rh(\tfrac{8}{r}) = 0$$
$$12\pi r = 8\pi h$$
$$h = \tfrac{3r}{2}.$$
Substituting this value of h into the third equation of the system, we find

$$3r^2\left(\tfrac{3r}{2}\right) = 64, \quad r^3 = \frac{128}{3\pi}, \ \text{or} \ r = \frac{4}{3}\sqrt[3]{\frac{18}{\pi}} \ \text{and} \ h = 2\sqrt[3]{\frac{18}{\pi}}.$$

23. Let the box have dimensions x' by y' by z'. Then $xyz = 4$. We want to minimize
$$C = 2xz + 2yz + \tfrac{3}{2}(2xy) = 2xz + 2yz + 3xy.$$
Form the Lagrangian function
$$F(x,y,z,\lambda) = 2xz + 2yz + 3xy + \lambda(xyz - 4).$$
Next, we solve the system
$$\begin{cases} F_x = 2z + 3y + \lambda yz = 0 \\ F_y = 2z + 3x + \lambda xz = 0 \\ F_z = 2x + 2y + \lambda xy = 0 \\ F_\lambda = xyz - 4 = 0. \end{cases}$$
Multiplying the first, second, and third equations by x, y, and z, respectively, we have
$$\begin{cases} 2xz + 3xy + \lambda xyz = 0 \\ 2yz + 3xy + \lambda xyz = 0 \\ 2xz + 2yz + \lambda xyz = 0. \end{cases}$$
The first two equations imply that $2z(x - y) = 0$. Since $z \neq 0$, we see that $x = y$. The second and third equations imply that $x(3y - 2z) = 0$ or $x = (3/2)y$. Substituting these values into the fourth equation in the system, we find
$$y^2\left(\tfrac{3}{2}y\right) = 4 \ \text{or} \ y^3 = \tfrac{8}{3}. \ \text{Therefore,} \ y = \frac{2}{3^{1/3}} = \frac{2}{3}\sqrt[3]{9} \ \text{and} \ x = \frac{2}{3}\sqrt[3]{9}, \ \text{and} \ z = \sqrt[3]{9}.$$
So the dimensions are $\dfrac{2}{3}\sqrt[3]{9} \times \dfrac{2}{3}\sqrt[3]{9} \times \sqrt[3]{9}$.

25. We want to maximize $f(x,y) = 100x^{3/4}y^{1/4}$ subject to $100x + 200y = 200{,}000$.
Form the Lagrangian function
$$F(x,y,\lambda) = 100x^{3/4}y^{1/4} + \lambda(100x + 200y - 200{,}000).$$
We solve the system
$$\begin{cases} F_x = 75x^{-1/4}y^{1/4} + 100\lambda = 0 \\ F_y = 25x^{3/4}y^{-3/4} + 200\lambda = 0 \\ F_\lambda = 100x + 200y - 200{,}000 = 0. \end{cases}$$
The first two equations imply that $150x^{-1/4}y^{1/4} - 25x^{3/4}y^{-3/4} = 0$ or upon multiplying by $x^{1/4}y^{3/4}$, $150y - 25x = 0$, which implies that $x = 6y$. Substituting

this value of x into the third equation of the system, we have

$$600y + 200y - 200{,}000 = 0$$

giving $y = 250$ and therefore $x = 1500$. So to maximize production, he should

spend 1500 units on labor and 250 units of capital.

27.   False.   See Example 1, Section 8.5.

8.6 Problem Solving Tips

1. To evaluate the double integral $\iint\limits_{R} f(x,y)\,dA = \int_{a}^{b}\left[\int_{g_1(x)}^{g_2(x)} f(x,y)\,dy\right]dx$, first

evaluate the inside y-integral, treating x as a constant, and then evaluate the outside

integral with respect to x.

To evaluate the double integral $\iint\limits_{R} f(x,y)\,dA = \int_{a}^{b}\left[\int_{h_2(y)}^{h_2(y)} f(x,y)\,dx\right]dy$, first evaluate

the inside x-integral, treating y as a constant, and then evaluate the outside integral with

respect to y.

8.6 CONCEPT QUESTIONS, page 604

1. It gives the volume of the solid region bounded above by the graph of f and below by
the region R.

3. $\iint\limits_{R} f(x,y)\,dA = \int_{a}^{b}\left[\int_{g_1(x)}^{g_2(x)} f(x,y)\,dy\right]dx$

5. The average value is $\dfrac{\iint\limits_{R} f(x,y)\,dA}{\iint\limits_{R} dA}$

EXERCISES 8.6, page 604

1. $\int_1^2 \int_0^1 (y+2x)\,dy\,dx = \int_1^2 \frac{1}{2}y^2 + 2xy \Big|_{y=0}^{y=1}\,dx = \int_1^2 (\frac{1}{2}+2x)\,dx = \frac{1}{2}x + x^2 \Big|_1^2 = 5 - \frac{3}{2} = \frac{7}{2}.$

3. $\int_{-1}^1 \int_0^1 xy^2\,dy\,dx = \int_{-1}^1 \frac{1}{3}xy^3 \Big|_0^1\,dx = \int_{-1}^1 \frac{1}{3}x\,dx = \frac{x^2}{6}\Big|_{-1}^1 = \frac{1}{6} - (\frac{1}{6}) = 0.$

5. $\int_{-1}^2 \int_1^{e^3} \frac{x}{y}\,dy\,dx = \int_{-1}^2 x\ln y \Big|_1^{e^3}\,dx = \int_{-1}^2 x\ln e^3\,dx = \int_{-1}^2 3x\,dx = \frac{3}{2}x^2 \Big|_{-1}^2$
$= \frac{3}{2}(4) - \frac{3}{2}(1) = \frac{9}{2}.$

7. $\int_{-2}^0 \int_0^1 4xe^{2x^2+y}\,dx\,dy = \int_{-2}^0 e^{2x^2+y}\Big|_{x=0}^{x=1}\,dy = \int_{-2}^0 (e^{2+y} - e^y)\,dy = (e^{2+y} - e^y)\Big|_{-2}^0$
$= [(e^2 - 1) - (e^0 - e^{-2}) = e^2 - 2 + e^{-2} = (e^2 - 1)(1 - e^{-2}).$

9. $\int_0^1 \int_1^e \ln y\,dy\,dx = \int_0^1 y\ln y - y\Big|_{y=1}^{y=e}\,dx = \int_0^1 dx = 1.$

11. $\int_0^1 \int_0^x (x+2y)\,dy\,dx = \int_0^1 (xy + y^2)\Big|_{y=0}^{y=x}\,dx = \int_0^1 2x^2\,dx = \frac{2}{3}x^3\Big|_0^1 = \frac{2}{3}.$

13. $\int_1^3 \int_0^{x+1} (2x+4y)\,dy\,dx = \int_1^3 2xy + 2y^2\Big|_{y=0}^{y=x+1}\,dx = \int_1^3 [2x(x+1) + 2(x+1)^2]\,dx$
$= \int_1^3 (4x^2 + 6x + 2)\,dx = (\frac{4}{3}x^3 + 3x^2 + 2x)\Big|_1^3$
$= (36 + 27 + 6) - (\frac{4}{3} + 3 + 2) = \frac{188}{3}.$

15. $\int_0^4 \int_0^{\sqrt{y}} (x+y)\,dx\,dy = \int_0^4 \frac{1}{2}x^2 + xy\Big|_{x=0}^{x=\sqrt{y}}\,dy = \int_0^4 (\frac{1}{2}y + y^{3/2})\,dy$
$= (\frac{1}{4}y^2 + \frac{2}{5}y^{5/2})\Big|_0^4 = 4 + \frac{64}{5} = \frac{84}{5}.$

17. $\displaystyle\int_0^2\int_0^{\sqrt{4-y^2}} y\,dx\,dy = \int_0^2 xy\Big|_0^{\sqrt{4-y^2}}\,dy = \int_0^2 y\sqrt{4-y^2}\,dy = -\tfrac{1}{2}(\tfrac{2}{3})(4-y^2)^{3/2}\Big|_0^2$

$\qquad = \tfrac{1}{3}(4^{3/2}) = \tfrac{8}{3}.$

19. $\displaystyle\int_0^1\int_0^x 2xe^y\,dy\,dx = \int_0^1 2xe^y\Big|_{y=0}^{y=x}\,dx = \int_0^1(2xe^x - 2x)\,dx = 2(x-1)e^x - x^2\Big|_0^1 = (-1)+2 = 1.$

21. $\displaystyle\int_0^1\int_x^{\sqrt{x}} ye^x\,dy\,dx = -\int_0^1\int_x^{\sqrt{x}} ye^x\,dy\,dx = \int_0^1 -\tfrac{1}{2}y^2e^x\Big|_{y=\sqrt{x}}^{y=x}\,dx = -\tfrac{1}{2}\int_0^1(x^2e^x - xe^x)\,dx$

$\qquad = -\tfrac{1}{2}[x^2e^x\Big|_0^1 - 2\int_0^1 xe^x\,dx - \int_0^1 xe^x\,dx] = -\tfrac{1}{2}[x^2e^x\Big|_0^1 - 3\int_0^1 xe^x\,dx]$

$\qquad = -\tfrac{1}{2}[x^2e^x - 3xe^x + 3e^x]\Big|_0^1 = -\tfrac{1}{2}[e - 3e + 3e - 3] = \tfrac{1}{2}(3-e).$

23. $\displaystyle\int_0^1\int_{2x}^2 e^{y^2}\,dy\,dx = \int_0^2\int_0^{y/2} e^{y^2}\,dx\,dy = \int_0^2 xe^{y^2}\Big|_{x=0}^{x=y/2}\,dy = \int_0^2 \tfrac{1}{2}ye^{y^2}\,dy$

$\qquad = \tfrac{1}{4}e^{y^2}\Big|_0^2 = \tfrac{1}{4}(e^4-1).$

25. $\displaystyle\int_0^2\int_{y/2}^1 ye^{x^3}\,dx\,dy = \int_0^1\int_0^{2x} ye^{x^3}\,dy\,dx = \int_0^1 \tfrac{1}{2}y^2e^{x^3}\Big|_{y=0}^{y=2x}\,dx = \int_0^1 2x^2e^{x^3}\,dx$

$\qquad = \tfrac{2}{3}e^{x^3}\Big|_0^1 = \tfrac{2}{3}(e-1).$

27. $\displaystyle V = \int_0^4\int_0^3 (6-x)\,dy\,dx = \int_0^4 (6-x)y\Big|_{y=0}^{y=3}\,dx = 3\int_0^4(6-x)\,dx$

$\qquad = 3(6x - \tfrac{1}{2}x^2)\Big|_0^4 = 3(24-8) = 48$, or 48 cu units.

29. $\displaystyle V = \int_0^2\int_0^{3-(3/2)z}(6-2y-3z)\,dy\,dz = \int_0^2 6y - y^2 - 3yz\Big|_{y=0}^{y=3-(3/2)z}\,dz$

$\qquad = \int_0^2[(6(3-\tfrac{3}{2}z) - (3-\tfrac{3}{2}z)^2 - 3(3-\tfrac{3}{2}z)z]\,dz$

$\qquad = -2(3-\tfrac{3}{2}z)^2 - \tfrac{2}{9}(3-\tfrac{3}{2}z)^3 - \tfrac{9}{2}z^2 + \tfrac{3}{2}z^3\Big|_0^2$

$\qquad = (-18+12) - (-18+6) = 6$, or 6 cu units.

31. $\displaystyle V = \int_0^1\int_0^{-2x+2}(4-x^2-y^2)\,dy\,dx = \int_0^1 (4y - x^2y - \tfrac{1}{3}y^3)\Big|_{y=0}^{y=2(1-x)}\,dx$

$$= \int_0^1 [8(1-x) - 2x^2 + 2x^3 - \tfrac{8}{3}(1-x)^3]\,dx$$

$$= [(8x - 4x^2 - \tfrac{2}{3}x^3 + \tfrac{1}{2}x^4) + \tfrac{2}{3}(1-x)^4]\Big|_0^1$$

$$= (8 - 4 - \tfrac{2}{3} + \tfrac{1}{2}) - \tfrac{2}{3} = \tfrac{19}{6}, \text{ or } \tfrac{19}{6} \text{ cu units.}$$

33. $V = \int_0^2 \int_0^2 5e^{-x-y}\,dx\,dy = \int_0^2 -5e^{-x-y}\Big|_{x=0}^{x=2}\,dy = \int_0^2 -5(e^{-2-y} - e^{-y})\,dy$

$$= -5(-e^{-2-y} + e^{-y})\Big|_0^2 = -5(-e^{-4} + e^{-2}) + 5(-e^{-2} + 1) = 5(1 - 2e^{-2} + e^{-4}) \text{ cu units.}$$

35. $V = \int_0^2 \int_0^{2x} (2x + y)\,dy\,dx = \int_0^2 2xy + \tfrac{1}{2}y^2\Big|_0^{2x}\,dx = \int_0^2 (4x^2 + 2x^2)\,dx$

$$= \int_0^2 6x^2\,dx = 2x^3\Big|_0^2 = 16, \text{ or } 16 \text{ cu units.}$$

37. $V = \int_0^1 \int_0^{-x+1} e^{x+2y}\,dy\,dx = \int_0^1 \tfrac{1}{2}e^{x+2y}\Big|_{y=0}^{y=-x+1}\,dx = \tfrac{1}{2}\int_0^1 (e^{-x+2} - e^x)\,dx$

$$= \tfrac{1}{2}(-e^{-x+2} - e^x)\Big|_0^1 = \tfrac{1}{2}(-e - e + e^2 + 1) = \tfrac{1}{2}(e^2 - 2e + 1) = \tfrac{1}{2}(e-1)^2 \text{ cu units.}$$

39. $V = \int_0^4 \int_0^{\sqrt{x}} \dfrac{2y}{1+x^2}\,dy\,dx = \int_0^4 \dfrac{y^2}{1+x^2}\Big|_0^{\sqrt{x}}\,dx = \int_0^4 \dfrac{x}{1+x^2}\,dx$

$$= \tfrac{1}{2}\ln(1+x^2)\Big|_0^4 = \tfrac{1}{2}(\ln 17 - \ln 1) = \tfrac{1}{2}\ln 17 \text{ cu units.}$$

41. $V = \int_0^4 \int_0^{\sqrt{16-x^2}} x\,dy\,dx = \int_0^4 xy\Big|_{y=0}^{y=\sqrt{16-x^2}}\,dx = \int_0^4 x(16-x^2)^{1/2}\,dx$

$$= (-\tfrac{1}{2})(\tfrac{2}{3})(16-x^2)^{3/2}\Big|_0^4 = \tfrac{1}{3}(16)^{3/2} = \tfrac{64}{3}.$$

43. $A = \dfrac{1}{\tfrac{1}{2}}\int_0^1 \int_0^x (x+2y)\,dy\,dx = 2\int_0^1 xy + y^2\Big|_0^x\,dx = 2\int_0^1 (x^2 + x^2)\,dx = 4\int_0^1 x^2\,dx$

$$= \dfrac{4x^3}{3}\Big|_0^1 = \tfrac{4}{3}.$$

45. The area of R is 1/2. The average value of f is

$$\frac{1}{1/2}\int_0^1\int_0^x e^{-x^2}\,dy\,dx = 2\int_0^1 e^{-x^2}y\Big|_{y=0}^{y=x}dx = 2\int_0^1 xe^{-x^2}\,dx = -e^{-x^2}\Big|_0^1 = -e^{-1}+1 = 1-\frac{1}{e}.$$

47. The area of the region is, by elementary geometry, $[4+\frac{1}{2}(2)(4)]$, or 8 sq units.
Therefore, the required average value is

$$A = \tfrac{1}{8}\int_1^3\int_0^{2x}\ln x\,dy\,dx = \tfrac{1}{8}\int_1^3 (\ln x)y\Big|_0^{2x}dx = \tfrac{1}{4}\int_1^3 x\ln x\,dx$$

$$= \tfrac{1}{4}(\tfrac{x^2}{4})(2\ln x - 1)\Big|_1^3 \quad \text{(Integrating by parts)}$$

$$= \tfrac{9}{16}(2\ln 3 - 1) - \tfrac{1}{16}(-1) = \tfrac{1}{8}(9\ln 3 - 4).$$

49. The average population density inside R is $\dfrac{43,329}{20} \approx 2166$ people/sq mile.

51. The average weekly profit is

$$\frac{1}{(20)(20)}\int_{100}^{120}\int_{180}^{200}(-0.2x^2 - 0.25y^2 - 0.2xy + 100x + 90y - 4000)dx\,dy$$

$$= \frac{1}{400}\int_{100}^{120}-\tfrac{1}{15}x^3 - 0.25y^2x - 0.1x^2y + 50x^2 + 90xy - 4000x\Big|_{x=180}^{x=200}dy$$

$$= \frac{1}{400}\int_{100}^{120}(-144,533.33 - 5y^2 - 760y + 380,000 + 1800y - 80,000)dy$$

$$= \frac{1}{400}\int_{100}^{120}(155,466.67 - 5y^2 + 1040y)dy$$

$$= \frac{1}{400}(155,466.67y - \tfrac{5}{3}y^3 + 520y^2)\Big|_{100}^{120}$$

$$= \frac{1}{400}(3,109,333.40 - 1,213,333.30 + 2,288,000)$$

$$\approx 10,460\text{ ,or }\$10,460/\text{wk}.$$

53. True. This result follows from the definition.

55. True. $\iint_R g(x,y)\,dA$ gives the volume of the solid bounded above by the surface $z = g(x,y)$. $\iint_R f(x,y)\,dA$ gives the volume of the solid bounded above by the surface $z = f(x,y)$. Therefore,

$$\iint\limits_{R} g(x,y)\,dA - \iint\limits_{R} f(x,y)\,dA = \iint\limits_{R} [g(x,y) - f(x,y)]\,dA$$

gives the volume of the solid bounded above by $z = g(x,y)$ and below by $z = f(x,y)$.

CHAPTER 8 CONCEPT REVIEW, page 608

1. xy; ordered pair; real number; $f(x,y)$
3. $z = f(x,y)$; f; surface   5. Fixed number; x
7. $\leq$; (a,b); $\leq$; domain
9. Scatter; minimizing; least-squares; normal
11. Volume; solid

CHAPTER 8 REVIEW EXERCISES, page 609

1. $f(0,1) = 0$; $f(1,0) = 0$; $f(1,1) = \dfrac{1}{1+1} = \dfrac{1}{2}$.

    $f(0,0)$ does not exist because the point $(0,0)$ does not lie in the domain of f.

3. $h(1,1,0) = 1 + 1 = 2$; $h(-1,1,1) = -e - 1 = -(e + 1)$;
    $h(1,-1,1) = -e - 1 = -(e + 1)$.

5. $D = \{(x,y)|y \neq -x\}$

7. The domain of f is the set of all ordered triplets (x,y,z) of real numbers such that $z \geq 0$ and $x \neq 1$, $y \neq 1$, and $z \neq 1$.

9. $z = y - x^2$

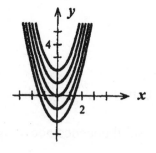

11. $z = e^{xy}$

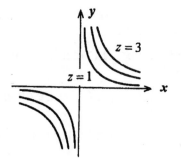

13. $f(x,y) = x\sqrt{y} + y\sqrt{x};\ f_x = \sqrt{y} + \dfrac{y}{2\sqrt{x}};\ f_y = \dfrac{x}{2\sqrt{y}} + \sqrt{x}$

15. $f(x,y) = \dfrac{x-y}{y+2x}.\ f_x = \dfrac{(y+2x)-(x-y)(2)}{(y+2x)^2} = \dfrac{3y}{(y+2x)^2}.$

$f_y = \dfrac{(y+2x)(-1)-(x-y)}{(y+2x)^2} = \dfrac{-3x}{(y+2x)^2}.$

17. $h(x,y) = (2xy+3y^2)^5;\ h_x = 10y(2xy+3y^2)^4;\ h_y = 10(x+3y)(2xy+3y^2)^4.$

19. $f(x,y) = (x^2+y^2)e^{x^2+y^2};$

$f_x = 2xe^{x^2+y^2} + (x^2+y^2)(2x)e^{x^2+y^2} = 2x(x^2+y^2+1)e^{x^2+y^2}.$

$f_y = 2ye^{x^2+y^2} + (x^2+y^2)(2y)e^{x^2+y^2} = 2y(x^2+y^2+1)e^{x^2+y^2}.$

21. $f(x,y) = \ln\left(1+\dfrac{x^2}{y^2}\right).\ f_x = \dfrac{\frac{2x}{y^2}}{1+\frac{x^2}{y^2}} = \dfrac{2x}{x^2+y^2};\ f_y = \dfrac{-\frac{2x^2}{y^3}}{1+\frac{x^2}{y^2}} = -\dfrac{2x^2}{y(x^2+y^2)}.$

23. $f(x,y) = x^4 + 2x^2y^2 - y^4;\ f_x = 4x^3 + 4xy^2;\ f_y = 4x^2y - 4y^3;$

$f_{xx} = 12x^2 + 4y^2,\ f_{xy} = 8xy = f_{yx},\ f_{yy} = 4x^2 - 12y^2.$

25. $g(x,y) = \dfrac{x}{x+y^2};\ g_x = \dfrac{(x+y^2)-x}{(x+y^2)^2} = \dfrac{y^2}{(x+y^2)^2},\ g_y = \dfrac{-2xy}{(x+y^2)^2}.$

Therefore, $g_{xx} = -2y^2(x+y^2)^{-3} = -\dfrac{2y^2}{(x+y^2)^3},$

$g_{yy} = \dfrac{(x+y^2)^2(-2x)+2xy(2)(x+y^2)2y}{(x+y^2)^4} = \dfrac{2x(x^2+y^2)[-x-y^2+4y^2]}{(x+y^2)^4}$

$= \dfrac{2x(3y^2-x)}{(x+y^2)^3}.$

and $\qquad g_{xy} = \dfrac{(x+y^2)2y - y^2(2)(x+y^2)2y}{(x+y^2)^4} = \dfrac{2(x+y^2)[xy+y^3-2y^3]}{(x+y^2)^4}$

$$= \frac{2y(x-y^2)}{(x+y^2)^3} = g_{yx}.$$

27. $h(s,t) = \ln\left(\frac{s}{t}\right)$. Write $h(s,t) = \ln s - \ln t$. Then $h_s = \frac{1}{s}$, $h_t = -\frac{1}{t}$.

Therefore, $h_{ss} = -\frac{1}{s^2}$, $h_{st} = h_{ts} = 0$, $h_{tt} = \frac{1}{t^2}$.

29. $f(x,y) = 2x^2 + y^2 - 8x - 6y + 4$; To find the critical points of f, we solve the
system $\begin{cases} f_x = 4x - 8 = 0 \\ f_y = 2y - 6 = 0 \end{cases}$ obtaining $x = 2$ and $y = 3$. Therefore, the sole critical
point of f is $(2,3)$. Next, $f_{xx} = 4, f_{xy} = 0, f_{yy} = 2$. Therefore,
$$D = f_{xx}(2,3)f_{yy}(2,3) - f_{xy}(2,3)^2 = 8 > 0.$$
Since $f_{xx}(2,3) > 0$, we see that $f(2,3) = -13$ is a relative minimum.

31. $f(x,y) = x^3 - 3xy + y^2$. We solve the system of equations $\begin{cases} f_x = 3x^2 - 3y = 0 \\ f_y = -3x + 2y = 0 \end{cases}$
obtaining $x^2 - y = 0$, or $y = x^2$. Then $-3x + 2x^2 = 0$, and $x(2x - 3) = 0$, and $x = 0$,
or $x = 3/2$ and $y = 0$, or $y = 9/4$. Therefore, the critical points are $(0,0)$ and $(\frac{3}{2}, \frac{9}{4})$.
Next, $f_{xx} = 6x, f_{xy} = -3$, and $f_{yy} = 2$ and $D(x,y) = 12x - 9 = 3(4x - 3)$. Therefore,
$D(0,0) = -9$ so $(0,0)$ is a saddle point. $D(\frac{3}{2}, \frac{9}{4}) = 3(6-3) = 9 > 0$,   and
$f_{xx}(\frac{3}{2}, \frac{9}{4}) > 0$ and therefore, $f(\frac{3}{2}, \frac{9}{4}) = \frac{27}{8} - \frac{81}{8} + \frac{81}{16} = -\frac{27}{16}$ is the relative minimum
value.

33. $f(x,y) = f(x,y) = e^{2x^2+y^2}$. To find the critical points of f, we solve the system
$$\begin{cases} f_x = 4xe^{2x^2+y^2} = 0 \\ f_y = 2ye^{2x^2+y^2} = 0 \end{cases}$$
giving $(0,0)$ as the only critical point of f. Next,
$$f_{xx} = 4(e^{2x^2+y^2} + 4x^2e^{2x^2+y^2}) = 4(1+4x^2)e^{2x^2+y^2}$$
$$f_{xy} = 8xye^{2x^2+y^2}$$
$$f_{yy} = 2(1+2y^2)e^{2x^2+y^2}.$$

Therefore, $D = f_{xx}(0,0)f_{yy}(0,0) - f_{xy}^2(0,0) = (4)(2) - 0 = 8 > 0$
and so $(0,0)$ gives a relative minimum of f since $f_{xx}(0,0) > 0$. The minimum value
of f is $f(0,0) = e^0 = 1$.

35. We form the Lagrangian function $F(x,y,\lambda) = -3x^2 - y^2 + 2xy + \lambda(2x + y - 4)$.
Next, we solve the system
$$\begin{cases} F_x = 6x + 2y + 2\lambda = 0 \\ F_y = -2y + 2x + \lambda = 0 \,. \\ F_\lambda = 2x + y - 4 = 0 \end{cases}$$
Multiplying the second equation by 2 and subtracting the resultant equation from
the first equation yields $6y - 10x = 0$ so $y = 5x/3$. Substituting this value of y into
the third equation of the system gives $2x + \frac{5}{3}x - 4 = 0$. So $x = \frac{12}{11}$ and consequently
$y = \frac{20}{11}$. So $(\frac{12}{11}, \frac{20}{11})$ gives the maximum value for f subject to the given constraint.

37. The Lagrangian function is $F(x,y,\lambda) = 2x - 3y + 1 + \lambda(2x^2 + 3y^2 - 125)$. Next, we
solve the system of equations
$$\begin{cases} F_x = 2 + 4\lambda x = 0 \\ F_y = -3 + 6\lambda y = 0 \\ F_\lambda = 2x^2 + 3y^2 - 125 = 0 \,. \end{cases}$$
Solving the first equation for x gives $x = -1/2\lambda$. The second equation gives
$y = 1/2\lambda$. Substituting these values of x and y into the third equation gives
$$2\left(-\frac{1}{2\lambda}\right)^2 + 3\left(\frac{1}{2\lambda}\right)^2 - 125 = 0$$
$$\frac{1}{2\lambda^2} + \frac{3}{4\lambda^2} - 125 = 0$$
$$2 + 3 - 500\lambda^2 = 0, \text{ or } \lambda = \pm\frac{1}{10}.$$
Therefore, $x = \pm 5$ and $y = \pm 5$ and so the critical points of f are $(-5,5)$ and $(5,-5)$.
Next, we compute
$$f(-5,5) = 2(-5) - 3(5) + 1 = -24.$$
$$f(5,-5) = 2(5) - 3(-5) + 1 = 26.$$
So f has a maximum value of 26 at $(5,-5)$ and a minimum value of -24 at
$(-5,5)$.

39. $\int_{-1}^{2}\int_{2}^{4}(3x-2y)dx\,dy = \int_{-1}^{2}\frac{3}{2}x^2-2xy\Big|_{x=2}^{x=4}dy = \int_{-1}^{2}[(24-8y)-(6-4y)]dy$

$$= \int_{-1}^{2}(18-4y)\,dy = (18y-2y^2)\Big|_{-1}^{2} = (36-8)-(-18-2) = 48.$$

41. $\int_{0}^{1}\int_{x^3}^{x^2}2x^2y\,dy\,dx = \int_{0}^{1}x^2y^2\Big|_{y=x^3}^{y=x^2}dx = \int_{0}^{1}x^2(x^4-x^6)dx$

$$= \int_{0}^{1}(x^6-x^8)dx = \frac{x^7}{7}-\frac{x^9}{9}\Big|_{0}^{1} = \frac{1}{7}-\frac{1}{9} = \frac{2}{63}.$$

43. $\int_{0}^{2}\int_{0}^{1}(4x^2+y^2)dy\,dx = \int_{0}^{2}4x^2y+\frac{1}{3}y^3\Big|_{y=0}^{y=1}dx = \int_{0}^{2}(4x^2+\frac{1}{3})dx$

$$= (\frac{4}{3}x^3+\frac{1}{3}x)\Big|_{0}^{2} = \frac{32}{3}+\frac{2}{3} = \frac{34}{3}.$$

45. The area of R is

$$\int_{0}^{2}\int_{x^2}^{2x}dy\,dx = \int_{0}^{2}y\Big|_{y=x^2}^{y=2x}dx = \int_{0}^{2}(2x-x^2)\,dx = (x^2-\frac{1}{3}x^3)\Big|_{0}^{2} = \frac{4}{3}.$$

Then

$$AV = \frac{1}{4/3}\int_{0}^{2}\int_{x^2}^{2x}(xy+1)dy\,dx = \frac{3}{4}\int_{0}^{2}\frac{xy^2}{2}+y\Big|_{x^2}^{2x}dx$$

$$= \frac{3}{4}\int_{0}^{2}(-\frac{1}{2}x^5+2x^3-x^2+2x)\,dx = \frac{3}{4}(-\frac{1}{12}x^6+\frac{1}{2}x^4-\frac{1}{3}x^3+x^2)\Big|_{0}^{2}$$

$$= \frac{3}{4}(-\frac{16}{3}+8-\frac{8}{3}+4) = 3.$$

47. $f(p,q) = 900-9p-e^{0.4q}$; $g(p,q) = 20{,}000-3000q-4p$.

We compute $\dfrac{\partial f}{\partial q} = -0.4e^{0.4q}$ and $\dfrac{\partial g}{\partial p} = -4$. Since $\dfrac{\partial f}{\partial q} < 0$ and $\dfrac{\partial g}{\partial p} < 0$

for all $p > 0$ and $q > 0$, we conclude that compact disc players and audio discs are complementary commodities.

49. a. We first summarize the data.

x	y	x^2	xy
0	19.5	0	0
10	20	100	200
20	20.6	400	412
30	21.2	900	636
40	21.8	1600	872
50	22.4	2500	1120
150	125.5	5500	3240

The normal equations are
$$6b + 150m = 125.5$$
$$150b + 5500m = 3240$$
The solutions are $b = 19.45$ and $m = 0.0586$. Therefore, $y = 0.059x + 19.5$.
b. The life expectancy at 65 of a female in 2040 is
$$y = 0.059(40) + 19.5 = 21.86 \quad \text{or } 21.9 \text{ years.}$$
The datum gives a life expectancy of 21.8 years.
c. The life expectancy at 65 of a female in 2030 is
$$y = 0.059(30) + 19.5 = 21.27 \quad \text{or } 21.3 \text{ years.}$$
The data give a life expectancy of 21.2 years.

51. Refer to the following diagram.

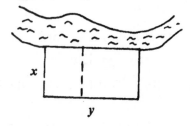

We want to minimize $C(x,y) = 3(2x) + 2(x) + 3y = 8x + 3y$ subject to
$xy = 303{,}750$. The Lagrangian function is
$$F(x,y,\lambda) = 8x + 3y + \lambda(xy - 303{,}750).$$

8 Calculus of Several Variables

Next, we solve the system

$$\begin{cases} F_x = 8 + \lambda y = 0 \\ F_y = 3 + \lambda x = 0 \\ F_\lambda = xy - 303{,}750 = 0 \end{cases}.$$

Solving the first equation for y gives $y = -8/\lambda$. The second equation gives $x = -3/\lambda$. Substituting this value into the third equation gives

$$\left(-\frac{3}{\lambda}\right)\left(-\frac{8}{\lambda}\right) = 303{,}750 \quad \text{or} \quad \lambda^2 = \frac{24}{303{,}750} = \frac{4}{50{,}625},$$

or $\lambda = \pm\frac{2}{225}$. Therefore, $x = 337.5$ and $y = 900$ and so the required dimensions of the pasture are 337.5 yd by 900 yd.

CHAPTER 8 BEFORE MOVING ON, page 610

1. We have the constraints $x \geq 0$, $y \geq 0$, $x \neq 1$ and $y \neq 2$. Therefore the domain of f is
   $D = \{(x, y) | x \geq 0,\ y \geq 0;\ x \neq 1 \text{ and } y \neq 2\}$.

2. $f_x = 2xy + ye^{xy}$, $f_y = x^2 + xe^{xy}$, $f_{xx} = 2y + y^2 e^{xy}$, $f_{xy} = 2x + (1 + xy)e^{xy} = f_{yx}$
   $f_{yy} = x \cdot xe^{xy} = x^2 e^{xy}$

3.
   $$\left.\begin{array}{l} f_x = 6x^2 + 6y = 6(x^2 - y^2) = 0 \\ f_y = 6y^2 - 6x = 6(y^2 - x) = 0 \end{array}\right\} \text{ gives } y = x^2 \text{ and } x = y^2. \text{ Therefore,}$$

   $x = x^4$, $x^4 - x = x(x^3 - 1) = 0$ giving $x = 0$ or 1. The critical points of f are $(0,.0)$ and $(1,1)$. $f_{xx} = 12x$, $f_{xy} = -6$, $f_{yy} = 12y$.

   $D(x, y) = 144x^2 + 144y^2 - 36$; $D(0,0) = -36 < 0$; and so $(0,0,-5)$ is a relative minimum.

   $f(x, y) = 2x^3 + 2y^3 - 6xy - 5$ and $f(1,1) = 2(1)^3 + 2(1)^3 - 6(1)(1) - 5 = -7$

4.

x	y	x^2	xy
0	2.9	0	0
1	5.1	1	5.1
2	6.8	4	13.6
3	8.8	9	26.4
5	13.2	25	66
11	36.8	39	113.1

The normal equations are
$$5b + 11m = 36.8$$
$$11b + 39m = 111.1$$
Solving, we find $m = 2.036$ and $b = 2.8797$. The least-squares equation is
$$y = 2.036x + 2.8797.$$

5. $F(x, y, \lambda) = 3x^2 + 3y^2 + 1 + \lambda(x + y - 1)$

$F_x = 6x + \lambda = 0$

$F_y = 6y + \lambda = 0$

$F_\lambda = x + y - 1 = 0$

Gives $\lambda = -6x - 6y$ so $y = x$. Substituting into the third equation gives
$2x = 1$ or $x = \frac{1}{2}$ and $y = \frac{1}{2}$. Therefore, $(\frac{1}{2}, \frac{1}{2}, \frac{5}{2})$ is the required minimum.

6. $\displaystyle\iint_R (1 - xy)\, dA = \int_0^1 \int_{x^2}^{x} (1 - xy)\, dy\, dx$

$= \int_0^1 [(y - \tfrac{1}{2}xy^2)|_{x^2}^{x}]\, dx$

$= \int_0^1 [x - \tfrac{1}{2}x^3 - x^2 + \tfrac{1}{2}x^5]\, dx$

$= \tfrac{1}{2}x^2 - \tfrac{1}{8}x^4 - \tfrac{1}{3}x^3 + \tfrac{1}{12}x^6 \Big|_0^1$

$= \tfrac{1}{2} - \tfrac{1}{8} - \tfrac{1}{3} + \tfrac{1}{12} = \tfrac{1}{8}$